MODELING
THE WORLD IN A
SPREADSHEET

ENVIRONMENTAL SIMULATION ON A MICROCOMPUTER

Timothy J. Cartwright

The Johns Hopkins University Press
Baltimore and London

The Johns Hopkins University Press
2715 North Charles Street
Baltimore, Maryland 21218-4319
The Johns Hopkins Press Ltd., London

Library of Congress Catalog Card Number 93-078894

ISBN 0-8018-4596-3
ISBN 0-8018-4597-1 (pbk)

A catalog record for this book is available from the British Library.

Lotus and 1-2-3 are registered trademarks of Lotus Development Corporation.

This book has been published from camera-ready copy prepared electronically by the author.

CONTENTS

Foreword   v

Introduction: Simulation Modeling and Spreadsheets   1

1. MODELING NATURAL SYSTEMS   17

1. Blowing Smoke: Atmospheric Dispersion of Air Pollution   19
2. Running Water: The Underground Transport of Pollutants   43
3. Preserving a Species: Finding the Minimum Viable Population   63
4. Sustainable Yield: Managing the Forest for the Trees   89
5. Here Comes the Sun: Solar Energy from a Flat-Plate Collector   121

2. MODELING SOCIAL SYSTEMS   145

6. Macroeconomic Policy: Econometrics and the Klein Model   147
7. Urban Form: The Lowry Model of Population Distribution   165
8. Affordable Housing: The Bertaud Model   183
9. Traffic on the Roads: Modeling Trip Generation and Trip Distribution   205
10. Throwing Things Away: A Model for Waste Management   229
11. Multi-Criteria Analysis: An Environmental Impact Assessment Model   249

3. MODELING ARTIFICIAL SYSTEMS   275

12. Life in a Spreadsheet: John Conway's Game   277
13. The Game of Choice: Neighborhood Tolerance   293
14. The Game Machine: Cellular Automata, Chaos, and Fractals   303

4. THE WORLD IN A SPREADSHEET   325

Conclusion: Modeling with Microcomputers   327

APPENDICES   335

A. Creating the Models   337
B. The Recipes   361

Index   419

Program Listings

Listing 1.1	SMOKE	26
Listing 2.1	WATER	49
Listing 3.1	BEARS	71
Listing 4.1	TREES	95
Listing 5.1	SOLAR	129
Listing 6.1	KLEIN	157
Listing 7.1	LOWRY	173
Listing 8.1	BERTAUD	191
Listing 9.1	TRAFFIC	215
Listing 10.1	WASTE	239
Listing 11.1	EIA	260
Listing 12.1	LIFE	284
Listing 13.1	CHOICE	296
Listing 14.1	GAMACH	308

Program Recipes

Recipe 1:	Blowing Smoke	362
Recipe 2:	Running Water	365
Recipe 3:	Preserving a Species	370
Recipe 4:	Sustainable Yield	373
Recipe 5:	Here Comes the Sun	378
Recipe 6:	The Klein Model	382
Recipe 7:	The Lowry Model	385
Recipe 8:	The Bertaud Model	387
Recipe 9:	Traffic on the Roads	393
Recipe 10:	Throwing Things Away	400
Recipe 11:	Multi-Criteria Analysis and EIA	402
Recipe 12:	The Game of Life	406
Recipe 13:	The Game of Choice	409
Recipe 14:	The Game Machine	414

FOREWORD

My aim in writing this book is not, of course, to prove that the world can be reduced to a spreadsheet. I am not so naive as to think that that is either desirable or possible. In fact, my purpose is quite the opposite. It is to say, look, if we can build all of these intriguing relationships into a two-dimensional spreadsheet using only a microcomputer and 640 Kb of memory, then think how wondrously complex the real thing must be!

Ultimately, the models in this book are all mechanical in nature. I do not for a moment believe that the real world is as simple as that. So why bother with such models? In a recent letter in the *Financial Times* (April 17, 1993), Andrew Britton of the National Institute of Economic and Social Research in London said,

> A model is not a "monster", but a convenient way of setting out in formal mathematical terms the lessons you think you have learned from experience. One advantage of doing this explicitly is that the basis on which you form your judgement is there for the public to see and to criticise. It makes economic forecasting less like soothsaying, and more like applied science.

We can debate how far the lessons of "experience" can be quantified and their relationships expressed in mathematical terms. But, in the end, that is relatively unimportant—and is just as likely to be decided by personal preferences, political expediency, or the availability of data as by any more scientific considerations. The real issue is whether humans are capable of creating models, or images, of the world that are *not* fundamentally mechanical in nature. In other words, the basic question is not so much what the world is like, but rather what are the limits to our capacity to understand it. Thus, the scientist who prefers informal, qualitative, "rules of thumb" to the more formal models presented in this book is not being any the less "mechanical" in his or her view of the world than I am. We are just acknowledging our common constraints in different ways.

I am always amused by people who ask if computers will ever be able to think. What is amusing is not the question, which is obviously a very serious one, but the implicit assumption that *thinking* means *thinking like us*. In fact, there may be other ways of thinking than the way humans think, and computers may turn out to be better at some of these other ways of thinking than we are. Computers will never be good humans but, judged on their own terms, they are already formidably capable.

For the time being, at least, computers (including so-called parallel computers, or transputers) and humans process information in different ways. For example, humans tend to process relatively few data in relatively elaborate ways, whereas computers tend to process a great many data in relatively simple ways. Naturally, we tend to regard our kind of processing as superior to the computer's kind. No doubt it is, at least in some respects. (After all, humans created computers; it was not, so far as we know, the other way round!) But it remains to be seen which kind of information-processing will be, so to speak, the "wave of the future"—which is the more capable of continued development, adaptation, and ultimately survival.

So the models in this book make no special claims. I do not believe that simulation models are better or worse than more qualitative kinds of models. But I do believe that the two kinds of models can complement each other. If computers are better than humans at "number crunching", then let us use computers for that kind of information-processing. If humans are better than computers at being creative and using judgement, then let us use humans for that kind of information-processing. And when quantitative models can be built as quickly and easily as they can in spreadsheets, there is little excuse (it seems to me) for even the most dedicated proponent of qualitative models not to take advantage of quantitative models as well. In the end, simulation models cannot replace human judgement; but they can certainly enhance it.

There are hundreds of formulas in this book, many of them containing dozens of variables and arithmetic operations. (Frankly, too, some of the models are at the outer limit of my own mathematical capacities!) I have tried to make sure there are no errors or misprints in the models. But inevitably there will be some I have missed (as readers of Michael Crichton's recent best-seller, *Jurassic Park*, will appreciate!), and for that I apologize.

Many people, colleagues and students, inspired and helped me with this book, and I would like to record my gratitude to a few of them: to Professor Iskandar Gabbour of the Université de Montréal, who has been my friend and "math coach" for years and who read the entire manuscript; to Tony Oliveri, who created a preliminary version of the grizzly-bear model in chapter 3 and helped debug many of my "recipes"; to John McLean (chapter 1), Alan Yoshioka (chapters 5 and 11), and Sutrisno (chapter 9) for substantial help with the chapters indicated; to Dr. Jean-Marc Choukroun, who showed me the Game of Life about 20 years ago, long before I had any idea what it was all about; and to Miriam Klieger, an editor for the Press, who did her job so well. Thanks to all; the faults that remain are mine alone.

I am also grateful to the Center for Urban Policy Research at Rutgers University for allowing me to use in this book revised versions (including revised versions of the models) of two chapters from R.E. Klosterman, E.G. Bossard, and R.K. Brail, eds., *Spreadsheet Models for Urban and Regional Analysis* (New Brunswick, N.J.: Center for Urban Policy Research, 1993).

Finally, I should like to record my gratitude to York University for creating the institutional setting in which people like me can work on multidisciplinary projects like this book. That institutional setting is the Faculty of Environmental Studies, which this year celebrates its 25th anniversary. Although multidisciplinary research is still regarded by many in academia with a mixture of envy and suspicion (especially when it comes to matters of promotion), the Faculty at least protects me from the usual disciplinary constraints on what I can teach and where I should publish. For these tender mercies, I am exceeding grateful.

Perhaps, in the same vein, I should also acknowledge the unwitting assistance of the Social Science and Humanities Research Council. On two occasions, the Council declined to provide funding for the work on which this book is based (in one case, on the rather bizarre grounds that there was insufficient emphasis on the sociology of knowledge!) Naturally, this just encouraged me to persevere.

One of the motives for writing a book like this is to share teaching materials with others. I hope that readers who develop any case studies related to the models in this book, or who create variations on or extensions of any of the models, will likewise feel impelled to share them with me (at the address given below).

At various points in this book, reference is made to certain companies and certain product names. Some of these names may be trademarks or registered trademarks of their respective companies. Such names are used here only for purposes of identification.

Introduction

SIMULATION MODELING AND SPREADSHEETS

"What do you consider the largest map that would be really useful?"
"About six inches to the mile."
"Only six inches!" exclaimed Mein Herr. We very soon got to
six yards to the mile. Then we tried a hundred yards to the mile.
And then came the grandest idea of all! We actually
made a map on the scale of a mile to the mile!"
"Have you used it much?" I enquired.
"It has never been spread out yet," said Mein Herr: "the farmers
objected; they said it would cover the whole country and shut out
the sunlight! So now we use the country itself as its own map."

Lewis Carroll, Sylvie and Bruno Concluded

This book has two principal subjects: simulation modeling and spreadsheets. The first (simulation modeling) is an exciting way of learning about and, to some extent, learning how to manage, environmental issues of the kind that now seem to confront us on scales ranging from the local to the global. The second (spreadsheets) is an easy-to-use and extraordinarily versatile type of computer software. Spreadsheets are often thought of as tools for financial analysis, but their range of application is much wider than that. Among other things, spreadsheets are very good for modeling all sorts of environmental phenomena.

Until recently, simulation modeling has been viewed as the domain of high-powered research centers and "think tanks"—organizations with access to heavy-duty computers and specialized software. Thanks to the spreadsheet, however, simulation modeling is now accessible to anyone with a microcomputer—students and teachers; public officials from municipal to national levels; ordinary professional people, such as engineers and planners; and employees of small businesses, non-governmental organizations, voluntary agencies, and the like. To show just how broadly spreadsheet simulation can be applied, this book presents 14 examples. They range from the physical dispersion of air and water pollution, to housing affordability and community waste management, to "cellular" decision making in various kinds of "neighborhoods."

Each model, moreover, is accompanied not by a printout of the entire model (which tends to overwhelm all but the most dedicated modeler) but by a concise "recipe." The recipe consists of instructions for using the commands of the spreadsheet to create the model as well as to run it.

More than most other kinds of software, spreadsheets are all broadly similar, and all work in more or less the same way. So it matters little which particular brand of spreadsheet you want to use. I have chosen to present the models in terms of one particular program (SuperCalc) rather than invent some artificial, generic version of a spreadsheet. This is not meant to be an endorsement of one product over others; it is merely a convenience. On the contrary, my aim has been to present models that can run under as many different spreadsheet programs as possible. Thus, none of the models requires features or commands that are found only in particular programs. All of the models should run in most commercial spreadsheet programs and in many of the public-domain versions as well.[1]

Simulation Modeling

At first sight, the term *simulation model* may seem redundant. After all, a model is by definition a representation or scaled-down version of the real thing. So why should we talk about "simulation models," if *all* models are in fact simulations of reality?

The reason for describing some models as simulation models is that the purpose of modeling is not always representation. Some models are intended to predict what is going to happen: these are often called *predictive models* or *forecasting models*. Some models are meant to determine what is the best policy to follow or the best action to take: these are often called *optimizing models*. *Simulation models* are different from both of these, in that their purpose is simulation. That is, they are

---

1.   The mechanics of entering and adjusting the models in various spreadsheet programs are discussed in detail in appendixes 1 and 2.

intended to represent what is happening now—not what is likely to happen or what ought to happen at some time in the future. Simulation models are meant to provide a kind of surrogate or laboratory for the real world, in which we can test the effect of changing some of the initial conditions or inputs, or of varying some of the assumptions or parameters of the model. Simulation models provide a vehicle for investigating what would happen or would have happened, if certain factors were or had been different. To put this another way, simulation models allow us to examine how sensitive a process is to changes in input variables or parameter settings. Thus, what-if modeling and sensitivity analysis are often the hallmarks of simulation models.

Of course, these distinctions are not entirely clear-cut. We often need to understand what is happening now, in order to be able to predict what is going to happen in the future. But this is by no means always true. On the one hand, there are successful "black box" types of forecasting techniques (such as the "technical" approach to analysis of stock prices in financial markets) in which the emphasis is almost entirely on behavior, with little effort made to understand fundamental or underlying causes. On the other hand, simulation models can, primarily through a process of trial and error, often suggest what might happen or what ought to happen in the future. Finally, optimizing models often depend on some, at least idealized, notion of how a system actually works, in order to suggest how to make the system work better. Nevertheless, the motivations are different; and it is the desire to represent the real world, albeit on a reduced or simplified scale, that forms the aim of simulation modeling.

If the term "simulation modeling" is not redundant, then is it not at least, by its very nature, doomed before it starts? That is, if simulation models are reduced or simplified versions of the real thing, does that not make them inherently unreliable? In one sense, this criticism is perfectly valid. No finite human perception of reality—whether a simulation model or any other kind of model, either computerized or worked out with pencil and paper—can ever duplicate the infinite complexity of the "real world." But representation need not mean duplication, as Lewis Carroll reminds us in the epigraph to this chapter.[2] The purpose of

---

2. In his book, *Dreamtigers* (Austin: University of Texas Press, 1964), Jorge Luis Borges writes about a country where "the Art of Cartography reached such Perfection that the map of one Province alone took up the whole of a City, and the map of the Empire the whole of a Province. In time, those Unconscionable Maps did not satisfy and the Colleges of Cartogra-

simulation modeling (like that of mapping) is to create a reasonably reliable guide, not a one-to-one analog. The best simulation model, therefore, is not necessarily the most complicated or the most realistic. The best simulation model is the one that proves most reliable for the uses to which it is put. Any model can be faulted in comparison with the real thing. The art lies in finding the model that is suitable for a given task—or, sometimes, the task that is suited to a given model! I hope that this book may help readers become more skilled at doing both.

Spreadsheets

Spreadsheets are essentially number-processing programs. In other words, spreadsheets are to numbers what word processors are to text. Spreadsheets are generic data-processing programs optimized for dealing with numbers rather than text. Like word processors, spreadsheets present the user with a blank page; but the page is organized into rows and columns instead of lines and pages. Like word processors, spreadsheets can manipulate both text and numbers, but spreadsheets are better at doing the kinds of things that are typically done with numbers (such as adding, averaging, randomizing, etc.). By contrast, spreadsheets are typically less effective at doing the kinds of things that are associated with text processing (such as reformating a paragraph, finding and replacing a word, checking spelling, etc.). It is possible to write a letter with a spreadsheet, just as it is possible to add up a column of numbers in some word-processing programs. But clearly, both of these tasks are more conveniently done with the appropriate program.

The spreadsheet is said by some to be one of the cleverest computer programs ever written. One reason for this is that spreadsheets did not offer merely to automate a task we had hitherto been doing manually. Instead, the spreadsheet offered a wholly new kind of working environment, something that could be described as a dynamic scratch pad. If ever there was one single reason for learning to use a computer as something other than a glorified typewriter, the spreadsheet was it. Indeed, many people have attributed the success of the microcomputer revolution as much to the first spreadsheet (Visicalc) as to the first computer (the Apple II) it

---

phers set up a Map of the Empire which had the size of the Empire itself and coincided with it point by point. Less addicted to the Study of Cartography, succeeding Generations understood that this widespread map was useless and not without Impiety they abandoned it." (Cited in an e-mail message from Duane Marble, Ohio State University.)

ran on. It is also notable that the spreadsheet was developed on and for *micro*computers; it was years before spreadsheet programs were written for minicomputers and mainframes.

A spreadsheet is deceptively simple. It looks like nothing more than a set of rows (horizontal) and columns (vertical), with each one labeled with a number or letter to identify it. But there is more to it than this. A word processor presents you with a blank screen, while "behind" it lies the power to edit, move, and process the text you enter in a far more flexible and convenient way than can be done manually or with a typewriter. In the same way, a spreadsheet presents a screen that is nearly blank, except for being divided into "cells" to keep data lined up in neat rows and columns; but the significant thing is what lies behind this simple face. Essentially, what a spreadsheet provides is the capacity to create a dynamic arithmetic and logical network of "links" among any or all of the cells in the spreadsheet.

For example, if we want to enter a column of numbers and find its total, it suffices to enter the data into the spreadsheet, one number to a cell, and then to tell the computer to calculate and display the sum in another cell. (The cells containing the data could be dispersed anywhere in the spreadsheet, although the normal procedure would be to enter them all together in a column, with the total at the bottom.) But the key point is this: if any or even all of the data are changed, the spreadsheet automatically recalculates the sum—and any other cells that depend on the changed data. In other words, you can organize a spreadsheet so that any change in the input data is automatically reflected in the output or results. You might almost say that the spreadsheet is a kind of automatic "answer machine."

Microcomputer spreadsheets are typically about 256 columns wide and perhaps 9,000 rows deep (although you will soon run out of memory if you try to put a complicated formula in every one of these cells!) Rows are usually numbered, and columns are usually labeled with letters (from A to Z, then from AA to AZ, BA to BZ, CA to CZ, and so on, as required). Thus, every cell in a spreadsheet can be referenced by a letter-number combination (e.g., A1, B7, GS4096, etc.). Naturally, only part of a large worksheet can be seen on the computer screen at one time; so the screen should be thought of as a movable "window" on the worksheet as a whole.

Spreadsheet cells can contain any of three different kinds of data: numerical constants, text (sometimes called "labels"), and formulas. Formulas in turn consist of numbers and/or references to other cells (e.g., B7) plus basic arithmetic operators

(+, -, *, /), exponentiation (^), Boolean operators (such as >, <, AND, OR, and NOT), and a variety of arithmetic, statistical, trigonometric, logical, string, calendar, data-management, and other functions. Thanks to such formulas, data can be entered into the cells of a spreadsheet and then linked mathematically and logically to other data. Essentially, cell references act like variable names in conventional algebra. Instead of writing something like

$$y = 2x^2 + 1$$

spreadsheets dictate a slightly different approach. Select a cell to show the value of x (say, cell A1) and a cell to hold the value of y (say, cell A2); then enter the following formula in cell A2:

```
2*A1+1
```

As soon as you do so, cell A2 displays not the formula you typed in but the answer you want: namely, the value of y.

Since the relationship between cells A1 (or x) and A2 (or y) in the spreadsheet is a dynamic one, any change you make in the value of cell A1 will be automatically reflected in the value of cell A2. Not only does this save time and reduce errors; it also permits users to "experiment" quickly and easily with their data by testing to see what would happen if certain data were changed. (Indeed, this feature has become so widely used that it is now referred to as "what-if" modeling.) Most spreadsheets also provide facilities for graphing, for sorting data in the spreadsheet, and for programming these and other operations in what are called "macros." Once a working spreadsheet (usually called a worksheet or a template) has been created, its contents—the numbers, labels, and formulas that have been entered—can be saved to disk for reuse or modification at a later time. In this way, a worksheet or set of worksheets can be used over and over again to solve similar or related problems.

By and large, spreadsheets were invented for financial analysis. In fact, they were often described as electronic versions of the accountant's columnar pad. But spreadsheets have grown far beyond this now and can be regarded less as an analytical tool and more as a kind of programming environment. As such, spreadsheets have a number of important advantages compared to conventional programming languages, such as BASIC, COBOL, Fortran, Pascal, or C. They also have a few disadvantages. Together, these define the applications for which spreadsheets are most suited.

The advantages of spreadsheets are essentially these:

- Spreadsheets are easy to "program". Anyone can learn how: it does not take an expert programmer to create a working model. It is true that writing macros can be a bit tricky, but you can do a great deal in a spreadsheet without ever using a macro.

- Once programmed, worksheets are easy to modify. The program "code" is not hidden from view the way it is in conventional programming environments.

- For both these reasons, a spreadsheet model is less likely to seem like a "black box" to its users. This in turn means that users are more likely to understand how the model works and what exactly it does to the input data in order to produce its answer.

- Spreadsheets allow construction of models that invite what-if modeling. If input data and results are side by side, the user is invariably tempted to "see what happens if" he or she adjusts some of the input data or some of the model parameters.

- Spreadsheets are extraordinarily flexible and powerful. Many of them have extensive built-in functions for statistical analysis, financial modeling, calendar-related functions, data management and string manipulation, and trigonometric calculations. The latest versions of some spreadsheets provide commands for linear regression and matrix multiplication, for creating user-defined functions, and for manipulating three-dimensional data structures.

- Spreadsheets have a built-in capability for creating graphs which goes far beyond that of any of the conventional programming languages. Of course, the user can always write programs in those languages to draw graphs. But spreadsheets have such programs built in.

However, spreadsheets have at least one big disadvantage: they are slow and inelegant compared to many of the conventional programming languages.[3] This is due primarily to their cellular—rather than command or procedural—orientation. In conventional programming, the same series of commands can be used to compute many different results. In a spreadsheet, however, if you wanted ten different answers, you would normally have to write the formula into ten different cells. Similarly, looping is awkward in a spreadsheet. You can write macros to achieve this effect, but you cannot write a loop directly into a formula in a cell. Nevertheless, spreadsheets are not inherently inferior to other programming environments. In fact, some experts argue that spreadsheets are better than conventional languages in some respects. For example, iteration works more efficiently in spreadsheets than in most conventional languages, because spreadsheets begin the iteration of a cell from its current value, rather than from zero or unity as is the case with most programming languages (Johansson 1965).

Spreadsheet Simulations

There have been several books as well as numerous articles written to illustrate the use of spreadsheets for environmental modeling. Among these are:

- Sipe and Hopkins 1984, which provides a set of templates for use in local government;

- Brail 1987, which presents a number of models for urban planners, including two that served as the basis for the models presented here in chapters 7 and 9;

- Gould and Tobochnik 1988, which presents models from the physical sciences;

- Gordon 1989, which includes (amongst others) a model for surface-water runoff;

---

3. There are compilers (such as @Liberty or Baler) which produce runtime versions of worksheets created in Lotus® 1-2-3®. These runtime versions run directly under DOS and do not require the spreadsheet program itself.

●   Misner and Cooney 1991, which also applies spreadsheets to physical models; and

●   Klosterman, Brail, and Bossard 1993, which includes 22 models on a variety of subjects related to urban and regional analysis.

Spreadsheet programming has its own logic and style, just as programming in more conventional languages has (Knuth 1973; Nevison 1978). In general, the objective is to produce worksheets that are efficient and user friendly. An efficient worksheet is one that minimizes the use of memory and contains as few and as simple formulas as possible. A user-friendly worksheet is one that is easy to understand and use. In some cases, of course, there has to be a trade-off between efficiency and user friendliness; but often the two objectives can coexist quite happily.

The first principle is to keep the worksheet as compact as possible. Most spreadsheet programs store worksheets as rectangular blocks of cells, from cell A1 in the upper left corner to the "last cell" in the lower right corner, even if much of the intervening space is unused. So a key design criterion is to keep the "last cell" as close as possible to cell A1 and minimize the number of blank cells in between.

A second principle of good spreadsheet programming is to design worksheets vertically rather than horizontally. Other things being equal, "tall, thin" worksheets take less space and run faster than "short, fat" ones. Thus, most of the models in this book are designed to fit, more or less, within the width of a standard computer screen. Not only does this make them slightly faster than they might otherwise be; it also makes them easier to read. Even with the biggest models, all you have to do to move around in them is to use the page-up and page-down keys. Narrow worksheets are also easier to print, since you do not have to resort to landscape mode and/or a "sideways" printing utility.

The main drawback with narrow worksheets is that they put more constraints on model design than wide worksheets do. This is because, while columns in the worksheet can be set to any width, the width cannot vary from row to row within a single column. In other words, if you choose a certain width for a particular column in (say) the top part of a model, you are stuck with that width for the rest of the model as well. For example, you cannot format a column to be, say, 10 characters wide for the top 20 rows and then 15 characters wide in succeeding rows. In a wider worksheet, this is not so much of a constraint, since you can always keep moving into new columns, which you can set to whatever width seems appropriate. In narrow worksheets, however, you have to find a set of column widths that will be suitable for the entire worksheet. This requires more careful planning and design.

A third principle of good worksheet design is to place model assumptions, parameters, and results *all* at the top of the worksheet—and, whenever possible, all within the confines of a single screen. This is in contrast to the more linear structure of conventional programs. In such programs, the user is typically presented with a sequence of different screens in which, first, the model is introduced; then data are entered; next, parameters for the run are specified; and, finally, the screen is cleared and the results are displayed. By then, however, the user may well have forgotten what input data were entered or what parameter values were used. Moreover, if the user wants to adjust the input data or parameter values, it may be necessary to go back to the beginning and go through the whole process all over again.

Most spreadsheet programs can duplicate this traditional style of programming, if that is what you want. For example, it is not difficult to write a spreadsheet macro that takes the user from, say, an introductory screen (with a menu of choices, if you like) to one or more data-entry screens, and then to a results screen.[4] But to me, this seems neither necessary nor desirable. One of the biggest appeals of the spreadsheet is that you can design your models so that input and output are side by side. Then it is a simple matter to see how results change in response to changes in input data and/or parameter values. This is the essence of simulation modeling.

As a result, macros are used sparingly in this book. Their use is confined to two distinct purposes. In some of the models (e.g., chapters 3, 4, and 11), macros are used to run a model over and over again, sometimes recording the results, and then to present some kind of summary of the results. The other use of macros is illustrated in the last three chapters in the book, where macros are used to initialize some or all of the 400 cells that need to be "set" before the model is run. Nevertheless, one of my prejudices about spreadsheet models is that excessive use of macros is prima facie evidence of poor design!

---

4.   See, for example, the style exhibited in Klosterman et al. 1993.

My fourth and last design principle is that spreadsheet models, like any models, should always make clear what is being modeled and how that is being done. Thus, input data and parameters should be clearly indicated, units of measurement should always be specified, procedures (especially macros) should be explained and annotated, and data and model sources should always be given. In the end, a good model is one that provides not just an answer but also a clear indication of how the answer was obtained.

Models in This Book

Fourteen models are presented in the following chapters. In order to illustrate the versatility of spreadsheets, the models vary a good deal in terms of the issues they deal with, as well as their size and nature. As shown in the tables on the following page, the models are divided into three groups: there are five models of natural systems, six models of social systems, and three models of artificial systems. Then, on the facing page is a summary of the size, structure, and features of the various models.

As shown below, the models range in size from the KLEIN model (which is only about 10 Kb and consists of only 387 cells) to the EIA model (which is about 170 Kb in size and consists of 4,278 cells). The computationally most intensive models (i.e., the ones with the largest number of calculations) are the forest management model (TREES) and the environmental impact assessment model (EIA). Each of these models performs more than 10,000 discrete arithmetic or logical operations each time it is run. In fact, if either of these models is iterated the number times suggested, the total number of calculations performed—approximately 250,000—is staggering. In the space of a few minutes, the computer does what we would need literally months to do by hand or even with a calculator.

According to our index of complexity, the most complex model is WATER, the underground-transport model, because it contains more calculations for its size (i.e., more operations per kilobyte) than any other model. The three artificial-system models are also noteworthy in this respect, in that they contain relatively few formulas (less than a dozen in each case), but some of them are quite long and complicated. Thus, for example, GAMACH contains only 11 different formulas but ranks as the third most complex of the 14 models.

Models of Natural Systems

SMOKE	A model of the spatial dispersion of a plume of smoke from a chimney stack, subject to atmospheric conditions, ambient temperature and wind, and temperature and velocity of the plume.
WATER	A model of the underground transport of pollution in groundwater arising from a momentary spill or continuous leaching.
BEARS	A cohort-survival model for small populations of grizzly bears, intended to identify minimum viable population levels necessary to ensure long-term survival.
TREES	A cohort-survival model of specific tree species, subject to varying rates of growth, management practices, and forest fires.
SOLAR	A model of the production of energy from a flat-plate solar collector.

Models of Social Systems

KLEIN	A simple version of the Klein Model, one of the first large-scale macroeconomic models of a national economy.
LOWRY	A version of the Lowry Model, which simulates interactions between economic activity and the distribution of population in an urban area.
BERTAUD	A model of housing costs and affordability, originally designed for use in developing countries.
TRAFFIC	A version of a standard traffic-planning model that simulates trip generation and trip distribution in urban areas.
WASTE	A model for the management of a waste disposal site under varying conditions of load and relief.
EIA	A model for making an environmental impact assessment of various development alternatives.

Models of Artificial Systems

LIFE	A spreadsheet version of John Conway's well-known "Game of Life".
CHOICE	A probabilistic version of "Life" to simulate the effects of intolerance.
GAMACH	A "Game Machine" which can simulate billions of different games.

Table I.1: Summary of Model Structure and Features

Model		Approx. Size	Layout Cols.	Layout Rows	Cells	Formulas No.	Formulas Size	Total Ops.	I.C.	Basis of Sim.	Uses RAND
1	SMOKE	29	16	65	1,040	21	110	3,200	110	Manual	No
2	WATER	65	11	121	1,331	23	149	9,050	139	Auto	No
3	BEARS	42	18	199	3,582	11	92	280	7	Macro	Yes
4	TREES	122	15	245	3,675	17	168	12,300	101	Macro	Yes
5	SOLAR	78	17	110	1,870	26	185	10,531	135	Manual	No
6	KLEIN	10	9	43	387	12	71	240	24	Auto	No
7	LOWRY	11	11	48	528	12	59	290	26	Auto	No
8	BERTAUD	16	10	77	770	50	81	250	16	Manual	No
9	TRAFFIC	46	14	172	2,408	63	98	2,650	58	Auto	No
10	WASTE	19	18	83	1,494	20	21	560	29	Manual	Yes
11	EIA	172	23	186	4,278	19	146	13,600	79	Macro	Yes
12	LIFE	46	26	86	2,236	5	109	3,200	70	Macro	Yes*
13	CHOICE	79	26	88	2,288	6	153	6,800	86	Macro	Yes
14	GAMACH	55	33	123	4,059	11	89	8,400	153	Macro	Yes*
All Models											
Minimum		10	9	43	387	5	21	240	7		
Maximum		172	33	245	4,278	63	185	13,600	153		
Mean		56	18	118	2,139	21	109	5,097	74		

Notes:

The approximate size (in thousands of bytes) refers to SuperCalc4 versions of the models, as they are saved to disk. Sizes may differ for other spreadsheet versions. The number of cells in each model is calculated as the product of the number of columns and the number of rows.

Models are characterized by the number of *different* formulas they use and the number of characters contained in the largest single formula.

The total number of operations in each model is the number of discrete arithmetic or logical operations plus the number of functions contained in all of the cells.

The Index of Complexity (I.C.) for each model is calculated as the total number of operations in the model divided by its size in kilobytes. The BEARS model has a very low index in part because more than 60% of its size is taken up by space for storing the results of runs.

The Basis of Simulation of each model is determined according to whether it relies on *manual* recalculation, built-in iteration (*automatic*), or a *macro*. The last column indicates whether a model uses the random number function (RAND); an asterisk means that the function is used in a macro but not in the model itself.

The biggest formula used in any of the models is found in SOLAR, the solar-energy model.[5] In part, this is due to the fact that SOLAR makes extensive use of trigonometric functions, which are by their nature more "verbose" than pure arithmetic operations. In any case, half the models contain formulas that are more than 100 characters long. The trip-distribution/trip-generation model (TRAFFIC) contains the largest number of different formulas (63) and LIFE the smallest number (5).

The last two columns of the Table I.1 show the basis for the simulation in each model and whether the model requires random numbers. From this, it can be seen that six of the models use macros to perform their simulation functions, four rely on the built-in iteration capability of the spreadsheet, and four require manual recalculation to test different scenarios. Random numbers are used in seven of the fourteen models. In two of these cases, however, the random numbers are used only in an initialization macro and not in the model itself.[6]

Using the Models

The 14 models are described, one at a time, in the next 14 chapters. Each chapter begins with a discussion of the purpose of the model and its conceptual basis. Next, there is a review of the data required to run the model and the results that can be expected from it. Then, the detailed working of the model is examined. Finally, there is an assessment of how the model might be used in practice, how reliable its results might be, and how it might be adapted or extended to other uses and other contexts. The remainder of the book consists of a concluding chapter and two appendixes. One appendix consists of a technical discussion of speadsheets and spreadsheet programming, while the other contains a complete set of "recipes" for building the models discussed in the book. The recipes are easy to follow and it should not take long to create even the more complex models, thanks in part to the spreadsheet's ability to "replicate" itself from one cell or cell-range to another. Besides there is no better way to learn how a model works than to build it yourself.

---

5.   Most spreadsheets have an upper limit on how many characters can be entered into a single cell (e.g., 241 characters in SuperCalc4). But it is usually possible to work around this limit by providing for interim calculations in a scratch-pad area of the worksheet.

6.   Spreadsheet randomization functions (which in fact generate only pseudo-random numbers) are discussed further in appendix A, below.

Bibliography and References

Arthur, J.L., J.O. Frendewey, P. Ghandfaroush, and L.P. Rees. 1986. "Microcomputer Simulation Systems." *Computers and Operations Research*, 13.2/3, pp. 167-83.

Biswas, Asit K., T.N. Khoshoo, and Ashok Khosla, eds. 1990. *Environmental Modelling for Developing Countries*. Natural Resources and the Environment Series, vol. 25. London: Tycooly.

Brail, Richard K. 1987. *Microcomputers in Urban Planning and Management*. New Brunswick, N.J.: Rutgers University Center for Urban Policy and Research.

Cartwright, T.J. 1993. "Geographic Information Technology as Appropriate Technology for Development". In *Diffusion and Use of Geographic Information Technologies*, edited by Harlan Onsrud and Ian Masser. Amsterdam: Kluwer.

———. 1991. "Planning and Chaos Theory." *Journal of the American Planning Association*, 57.1 (Winter), pp. 44-56.

———. 1989. "Urban Management as a Process, Not a Result". *Review of Urban and Regional Studies*, 2 (July), pp. 107-13.

———, M.R. Brown, and H.V. Seaforth. 1988. "Urban Data Management Software: The Growth and Development of Microcomputer Software for Planning." *Habitat International*, 12.4, pp. 171-93.

Forrester, Jay W. 1973. *World Dynamics*. Cambridge, Mass.: MIT Press.

———. 1969. *Urban Dynamics*. Cambridge, Mass.: MIT Press.

———. 1968. *Principles of Systems*. Cambridge, Mass.: MIT Press.

———. 1961. *Industrial Dynamics*. Cambridge, Mass.: MIT Press.

Gordon, Steven I. 1989. *Microcomputer Applications in City Planning and Management*. New York: Praeger.

Gould, Harvey, and Jan Tobochnik. 1988. *An Introduction to Computer Simulation Methods: Applications to Physical Systems*. 2 vols. Reading, Mass.: Addison-Wesley.

Kim, Tschango John. 1989. *Integrated Urban Systems Modeling: Theory and Applications*. Dordrecht: Martinus Nijhoff.

Klosterman, Richard, Richard Brail, and Earl Bossard, eds. 1993. *Spreadsheet Models for Urban and Regional Analysis*. New Brunswick, N.J.: Center for Urban Policy Research.

Knuth, Donald. 1973. *The Art of Computer Programming*. Reading, Mass.: Addison-Wesley.

Ludwig, J.A., and J.F. Reynolds. 1988. *Statistical Ecology: a Primer on Methods and Computing*. Includes a disk. New York: Wiley.

Meadows, Dennis L., W.W. Behrens III, D.H. Meadows, R.F. Naill, Jorgen Randers, and E.K.O. Zahn. 1974. *Dynamics of Growth in a Finite World*. Cambridge, Mass.: MIT Press.

Meadows, Dennis L., and Donella H. Meadows. 1973. *Towards Global Equilibrium: Collected Papers*. Cambridge, Mass.: MIT Press.

Meadows, Donella H., D.L. Meadows, Jorgen Randers, and W.W. Behrens III. 1973. *The Limits to Growth*. New York: Universe Books.

Misner, C.W., and P.J. Cooney. 1991. *Spreadsheet Physics*. Readin, Mass.: Addison-Wesley.

Nevison, J.M. 1978. *The Little Book of BASIC Style*. Reading, Mass.: Addison-Wesley.

Roberts, Nancy, D.F. Andersen, R.M. Deal, M.S. Garet, and W.A. Shaffer. 1983. *Introduction to Computer Simulation: The System Dynamics Approach*. Reading, Mass.: Addison-Wesley.

Sipe, Neil, and R.W. Hopkins. 1984. *Microcomputers and Economic Analysis: Spreadsheet Templates for Local Government*. Bureau of Economic and Business Research (BEBR) Monograph No. 2. Miami: University of Florida.

Swartzman, Gordon L., and S.P. Kaluzny. 1987. *Ecological Simulation Primer*. New York: Macmillan.

Part One

MODELING
NATURAL SYSTEMS

Chapter 1

BLOWING SMOKE: ATMOSPHERIC DISPERSION OF AIR POLLUTION

The owner of every alkali work shall use the
best practicable means, within a reasonable cost,
of preventing the discharge into the atmosphere
of all noxious gases arising from such work,
or of rendering such gases harmless when discharged.

Alkali Act (1863) Amendment Bill, Great Britain[1]

One of the more dubious achievements of the twentieth century has been to make air pollution an almost universal problem. There was a time when belching smokestacks were regarded as symbols of progress and development (Brimblecombe 1987). Not so now, as the effects of air pollution become daily more apparent through acid rain, ozone depletion, and even climate change (Shen 1986). In this century, air pollution has become a problem of truly global proportions (Berlyand 1973; Sharma et al. 1990).

For environmental planners, one of the main difficulties in dealing with air pollution is that its effects are often invisible beyond the immediate area of the source. Furthermore, the effects are almost always widely dispersed in space and can vary a lot from day to day, depending on the ambient wind and weather. This means that even the worst cases of air pollution can take a relatively long time to affect a lot of people to a significant degree. Until they do, they are unlikely to get on the political agenda (Cook 1988; Crandall 1983; Haskell 1982; Ashby and

---

1. The words "within a reasonable cost" were eventually deleted, and the bill became law (Alkali Act, 1874, 37 & 38 Vict., c. 43) without them. See also Cohen and Ruston 1912.

Anderson 1981; Jones 1975; and Crenson 1971). For this reason, planners can find it difficult to know what to do about air pollution, short of closing down the source or re-designing our cities—neither of which may be realistic or even desirable (Rydell and Stevens 1968).

To estimate the impact of airborne pollutants, there are mathematical models that predict the dispersion of a plume of smoke downwind from a point source under different conditions of emission, wind, and weather (OECD 1971; Weber 1982; Grandell 1985; Sharma et al. 1990). Such models can help planners to see both where pollution is most likely to occur and how its dispersion might best be managed through various mechanisms of control (Kneese 1984). Up to now, most of these models have been written in conventional high-level programming languages like Fortran and Pascal. This is certainly a great deal more convenient than relying on pencil and paper, or even a calculator. This chapter presents a spreadsheet version of a simple Gaussian dispersion model, which takes advantage of the spreadsheet's facility for "what-if" modeling and its built-in graphing capabilities.

Purpose of the Model

The purpose of the model presented here is to simulate the dispersion of a plume of smoke downwind from an elevated point source—such as a chimney stack—under various conditions of emission, wind, and weather (Jakeman and Simpson 1985). The model provides estimates of the concentration of pollution at ground level at selected distances downwind of the source along the centerline of the plume.

Naturally, models like these are designed for somewhat idealized conditions of airflow and pollution. In particular, the model discussed here assumes moderately dense pollutants (e.g., particulates), positive and consistent wind conditions, no overwhelming local disturbances of airflow due to nearby buildings or other obstructions, no excessive channeling of airflow due to hilly terrain, and no significant chemical or physical interaction between the pollutants and the ground (Pasquill 1962, p. 179; Davison and Leavitt 1981, p. 10).

Conceptual Basis

The dispersion of air pollution from a point source normally occurs in three dimensions:

- longitudinally along a centerline of maximum concentration running downwind from the source;

- laterally on either side of that centerline, as the pollution spreads out sideways (so to speak); and

- vertically above and below a horizontal axis drawn through the source (and, thus, down towards and up away from the surface of the earth).

The most commonly used model for predicting this kind of dispersion is the Gaussian plume model. This model was originally proposed more than fifty years ago by O.G. Sutton (Sutton 1932) and then refined by others in the 1960s (e.g., Pasquill 1961 and 1975; Gifford 1961 and 1968). According to the Gaussian plume model (see Zannetti 1990, Fig. 7-1, for a good diagram), emissions from a point source are carried downwind from the source in a continuous plume; at any point in this plume, the vertical and lateral dispersion of the emission (i.e., its cross-section) will correspond to a normal, or Gaussian, distribution in each dimension.

The fundamental equation of the Gaussian plume model can be written thus:[2]

$$C(x,y,z) = \frac{Q}{2 * \pi * u * \sigma y * \sigma z} * \exp(\frac{-y^2}{2 * \sigma y^2}) * (\exp(\frac{-(z-h)^2}{2 * \sigma z^2}) + \exp(\frac{-(z+h)^2}{2 * \sigma z^2}))$$

where $C(x,y,z)$ is the concentration of emission (in micrograms per cubic meter) at any point x meters downwind of the source, y meters laterally from the centerline of the plume, and z meters above ground level;

              Q is the quantity or mass of the emission (in grams) per unit of time (seconds);

              u is the wind speed (in meters per second);

              h is the height (in meters) of the source above ground level; and

---

2.   For further technical discussion of this model, including a note on exponents, see the Technical Note beginning on page 37 at the end of the chapter.

σy and σz are the standard deviations of a statistically normal plume in the lateral and vertical dimensions respectively.

This form of the model assumes that the plume is reflected back into the atmosphere whenever it strikes the ground. If a temperature inversion creates a similar barrier in the atmosphere, the model can be modified accordingly (see Dobbins 1979; or Clark et al. 1984).

The Gaussian plume model has been tested in numerous studies (Weil and Jepsen 1977; Davison and Leavitt 1981; Harrison and McCartney 1980) and has been found to give satisfactory results under normal conditions (Dop and Steyn 1991; Johnson et al. 1976; Hanna et al. 1981; Clark et al. 1984). Although the model was originally intended to apply to fairly short periods (up to several hours, during which time wind and atmospheric conditions could reasonably be assumed to remain constant), it has nevertheless been adapted to seasonal and even annual time frames (Clark et al. 1984, p. 209). Finally, studies (cited in ibid., pp. 210-11) have shown that the Gaussian plume model can be adapted to both multiple-point emission sources (e.g., automobile pollution along a highway) and area sources (e.g., air pollution in urban areas). In short, the Gaussian plume model provides a simple, robust, and flexible approach to modeling the dispersion of air pollution under a variety of conditions (Johnson et al. 1976, p. 515).

One critical component of the Gaussian plume model is the pair of standard deviations (σy and σz). The reason for using these values is that they can be specified "monotonically increasing functions of distance downwind from the source for a series of different atmospheric stability conditions corresponding to different intensities of turbulent diffusion" (Clark et al. 1984, p. 207). Various approaches to determining the value of the standard deviations have been tried. In some cases, values are derived from theoretical calculations of the effect of wind, insolation, nighttime cloud cover, the nature of the terrain, and other factors. In other cases, empirical observations at the site are used to calibrate estimated standard deviations.[3]

A second critical factor in applying the Gaussian plume model to air pollution is that emission from a smokestack is rarely just released into the atmosphere.

---

3.   See Pasquil 1962, p. 209 (Figure 5.7); or Clark et al. 1984, p. 208. A study carried out in Mol, Belgium (Kretzchmar and Mertens 1984), tested 12 different approaches.

Instead, smoke is more likely to "shoot" out of the stack at an (often fairly high) velocity and/or temperature. Because of this momentum and buoyancy, the *effective* height at which emissions are released into the atmosphere can be a good deal higher (e.g., as much as ten times higher) than the *physical* height of the stack above ground. Clearly, the model must make provision for this kind of effect.

Finally, it should be noted that the general form of the Gaussian model is often simplified (as it is here) to a form that yields concentrations of pollution at ground level. This is achieved by setting the parameter z to zero. If all you are interested in is *maximum* ground-level pollution (which is assumed to occur along the centerline of the plume), the Gaussian model can be further simplified by setting the parameter y to zero as well.

Data Requirements

There are seven input parameters required to use the model. Five of these pertain to the source of the pollution itself and should be more or less self-evident. These five parameters are

- the height of the stack above the ground (in meters),
- the diameter of the opening of the stack (in meters),
- the velocity of the gas emitted from the stack (in meters per second),
- the temperature of the gas as it exits the stack (in degrees Celsius), and
- the rate at which pollution is emitted from the stack (in grams per second).

The other two parameters are the ambient air temperature at the source (in degrees Celsius), which is also quite straightforward, and an index of atmospheric stability, which may need a bit more explanation.

Atmospheric stability is defined in terms of the environmental lapse rate: that is, the rate at which the temperature of an air mass varies with altitude (see, e.g., Sutton 1960; Johnston 1961). In neutral or *adiabatic* conditions (i.e., conditions in which an air mass neither absorbs heat from, nor gives up heat to, its environment), the temperature of an air mass cools by about 2° Celsius per 1,000 feet of altitude.[4]

---

4.   This is the adiabatic lapse rate for "standard" air. In moist air, the rate is lower (about 1.5°C per 1,000 feet); in dry air, it is higher (about 3°C per 1,000 feet).

If the temperature of an air mass cools at a higher rate than this, it is said to be *superadiabatic*; if it cools at a lower rate, it is said to be *subadiabatic*. According to one textbook,

> A rising air parcel, cooling at the adiabatic rate, becomes warmer and less dense than its environment and therefore buoyancy tends to accelerate it upward. . . . In such an environment, the parcel is in unstable equilibrium. Vertical motions upward or downward are reinforced. When the environmental lapse rate is less than adiabatic ("subadiabatic") or negative, a rising air parcel becomes cooler and more dense than its environment and tends to return to its starting point. The parcel is in stable equilibrium [Wanta and Lowry 1976, pp. 339-40].

For this reason, the model requires the user to define atmospheric stability in terms of one of the following six categories:

- very unstable (superadiabatic),
- moderately unstable,
- slightly unstable,
- neutral (adiabatic),
- somewhat stable, and
- stable (subadiabatic).

In effect, what these categories reflect is the tendency of a plume to disperse downwards towards the surface of the earth (Pasquill 1962, pp. 181-82). In unstable conditions, plumes tend to cycle down to the surface and then bounce back up into the atmosphere; this is often described as "looping". In neutral conditions, plumes tend to spread out equally in the vertical and horizontal dimensions; this is called "coning". Finally, in stable conditions, plumes tend to spread out more in the horizontal than in the vertical dimension; this is called "fanning".

Using this information, the model presented here provides an estimate of the hourly accumulation of pollution at ground level at various points along the centerline of the plume, subject to ten different windspeeds in any of the six different atmospheric conditions discussed above. The model calculates dispersion for every odd wind speed from 1 to 19 meters per second (or from 3.6 to 68.4 kilometers per hour). Estimates of accumulation are provided for various distances from the source, ranging from 500 meters to 100 kilometers.[5]

Overview

The entire model consists of 16 columns and 65 rows (see Listing 1.1). The top 25 rows are for entering the parameters of the model and displaying its results, while the rest contains labels for the graphic display and otherwise serves as a "scratch pad" for some of the longer and more complicated formulas in the model.

The model makes extensive use of three important features of spreadsheets: branching, lookup tables, and graphs. The first is used for choosing among various formulas (according to the stability of the atmosphere) and for abandoning calculations when results become so small as to be insignificant. The second (lookup tables) is used for accessing a matrix (located in the lower half of the model) of Gaussian coefficients of dispersion at selected distances from the source under each of the six states of atmospheric stability. Graphing capabilities are used to provide a graph of the ten dispersion curves (according to wind speed) for any given set of emission and atmospheric conditions. However, the model uses no menus or macros. Instead, data entry and display of the results are on the same screen (although some up-and-down scrolling is necessary to see the results for the highest wind speeds). In this way, the user can more easily relate the output of the model to its input parameters.

The model has three main parts. First, it calculates what is called "effective stack height". This is the height above ground (subject to ambient temperature and wind speed) at which the gas (subject in turn to its own temperature and velocity) comes out of the stack and begins to "bend over" and disperse downwind.[6] Hot gases emitted at high exit velocities in neutral or unstable atmospheric conditions will naturally rise some considerable distance above their stacks before beginning to disperse downwind. Under these conditions, the effective stack height will be much higher than the height of the physical stack. In stable conditions, on the other hand,

---

5.   These wind speeds and distances downwind can all be set to whatever values a user may desire, just by entering new values in the appropriate cells.

6.   To make these calculations, temperatures are first converted to the Kelvin scale, where degrees are equivalent to Celsius degrees above absolute zero (which is -273°C).

Listing 1.1

```
   | A||B|| C|| D|| E || F || G || H ||  I || J || K || L || M || N || O || P |
1  DIFFUSION OF POINT-SOURCE POLLUTION                    © 1993 by T.J. Cartwright
2
3  Stack height (m)          30        Gas exit velocity (m/s)         5
4  Stack diameter (m)         2        Gas exit temperature (°C)     200
5  Emission rate (g/s)       10        Ambient temperature (°C)       20
6
7  Atmospheric                         1 = Very Unstable      4 = Neutral
8     Condition               4        2 = Moderately Unstable 5 = Somewhat Stable
9     Category:                        3 = Slightly Unstable  6 = Stable
10
11 Wind    Stack   Estimated Concentration of Ground-Level Pollution (mmg/m3)
12 Veloc   Effect  on Plume Centerline at Selected Distances (km) from Source
13 (m/s)   Ht (m)     0   .5   .8  1.5   3    5   10   20   35   60  100
14
15    1      222      0    0    0    0    3    9   13    9    6    4    2
16    3       94      0    0    8   32   31   21   10    5    3    2    1
17    5       68      0    8   34   45   26   15    7    3    2    1    1
18    7       57      0   21   47   42   21   12    5    2    1    1    0
19    9       51      0   31   50   37   18    9    4    2    1    1    0
20   11       47      0   37   50   33   15    8    3    1    1    0    0
21   13       45      0   39   47   29   13    7    3    1    1    0    0
22   15       43      0   40   44   26   11    6    2    1    1    0    0
23   17       41      0   40   42   24   10    5    2    1    0    0    0
24   19       40      0   40   39   22    9    5    2    1    0    0    0
25
26
27 Graph Labels
28
29 Head       DIFFUSION OF POINT-SOURCE POLLUTION USING A GAUSSIAN DISPERSION
30 Subhead    Centerline Pollution at Ground Level for Selected Wind Condition
31 X-axis     Distance from Source (km)
32 Y-axis     Concentration (mmg/m3)
33
34 Calculation Scratchpad
35
36       >>     gas exit temperature in degrees Kelvin
37       >>     ambient temperature in degrees Kelvin
38       19    }
39       87    }    interim values for calculating effective stack height
40       >>    }
41        0    }
42
43 Lateral and Vertical Dispersion Coefficients for Each Stability Category
44
45 Lat-   1       0  107  169  308  579  898 1556 2540 3630 4989 6633
46 eral   2       0   78  123  224  421  653 1131 1848 2640 3628 4824
47        3       0   54   85  154  289  449  778 1270 1815 2495 3317
48        4       0   39   62  112  210  327  566  924 1320 1814 2412
49        5       0   29   46   84  158  245  424  693  990 1361 1809
50        6       0   20   31   56  105  163  283  462  660  907 1206
51
52 Current lookup  0   39   62  112  210  327  566  924 1320 1814 2412
53
54 Ver-   1       0  100  160  300  600 1000 2000 4000 7000 >>>> >>>>
55 ti-    2       0   60   96  180  360  600 1200 2400 4200 7200 >>>>
56 cal    3       0   38   59  105  190  283  462  716  990 1331 1746
57        4       0   23   32   50   77  103  150  216  287  377  488
58        5       0   13   19   31   47   60   75   86   91   95   97
59        6       0    7   10   17   25   32   40   46   49   51   52
60
61 Current lookup  0   23   32   50   77  103  150  216  287  377  488
62
63
64 Source:  Adapted from a program (GAUSS) in Pascal written by an anonymous
65    programmer.
```

effective stack height may be little greater than physical height. In any case, the results are displayed at the left side of the top part of the model (column A).

The second step in the operation of the model is calculation of the table of Gaussian dispersion coefficients for lateral and vertical dispersion at the predetermined distances downwind of the source. As noted, these results are stored in the lower portion of the spreadsheet. The third step is calculation of the concentration of pollution at ground level at each point along the plume centerline under each of the ten different wind conditions; the results are displayed in the top part of the model.

When all this is done, the model can provide a graph of the distribution of ground-level pollution along the centerline of the plume for various wind speeds, subject to the various conditions of atmospheric stability provided in the model.

Operation

Row 1 contains the title of the spreadsheet. Rows 3, 4, and 5 provide for entering the model's parameters. Variable names go into columns A and I, while the data go into columns F and N. Rows 7, 8 and 9 describe the basis for categorizing atmospheric stability. The category selected is entered at cell E8. Rows 11, 12, and 13 contain headers for the table of results. The various distances shown in row 13 are usually treated as constants; however, they can be varied by the user, if desired, in order to explore accumulations over a particular range of distances from the source. Similarly, column A in rows 15-24 gives a list of ten wind speeds, but these can be varied by the user, if desired, to cover a wider or narrower range.

The rest of rows 15-24 shows the results of two key calculations: i.e., effective stack height above ground level in column C, and centerline pollution concentrations in columns E to O (at selected distances from the source for each of the wind velocities shown in column A). Details of all the calculations described here are given in a technical note at the end of this chapter.

Rows 27-61 constitute the "hidden" part of the model. Rows 29-32 provide the labels for the graphic. Rows 36 and 37 convert the input temperatures from degrees Celsius to degrees Kelvin. Rows 38-41 contain some interim results in the calculation of effective stack height. Finally, rows 45-50 and 54-59 contain (respectively) the lateral and vertical dispersion coefficients referred to above. Note that the columns may be too narrow to display some of the results, with the result that the overflow symbol appears in some cells. Since these are only interim calculations, the display does not matter; the results are still available for use elsewhere in the model.

The model is designed to produce simple graphs, like those shown in Figure 1.1, which are based on the data in Listing 1.1. Here the concentration of pollution (on the y-axis) is graphed over the distance from the source (on the x-axis) as a set of curves, with each curve representing the effect of a different wind speed. Appropriate headings and labels are provided for the graph. Initial values are provided for wind speeds (in column A) and distances downwind from the source (in row 13). However, the user can alter these to other values, if desired, to obtain results for different distances.[7]

Interpretation and Use

Of course, the model shown here provides only an approximation to real life. As one expert admitted, after a detailed, technical discussion of pollution modeling:

> It is unlikely that the assessment of the distribution of an air pollutant over a region extending up to, say, 100 km from the point of release can always be confined to the substitution of a single set of parameters in one simple equation. This is so because the growth of a cloud or plume in the vertical will not necessarily follow a simple (e.g., power-law) variation with distance indefinitely. Even when a simple formula is applicable, estimates of the parameters to be used in it will vary according to the quality of the meteorological data available. . . . Moreover, there are aspects of diffusion which have so far been examined only in a very preliminary way, and generalization from the data available is then essentially speculative and should be reappraised frequently [Pasquill 1962, p. 297].

---

7.   Note that the x-axes in Figure 1.1 are not scaled consistently, as nonlinear scaling (e.g., a logarithmic scale) is not yet widely available in spreadsheets. Thus, a bar graph is really more appropriate than a line graph; however, the line graphs shown here are easier to read. Note, too, that the y-axes have been "manually" scaled to a consistent range, in order to facilitate comparisons between related graphs.

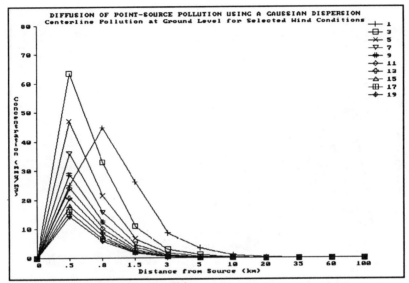

Fig. 1.1a. Very Unstable Conditions

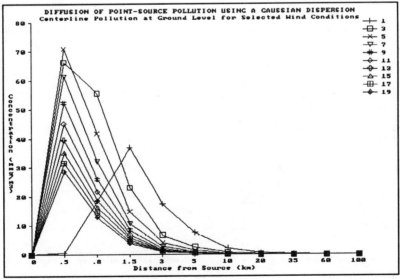

Fig. 1.1b. Moderately Unstable Conditions

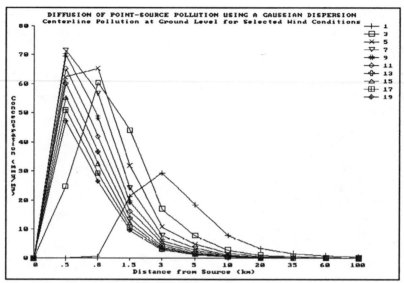

Fig. 1.1c. Slightly Unstable Conditions

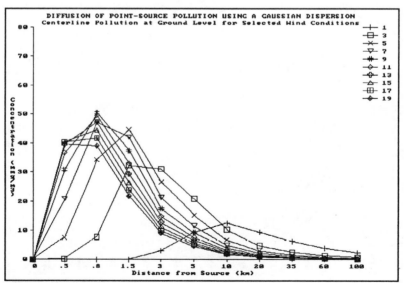

Fig. 1.1d. Neutral Conditions

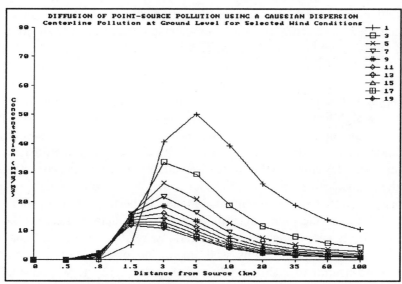

Fig. 1.1e. Somewhat Stable Conditions

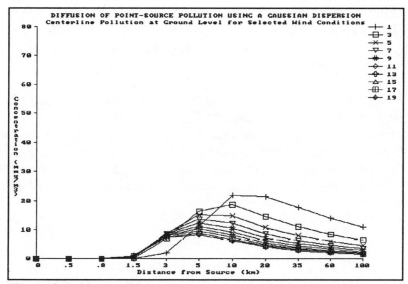

Fig. 1.1f. Stable Conditions

However, the use of even approximate simulation models like these can do a lot to help planners and even ordinary citizens to understand the nature of air pollution and to explore possible solutions and tradeoffs.[8]

To illustrate how a model like this can help, consider Figure 1.1 again. From this example, it is clear that atmospheric stability affects the "footprint" of the pollution in a significant way. As noted previously, unstable conditions promote "looping" by the plume. This has two principal effects: (a) higher centerline concentrations of pollution at the expense of lateral and vertical diffusion, and (b) higher concentrations closer to the source. As atmospheric turbulence becomes more neutral, the plume "cones" out from the source in a more stable fashion, resulting in a smoother distribution in the longitudinal plane. In perfectly stable conditions, the cone flattens out and the plume spreads more laterally than vertically ("fanning"). This results in lower centerline concentrations relative to lateral and vertical diffusion, and peak concentrations well downwind of the source. Indeed, the transition from neutral to stable conditions (Figures 1.1d, 1.1e, and 1.1f) has a particularly dramatic effect on concentrations close to the stack.

The effect of wind is also interesting. In general, the greater the wind speed, the more the pollution will be dispersed away from the stack. However, in unstable and especially neutral conditions (see Figure 1.1), moderate winds actually cause *higher* pollution concentrations close to the source than do light winds. Presumably, this is because moderate winds promote more rapid "coning" than occues in either very light or very strong winds.

Similarly, the effect of changes in ambient temperature is to raise or lower the profiles slightly but not to alter them. In Figure 1.2, for example, we can see the effect in neutral atmospheric conditions of first lowering (Figure 1.2a) and then raising (Figure 1.2b) the ambient temperature relative to what it was in Figure 1.1d. Essentially, pollution concentrations tend to be generally higher in hot weather than in cold weather. But there is no fundamental change in the overall pattern of dispersion.

---

8.   The value of rough approximations for helping to "frame" planning problems is one implication of recent research on "chaos" in planning: see Lee 1973 and Cartwright 1991.

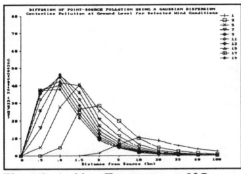

Fig. 1.2. Ambient Temerature = 0°C

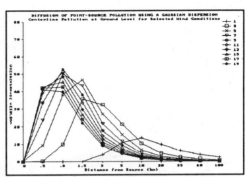

Fig. 1.2b. Ambient Temperature = 30°C

So what can be done to manage the footprint of pollution from a point source such as a smoke stack? The model presented here allows planners to simulate the effect of two basic approaches: managing the conditions in which the plume exits the stack and changing the design of the stack itself. This is not the place to make a comprehensive investigation of these options. But for purposes of illustration, let us once again take the stack described in Listing 1.1 and examine the effect of various incremental changes, under the neutral atmospheric conditions of Figure 1.1d.

To start with, suppose we try varying the temperature and exit velocity of the gas as it leaves the stack. Such effects can be achieved through a variety of means: exhaust gases can be cooled and slowed by means of settling chambers, baffles, or scrubbers (which also help remove larger particulates); similarly, exhaust gases can be heated and accelerated by the addition of heated air (Stern 1976, p. 253).

The results are illustrated in Figures 1.3 and 1.4 on the next page. Figure 1.3 shows the effect of first lowering (Fig. 1.3a) and then raising (Fig. 1.3b) the temperature of the smoke as it leaves the stack. Similarly, Figure 1.4 shows the effect of first decreasing (Fig. 1.4a) and then increasing (Fig. 1.4b) its exit velocity.

In general, it appears that increasing the temperature and/or the exit velocity of the smoke can significantly reduce concentrations of pollution near the stack. For that reason, these strategies often appear attractive both to those who create the pollution as well as to those who suffer from its effects. However, these practices have a much less discernible effect on pollution further away (Harrison and Perry 1986).

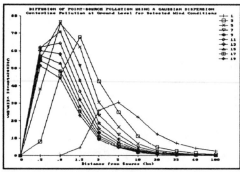

Fig. 1.3a. Exit Temperature = 100°C

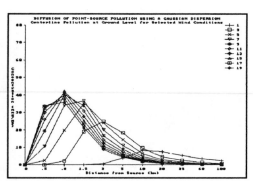

Fig. 1.3b. Exit Temperature = 300°C

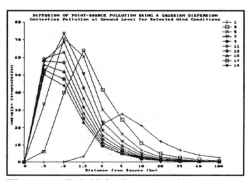

Fig. 1.4a. Exit Velocity = 3 m/sec

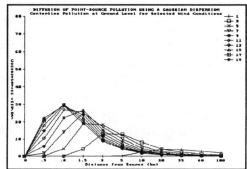

Fig. 1.4b. Exit Velocity = 10 m/s

Now suppose we consider the more drastic step of altering the height or diameter of the stack itself.[9] Again, it is a simple matter to change the input parameters in the model and view the effect. Figure 1.5 shows the effect of first lowering (Fig. 1.5a) and then raising (Fig. 1.5b) the height of the stack, while Figure 1.6 shows the effect of first reducing (Fig. 1.6a) and then expanding (Fig. 1.6b) the diameter of the stack. In these examples, all the other parameters are as shown in Listing 1.1, and atmospheric conditions are still presumed neutral, as in Figure 1.1d.

---

9.   The same effect can be achieved by multiplying the number of stacks instead of increasing the capacity of a single stack. In general, multiple stacks close together tend to create a plume higher than any single one of them would but lower than the plume created by a single combined stack. The effect of a line of multiple stacks is also related to its orientation (parallel or perpendicular) to the wind direction (Strom 1976, p. 250).

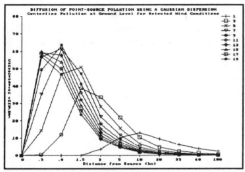

Fig. 1.5. Stack Height = 25 m

Fig. 1.5b. Stack Height = 40 m

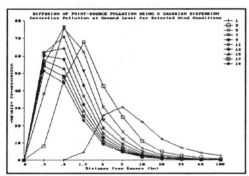

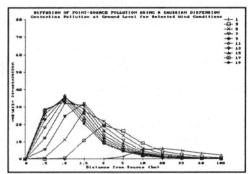

Fig. 1.6a. Stack Diameter = 1.5 m

Fig. 1.6b. Stack Diameter = 2.5 m

In general, it appears from these examples that there are things we can do to reduce the impact of a plume *close to its source*. For example, we can build the stacks taller and wider.[10] However, it is rather more difficult to do much to attenuate the effects of air pollution further away from the source (Frankenburg 1968; and Lee and Stern 1973).

In summary, what our model suggests is that there is considerable scope for managing the impact of air pollution locally, but not at long distance. Given adequate tools for monitoring current meteorological conditions and for controlling

---

10. Note that, for these runs, the exit temperature and velocity have been maintained at their original values. In practice, you would probably need to take additional steps, if you wanted to maintain the same temperature and velocity with a taller or wider stack.

the output of emissions at the stack, we should be able to manage the local impact of air pollution quite effectively.   On a regional scale, however, there is little evidence that management of individual plumes can substantially affect the dispersion of air pollution (Haaugen 1975).   Moreover, there is also the risk that the ability to manage and thus hide the more visible (local) impacts of air pollution may create a false sense of security about the broader impacts of air pollution—such as acid rain, ozone depletion, and even climate change.

Evaluation and Extension

Air pollution is both more and less complicated than we tend to think. On the one hand, turbulent flows over smooth surfaces—let alone over the features typical of urban and even regional landscapes—are so complex that simulating them with any degree of accuracy over more than a few hundred meters of distance may in reality be impossible. On the other hand, as this chapter shows, we can still do some surprisingly useful things with only a microcomputer and a standard off-the-shelf software package like a spreadsheet.

Thus, the real question is not so much whether we can use a model like the one presented here to get an accurate prediction of pollution concentrations downwind of a real stack—because the answer is probably that we cannot (Dobbins 1979). The important question is whether—and, if so, how—models like these can help planners to do their jobs more effectively.

There are at least three important ways in which simulation models like these can help with planning:

●   Simulation models can help to frame air-pollution problems more effectively. For example, using this model has given us a better idea of the relationship between centerline and ambient pollution, between peak concentrations and overall footprints, and between local and long-distance impacts.

●   Simulation models can help us identify what kind of payoff we can expect from different kinds of abatement strategies. For example, we have looked at the potential gains to be realized from managing the exit flow as opposed to reconstructing the stack.

- Simulation models can provide a basis for sensitivity analysis—for finding out whether there are any situations in which a relatively small change in one or more parameters can have a relatively major impact on dispersion. For example, in Figure 1.1, we noted how the pollution footprint of that particular stack seemed particularly sensitive to the transition from neutral to stable atmospheric conditions.

From a planner's point of view, the search for useful models is often as important as the search for exact models (Sharma et al. 1990). Because of their accessibility, their flexibility, and their transparency, spreadsheets provide an ideal vehicle for creating useful planning models over a broad range of applications—including the evaluation of pollution due to blowing smoke.

Technical Note

1. Exponents

An exponent is a number placed as a superscript to the right of a quantity to indicate the power of that quantity, or the number of times it is to be multiplied by itself. Thus, 2 is an exponent in the expression 3^2, which means the quantity 3 multiplied by itself twice, which equals 9.

Taking the exponent of an expression in an equation is just a way of magnifying the influence of that expression in the equation in a non-linear way (specifically, in an exponential way). By convention, the natural number e (which has a value of approximately 2.718282) is used for this purpose. Thus, exp(x), or the exponent of x, equals the value of e raised to the power of x; in fact, exp(x) can also be written as e^x. So, for example, exp(2) means e raised to the power of 2, or e-squared (e^2).

You can, incidentally, achieve the opposite effect (reducing the influence of an expression in an equation) by using the logarithm of the expression, instead of just the expression itself. In fact, if you combine a *natural* logarithm (which is a logarithm on the base e rather than the base 10) with an exponent, you end up where you started. That is, the natural log of the exponent of a number is the number itself: i.e., $\log_e(\exp(n)) = n$ or, if you prefer, $\log_e(e^n) = n$.

You can verify all this for yourself, if you want to, since most spreadsheets have both the EXP() function and the LN() function.

2. Effective Stack Height

Effective stack height (H_{eff}) is given by the formula:

$$H_{eff} = H + \frac{(1.6*\exp(\frac{\ln f_0}{3})*\exp(\frac{2*\ln (3.5*x_0)}{3})}{u}$$

where   H = physical height of the stack (in meters),
u = wind speed (in meters per second), and
f_0 and x_0 are given by the formulas

$$f_0 = 3.12*.785*v_0*d^2*\frac{t_0-t_1}{t_0}$$

and

$$\text{if } f_0 > 55, \quad \text{then } x_0 = 34*\exp(.4*\ln(f_0))$$
$$\text{if } f_0 \leq 55, \quad \text{then } x_0 = 14*\exp(.625*\ln(f_0))$$

where v_0 = gas exit velocity (in meters per second),
d = stack diameter (in meters),
t_0 = gas exit temperature (in degrees Kelvin), and
t_1 = ambient temperature (in degrees Kelvin);

A slightly modified formula is used for stable and somewhat stable atmospheric conditions, in order to reflect the tendency towards horizontal fanning in such conditions (see page 24 above).

3. Lateral and Vertical Dispersion Coefficients

For any value of x (the horizontal distance downwind from the source), the standard deviations, σy (laterally) and σz (vertically), for each of the six different conditions of atmospheric stability defined above are calculated as described below.

For lateral dispersion (σy),

$$\sigma y = x*\alpha*\frac{1}{\sqrt{1+.0001*x}}$$

where,   for Very Unstable Conditions, $\alpha = .22$,
for Moderately Unstable Conditions, $\alpha = .16$,

for Slightly Unstable Conditions,	$\alpha = .11,$
for Neutral Conditions,	$\alpha = .08,$
for Somewhat Stable Conditions,	$\alpha = .06,$ and
for Stable Conditions,	$\alpha = .04.$

For vertical dispersion (σz), different formulas are used for different atmospheric conditions:

for Very Unstable Conditions,

$$\sigma z = x * .20$$

for Moderately Unstable Conditions,

$$\sigma z = x * .12$$

for Slightly Unstable Conditions,

$$\sigma z = x * .08 * \frac{1}{\sqrt{1 + .0002 * x}}$$

for Neutral Conditions,

$$\sigma z = x * .06 * \frac{1}{\sqrt{1 + .00015 * x}}$$

for Somewhat Stable Conditions,

$$\sigma z = x * .03 * \frac{1}{1 + .0003 * x}$$

for Stable Conditions,

$$\sigma z = x * .016 * \frac{1}{1 + .0003 * x}$$

4.   Ground-Level Concentrations

The ground-level centerline concentration of pollution (C_{max}) is derived from the Gaussian formula given in the text by setting $y = 0$ and $z = 0$. Thus,

$$C_{max} = 1,000,000 * \frac{Q}{2 * \pi * u * \sigma y * \sigma z} * (\exp(\frac{-(H_{eff}/\sigma z)^2}{2}) + (\exp(\frac{-(H_{eff}/\sigma z)^2}{2})))$$

If either standard deviation (σy or σz) is less than .01, C_{max} is set to zero.

Bibliography and References

Ashby, Eric, and Mary Anderson. 1981. *The Politics of Clean Air*. Oxford: Clarendon.

Benarie, Michel M. 1980. *Urban Air Pollution Modeling*. Cambridge, Mass.: MIT Press.

Berlyand, M.E. 1991. *Prediction and Regulation of Air Pollution*. Dordrecht: Kluwer Academic Publishers.

————, ed. 1973. *Air Pollution and Atmospheric Diffusion*. Translated from the Russian by A. Baruch and D. Slutzkin. New York: Wiley.

Biswas, Asit K., T.N. Khoshoo, and Ashok Khosia, eds. 1990. *Environmental Modelling for Developing Countries*. London: Tycooly.

Brimblecombe, Peter. 1987. *The Big Smoke: A History of Air Pollution in London since Medieval Times*. London; New York: Methuen.

Cartwright, T.J. 1993. "SMOKE: Air Pollution Dispersion". In *Spreadsheet Models for Urban and Regional Analysis*, edited by R. Klosterman, R. Brail, and E. Bossard, chapter 13. New Brunswick, N.J.: Center for Urban Policy Research Press. Contains a slightly earlier version of the model presented here.

————. 1991. Planning and Chaos Theory." *Journal of the American Planning Association*, 57.1 (Winter), pp. 44-56.

Clark, A.I., A.E. McIntyre, J.N. Lester, and R. Perry. 1984. "Air Quality Impact Assessment." *Environmental Monitoring and Assessment*, 4, pp. 205-32.

Cohen, J.B., and A.G. Ruston. 1912. *Smoke: A Study of Town Air*. London: Arnold.

Cook, Brian J. 1988. *Bureaucratic Politics and Regulatory Reform: The EPA and Emissions Trading*. Contributions in Political Science, no. 196. New York: Greenwood Press.

Crandall, Robert W. 1983. *Controlling Industrial Pollution: The Economics and Politics of Clean Air*. Washington, D.C.: Brookings Institution.

Crenson, Matthew A. 1971. *The Un-Politics of Air Pollution: A Study of Non-Decision-Making in the Cities*. Baltimore: Johns Hopkins Press.

Davison, D.S., and E.D. Leavitt. 1981. *Airshed Management System for the Alberta Oil Sands*. Vol. 1, *A Gaussian Frequency Distribution Model*. Report nos. 119 and 120. Edmonton: Alberta Oil Sands Environmental Research Program (AOSERP).

Dobbins, R.A. 1979. *Atmospheric Motion and Air Pollution*. New York: Wiley.

Dop, Han van, and D.G. Steyn. 1991. *Air Pollution Modeling and Its Applications VIII*. New York: Plenum.

Frankenburg, T.T. 1968. "High Stacks for Diffusion of Sulfur Dioxide and Other Gases Emitted by Electric Power Plants." *Journal of the American Industrial Hygiene Association*, 29, pp. 181-85.

Grandell, Jan. 1985. *Stochastic Models of Air Pollutant Concentration*. Lecture Notes in Statistics, no. 30. Berlin: Springer-Verlag.

Harrison, Roy M., and Roger Perry, eds. 1986. *Handbook of Air Pollution Analysis*. 2nd ed. London: Chapman and Hall.

Haskell, Elizabeth H. 1982. *The Politics of Clean Air: Standards for Coal-Burning Power Plants*. New York, N.Y.: Praeger.

Haugen, Duane A., ed. 1975. *Lectures on Air Pollution and Environmental Impact Assessment*. Sponsored by the American Meteorological Association. Boston: AMS.

Jakeman, A.J., and R.W. Simpson. 1985. "Assessment of Air Quality Impacts from an Elevated Point Source." *Journal of Environmental Management*, 20.1, pp. 63-72.

Johnson, Warren B., R.C. Sklarew, and D.B. Turner. 1976. "Urban Air Quality Simulation Modeling." In Stern 1976.

Johnston, Harvey. 1961. *Weather Ways*. Meteorological Branch. Ottawa: Department of Transport.

Jones, Charles O. 1975. *Clean Air: The Policies and Politics of Pollution Control*. Pittsburgh: University of Pittsburgh Press.

Kneese, Allen V. 1984. *Measuring the Benefits of Clean Air and Water*. Baltimore: Johns Hopkins University Press for Resources for the Future.

Kohn, Robert E. 1978. *A Linear Programming Model for Air Pollution Control*. Cambridge, Mass.: MIT Press.

Lee, Douglass B., Jr. 1973. "Requiem for Large-Scale Models." *Journal of the American Institute of Planners*, 39.3 (May), pp. 163-78.

Lee, W.L., and A.C. Stern. 1973. "Stack Height Requirements Implicit in Federal Standards of Performance for New Stationary Sources." *Air Pollution Control Association Journal*, 23.6, pp. 503-13.

Organisation for Economic Cooperation and Development (OECD). 1971. *Models for Prediction of Air Pollution*, edited by J.R. Mahoney. Paris: OECD.

Pasquill, F. 1962. *Atmospheric Diffusion: The Dispersion of Windborne Material from Industrial and Other Sources*. London: Van Nostrand.

——. 1975. "The Dispersion of Materials in the Atmospheric Boundary Layer—The Basis for Generalization." In *Lectures on Air Pollution and Environmental Impact Analysis*, edited by D.A. Haugen. Boston: American Meteorological Society (AMS).

Rydell, C. Peter, and Gretchen Schwarz. 1968. "Air Pollution and Urban Form—A Review of Current Literature." *Journal of the American Institute of Planners*, 34.2 (March), pp. 115-20.

Rydell, C. Peter, and B.H. Stevens. 1968. "Air Pollution and the Shape of Urban Areas." *Journal of the American Institute of Planners*, 34.1 (January), pp. 50-51.

Sharma, M., M.N. Mohanan, C. Prakash, and N. Bagchi. 1990. "Development and Validation of an Air Quality Model for Sulphur Dioxide (SO_2) for Delhi City." In Biswas et al. 1990, chapter 4.

Shen, T.T. 1986. "Assessment of Air Pollution Impact." *Atmospheric Environment*, 20.10, pp. 2039-45.

Stern, Arthur C., ed. 1976. *Air Pollution*. Vol. 1, *Air Pollutants, Their Transformation and Transport*. New York: Academic Press.

Strom, Gordon H. 1976. "Atmospheric Dispersion of Stack Effluents." In Stern 1976.

Sutton, O.G. 1960. *Understanding the Weather*. Baltimore: Penguin.

Wanta, R.C., and W.P. Lowry. 1976. "The Meteorological Setting." In Stern 1976.

Wark, K., and C.P. Warner. 1981. *Air Pollution, Its Origin and Control*. New York: Harper and Row.

Weber, Erich, ed. 1982. *Air Pollution: Assessment Methodology and Modeling*. New York: Plenum Press for the NATO Committee on the Challenges of Modern Society.

Zanetti, Paolo. 1990. *Air Pollution Modeling: Theories, Computational Methods and Available Software*. New York: Van Nostrand.

Chapter 2

RUNNING WATER: THE UNDERGROUND TRANSPORT OF POLLUTANTS

I will show you how any hydrologist can build his own groundwater models using the same general software he may have been using to do the bookkeeping for his local golf club.

T.N. Olsthoorn, "The Power of the Electronic Worksheet"

Just as smoke disperses in a plume downwind of a stack, so pollution beneath the earth's surface may be transported downstream of its source by the flow of the ambient groundwater. The physical and chemical processes are quite different, of course, and time is measured in weeks and months rather than hours and days; but the environmental effects are remarkably similar. A single source can pollute a vast area and much of the impact can remain invisible until long after it has begun to accumulate.

For groundwater pollution, the causes can be numerous and diverse:

Groundwater pollution due to human activities occurs in a number of ways. . . . Polluted water may infiltrate into the aquifer from a polluted surface water body, from leaking waste-water pipes, or from ponds or cesspools (bacterial pollution). The pollutant may also be leached from the top soil by rain and be carried down into the saturated zone by seepage (nitrates, pesticides). Similarly, seepage through landfills and waste deposits is a source of pollution. Pollutants may enter the soil in a form immiscible with water (e.g., oil, chlorinated hydrocarbons). Their gradual dissolution by seepage water or the passing groundwater flow causes groundwater pollution [Kinzelbach 1986, p. 188].

Once a pollutant enters the soil, its transport is assumed to have two distinct components. First, the pollutant travels in a more or less vertical direction downwards through what is essentially an *unsaturated* medium. Then, when it reaches the water table, the pollutant travels in a more or less horizontal direction through a *saturated* medium, as it is carried along by the groundwater flow.

Predicting the course of groundwater pollution is more difficult than predicting the dispersion of airborne pollution, because we can usually only guess at the nature of the underground "atmosphere" through which the groundwater flows. However, groundwater travels relatively slowly compared to air flows above the surface. This means there is often time to take remedial measures, even after a spill has occurred, as long as we know where to act. Thus, it can be very helpful for environmental planners and managers to know, even if only roughly, how far and how fast the effects of pollution may be likely to spread in a given period of time.

The purpose of the model presented here is to show, as has been shown for air pollution (see chapter 1), how far and how fast a pollutant can travel, given various ambient parameters. However, unlike the air-pollution model, which accounted for dispersion in three dimensions, the model here deals only with horizontal dispersion. That is, the model simulates downstream and lateral dispersion; vertical dispersion is effectively ignored. This is not considered a major limitation, however, since vertical dispersion in groundwater flows is typically a matter of a few meters, whereas vertical dispersion within the atmosphere may involve thousands of meters.

Purpose of the Model

The present state of groundwater modeling is summarized in a recent report (1990) by the Committee on Ground Water Modeling Assessment of the National Research Council in the United States. In its view,

The processes that control saturated ground water flow [i.e., below the water table] are reasonably well understood, and standard models of these processes are generally believed to be able to give reliable predictions if provided with adequate amounts of data. Nevertheless, the impacts of field-scale heterogeneity are still widely debated, and there are few clear guidelines on how model inputs should be estimated from limited databases or on how hydrologic monitoring programs should be designed. . . .

Unsaturated flow [e.g., between the surface and the water table] is less well understood. The basic "laws" that govern such flow are still questioned by some investigators. . . . There have been very few field studies . . . and most of these have focused on one-dimensional transport in the vertical direction. Some investigators believe that unsaturated flow can move horizontally over significant distances, although available evidence is insufficient to either confirm or reject this hypothesis [p. 3].[1]

The model presented here deals with only the first kind of flow: namely, horizontal transport within a saturated medium. Nevertheless, there are at least four different ways in such transport can be initiated and sustained. These are:

- A pollutant may be discharged into the groundwater either (a) as a momentary spill or injection that can be measured in terms of its mass (e.g., kilograms), or (b) as a continuous leaching or flow that can be measured in terms of its mass per unit of time (e.g., kilograms per day).

- Once it traverses the water table and enters the saturated zone of groundwater flow, a pollutant may then disperse in either one dimension or two: that is, the pollutant may disperse either (a) in only a longitudinal direction, parallel to the groundwater flow, or (b) in both a longitudinal and a transverse or lateral direction (i.e., at right angles to the groundwater flow).

Thus, there are at least four different profiles of groundwater transport that can be simulated. These are:

- a momentary injection that disperses in one dimension,
- a momentary injection that disperses in two dimensions,
- a continuous injection that disperses in one dimension, and
- a continuous injection that disperses in two dimensions.

There are many factors that can influence the regional transport of pollution in the groundwater. To begin with, any infiltration from the surface passes through an unsaturated zone before it traverses the water table and enters the saturated,

---

1. The report refers to a third kind of flow—"through fractured media [which] may be either saturated or unsaturated"—but concludes (p. 4) that this kind of flow is so complicated that "true predictive modeling is not yet a reality."

groundwater zone, or aquifer. Although complicated processes may be at work in the unsaturated zone (e.g., Nielsen et al. 1989), we shall assume here for the purposes of this model that transport through this zone is confined to the vertical. Once the pollutant enters the aquifer, however, we assume that vertical flow becomes negligible. In most cases, the thickness of the aquifer (and thus dispersion in the vertical dimension) is so insignificant relative to dispersion in the two horizontal dimensions (a few meters compared to possibly hundreds of meters) that we can allow for it by modeling averages of vertical concentrations.[2] Thus, we ignore the vertical dimension and confine our attention to the horizontal and transverse (or lateral) dimensions.

In addition, standard models of groundwater transport make two other important assumptions (Nielsen et al. 1989):

- that the flow within the saturated layer is uniform and/or the porous medium is homogeneous; and

- that the aquifer is fully confined in the vertical dimension but of infinite extent in the horizontal dimension.[3]

Finally, it should be noted that standard groundwater models usually ignore any changes in the groundwater flow that might be induced by the quantity or density of the pollutant itself. As Clarke 1987 explains,

> Depending on the density of the solute, the pollutant may influence the flow field ([in which case, it is called a] hydrodynamically active solute). We consider only hydrodynamically inactive solutes; this means pollutant concentrations that are [assumed to be] so small that density-induced flows can be ignored [p. I-2].

---

2. Clarke 1987, p. I-2. Of course, there may be cases where this assumption is not justified. For example, vertical dispersion may be significant in the vicinity of the infiltration itself (especially if it is under pressure from some source) and in strongly stratified aquifers.

3. Aral 1990 discusses the case of unsteady flows in multilayer aquifers; his book includes a Fortran program (supplied on a disk) for modeling such cases.

Conceptual Basis

Simulation models like the one presented here have been available for many years (e.g., Bear and Jacobs 1965). General reviews of the literature on groundwater modeling are also available (e.g., Bouwer 1978; Marino and Lutkin 1982). Clarke 1987 presents two studies dealing with discharges from wells directly into the aquifer. Das Gupta (1990) reports on an extensive case study carried out in Bangkok (Thailand), where groundwater exhaustion has reached such a point that parts of the city are reported to be sinking at an annual rate of tens of centimeters! There are numerous other cases where groundwater flows are known to be having major environmental impacts: for example, Whittington and Guariso 1983 deal with management of the reservoir behind the Aswan Dam, including the effect of underground leaks.[4]

The model to be presented here is derived from one written in BASIC by Wolfgang Kinzelbach (1986, chapter 7, program 14). Kinzelbach assumes that there are four main processes affecting the transport of a pollutant, and all of them are included in the spreadsheet version too. These four processes are:

- *convective transport* caused by the physical flow of groundwater, which is assumed to cause movement of a homogeneous distribution of the pollutant within the groundwater;

- *molecular diffusion or dispersion*, which causes the concentration of the pollutant to spread out in both horizontal dimensions (as well as the vertical, but that is ignored here);

- *adsorption* of the pollutant by the medium through which it travels, which is assumed to cause a delay in the transport process in the horizontal dimension (or dimensions) but not to alter the concentration of the pollutant; and

- *chemical decay*, which is assumed to reduce the mass (and thus the concentration) of the pollutant in the groundwater flow.

---

4. The report of the Committee on Ground Water Modeling Assessment (1990, especially chapter 2) contains a useful bibliography; see also Huyakorn and Pinder 1983, and Celia 1988.

Data Requirements

Naturally, the critical input data required for groundwater modeling is almost always the pore velocity, or Darcy velocity, of the ambient groundwater flow (which is usually measured in meters per day).[5] In addition to the pore velocity of the water, the model requires the following data on the aquifer through which the groundwater flows:

- the coefficient of retardation
- the decay constant (in liters per day)
- the thickness of the aquifer (in meters)
- the effective porosity of the aquifer
- the coefficient of longitudinal dispersivity
- the coefficient of transverse or lateral dispersivity

The last parameter is not required if the model is to be run only for one-dimensional scenarios. As noted previously, the model is based on the assumption that all these parameters are uniform throughout the aquifer.

Overview

The model consists of 11 columns and 121 rows (see Listing 2.1). The top 20 rows are for entering the parameters of the model and displaying its results, while the rest of the model serves essentially as a "scratch pad" for the calculations of the model. The formulas used in the model are quite complex; so they are discussed briefly here and in more detail in a technical note beginning on page 59 below. Rows 3 to 9 are used for entering the required input data for the model. The data on the left are entered in the cells shown. On the right, note that the model can simulate four different cases, based on the four possible combinations of momentary or continuous injection and one- or two-dimensional dispersion. If any positive value is entered in cell K4, the model assumes a momentary mass and ignores any data that may be entered in cell K5. (To make this quite clear, an asterisk appears beside the

---

5.   The pore velocity is also called the Darcy velocity, since it is derived from Darcy's law, which has been described as the "cornerstone of all groundwater modeling" (e.g., Bouwer 1978, chapter 3).

Listing 2.1

```
  | A || B| | D || E || F || G || H || I || J || K |
1 GROUNDWATER CONCENTRATION DISTRIBUTION        ©1993 by T.J. Cartwright
2
3 Pore velocity (m/day)          1    Source (enter one value only):
4 Retardation factor             1    Momentary mass (kg)            0
5 Decay constant (l/day)         0  * Continuous flow (kg/day)       1
6 Aquifer thickness (m)         10
7 Effective porosity            .1    Select 1 or 2 dimensions       2
8 Long. dispersivity (m)       4.5
9 Trans. dispersivity (m)    1.125    Elapsed time (days)          100
10
11 Lateral            Longitudinal Transport (meters)
12 Transport      5     25    50     75    100    125    150    200
13
14                    Concentration (milligrams/m3)
15   C/line      71.5  47.0  36.2  30.0  26.3   7.3   5.9    .0
16   10 m         8.8  19.8  22.8  22.1  21.0   6.0   5.1    .0
17   20 m          .8   3.2   6.8   9.3  10.9   3.5   3.2    .0
18   30 m          .1    .4   1.3   2.6   3.9   1.4    .0    .0
19   40 m          .0    .0    .2    .5    .3    .4    .0    .0
20   50 m          .0    .0    .0    .0    .1    .1    .0    .0
21
22
23
24 Momentary injection in one dimension
25
26              0     0     0     0     0     0     0     0
27
28 Momentary injection in two dimensions
29
30    0         0     0     0     0     0     0     0     0
31   10         0     0     0     0     0     0     0     0
32   20         0     0     0     0     0     0     0     0
33   30         0     0     0     0     0     0     0     0
34   40         0     0     0     0     0     0     0     0
35   50         0     0     0     0     0     0     0     0
36
37
38 Continuous injection . . .
39
40 C0         1000
41 GA            1
42
43 . . . in one dimension
44
45 A1       -2.2392 -1.7678 -1.1785 -.58926  -7e-16  .589256 1.17851 2.35702
46 HH1       .612279 .672156 .765766 .889668 1.06141 .889668 .765766 .598940
47 HH2      -.48506 -.49458 -.49592 -.47231 -.39175 -.47231 -.49592 -.48202
48 HH3       .540140 .586935 .667708 .795536 1.02967 .795536 .667708 .530088
49 HH4       .147470 .191525 .276473 .428353 .745170 .428353 .276473 .138585
50 HH5       .232069 .282663 .383316 .572642 1.00000 .572642 .383316 .222000
51 F1       1.99846 1.98758 1.90442 1.59534 1.00000 .404657 .095581 .000858
52
53 A2       2.47487 2.94628 3.53553 4.12479 4.71405 5.30330 5.89256 7.07107
54 HH1       .586170 .540107 .491799 .451423 .417173 .387754 .362211 .320046
55 HH2      -.47880 -.46461 -.44544 -.42604 -.40718 -.38921 -.37229 -.34167
56 HH3       .520567 .486879 .452215 .423338 .398633 .377085 .358020 .325577
57 HH4       .130372 .102984 .077712 .059050 .044860 .033825 .025090 .012387
58 HH5       .212731 .182077 .154082 .133495 .117790 .105452 .095524 .080574
59 F2        .000465 .000031 .000001 5.45e-9 2.6e-11 6.4e-14 8.0e-17 1.6e-23
60
61 C         998.522 989.790 932.999 750.505 441.105 165.070 35.8807 .273371
62
```

		A || B ||	D ||	E ||	F ||	G ||	H ||	I ||	J ||	K |
63	... . in two dimensions									
64	A1									
65		0	.013889	.347222	1.38889	3.125	5.55556	8.68056	12.5	22.2222
66		10	.236111	.569444	1.61111	3.34722	5.77778	8.90278	12.7222	22.4444
67		20	.902778	1.23611	2.27778	4.01389	6.44444	9.56944	13.3889	23.1111
68		30	2.01389	2.34722	3.38889	5.125	7.55556	10.6806	14.5	24.2222
69		40	3.56944	3.90278	4.94444	6.68056	9.11111	12.2361	16.0556	25.7778
70		50	5.56944	5.90278	6.94444	8.68056	11.1111	14.2361	18.0556	27.7778
71	A2									
72		0	.555556	2.77778	5.55556	8.33333	11.1111	13.8889	16.6667	22.2222
73		10	2.29061	3.55729	5.98352	8.62454	11.3312	14.0655	16.8142	22.3331
74		20	4.47903	5.24110	7.11458	9.44444	11.9670	14.5827	17.2491	22.6623
75		30	6.68977	7.22222	8.67806	10.6719	12.9577	15.4060	17.9505	23.2007
76		40	8.90623	9.31281	10.4822	12.1843	14.2292	16.4898	18.8889	23.9341
77		50	11.1250	11.4531	12.4226	13.8889	15.7135	17.7865	20.0308	24.8452
78	S4									
79		0	0	0	0	0	0	0	0	0
80		10	0	0	0	0	0	0	0	0
81		20	0	0	0	0	0	0	0	0
82		30	0	0	0	0	0	0	0	0
83		40	0	0	0	0	0	0	0	0
84		50	0	0	0	0	0	0	0	0
85		0	1.18059	.097290	.004856	.000255	.000014	0	0	0
86		10	.165141	.041837	.003070	.000188	.000012	0	0	0
87		20	.015462	.006806	.000920	.000080	.000006	0	0	0
88		30	.001445	.000821	.000178	.000023	.000002	0	0	0
89		40	.000140	.000092	.000028	.000005	0	0	0	0
90		50	.000014	.000010	.000004	0	0	0	0	0
91	S2									
92		0	0	0	0	0	0	0	0	0
93		10	0	0	0	0	0	0	0	0
94		20	0	0	0	0	0	0	0	0
95		30	0	0	0	0	0	0	0	0
96		40	0	0	0	0	0	0	0	0
97		50	0	0	0	0	0	0	0	0
98		0	.897391	.077564	.004537	.000260	.000014	.000001	3.93e-8	0
99		10	.128379	.034892	.002927	.000192	.000011	.000001	3.36e-8	0
100		20	.013621	.006257	.000916	.000082	.000006	3.63e-7	2.11e-8	0
101		30	.001418	.000820	.000182	.000023	.000002	1.51e-7	0	0
102		40	.000143	.000094	.000028	.000005	.000001	4.75e-8	0	0
103		50	.000014	.000010	.000004	.000001	1.09e-7	1.19e-8	0	0
104	W									
105		0	1.15921	.082712	.003955	.000204	.000011	1.91e-7	9.72e-9	2.9e-17
106		10	.142819	.034885	.002494	.000150	.000009	1.58e-7	8.29e-9	2.4e-17
107		20	.012712	.005556	.000743	.000063	.000005	9.09e-8	5.20e-9	1.3e-17
108		30	.001169	.000662	.000142	.000018	.000002	3.76e-8	5.6e-14	5.1e-18
109		40	.000111	.000073	.000021	.000004	1.33e-7	1.18e-8	1.1e-14	1.3e-18
110		50	.000011	.000008	.000003	1.91e-7	2.70e-8	2.92e-9	1.4e-15	2.6e-19
111	C									
112		0	71.4571	47.0492	36.1831	29.9858	26.3380	7.26949	5.94918	.000005
113		10	8.80375	19.8435	22.8158	22.0518	20.9635	6.01801	5.07741	.000004
114		20	.783588	3.16042	6.79576	9.31310	10.8560	3.45994	3.18194	.000002
115		30	.072073	.376746	1.29591	2.58479	3.90974	1.43158	.000034	.000001
116		40	.006867	.041335	.195795	.539336	.314087	.447291	.000007	2.14e-7
117		50	.000676	.004441	.026225	.028103	.064001	.111050	.000001	4.11e-8
118										
119										

120 Source: Adapted from a program written in BASIC in Walter Kinzelbach,
121   Groundwater Modeling (Amsterdam: Elsevier, 1986).

words "Momentary mass" whenever that option is selected.) Otherwise, the model assumes a continuous flow of pollutant. Similarly, the user must indicate in cell K7 whether the model is to compute dispersion in one dimension (parallel to the ambient groundwater flow) or in two dimensions (parallel and at right angles to the ambient groundwater flow). Finally, cell K9 contains the number of days that have elapsed between the time when the pollution occurred (or, in the case of a continuous flow, started to occur) and the time for which the dispersion profile is to be simulated.

The data in row 12 and in cells A15:A20 define the points in space where the concentration of the pollutant is to be calculated, in meters downstream of the flow and (in the case of two-dimensional dispersion) laterally from its centerline. The locations shown here are typical of what might be required, but they can be adjusted at any time according to need. The rest of rows 15-20 (cells D15:K20) is used to present the results of the simulation. In effect, this part of the model is a "scoreboard" that displays the results of the calculations appropriate for whichever of the four types of transport is being modeled—momentary or continuous injection, and one- or two-dimensional flow.

Thus, cell D15 (for example) reads as follows:

```
IF($K$4>0,IF($K$7=1,D26,D30),IF($K$7=1,D61,D112))
```

This amounts to saying that, if the injection of the pollutant is momentary (i.e., K4>0), then (a) if the dispersion is one-dimensional (i.e., K7=1), display the value of cell D26; otherwise (b) display the value of cell D30. But, if the injection is not momentary but continuous (i.e., K4= <0), then (a) if the dispersion is one-dimensional (i.e., K7=1), display the value of cell D61; otherwise (b) display the value of cell D112. The formulas in the rest of row 15 are identical to this one, subject to adjustment of the cell references.

The formulas in the other rows (rows 16-20) are slightly different, in that they display a blank (" "), rather than a cell reference, in the event that dispersion is to be one-dimensional. Thus, the formula in cell D16, immediately below the one shown above, is

```
IF($K$4>0,IF($K$7=1," ",D31),IF($K$7=1," ",D113))
```

The rest of the model consists of four separate algorithms, one for each of the four cases reflected in the above formulas and each corresponding to one of the four transport profiles described in the previous section. Specifically, there is an algorithm for each of the following scenarios:

- **Momentary Injection and One-Dimensional Dispersion (Row 26)**

    The contents of cell D26, for example, are as follows:

    ```
    1000*$K$4/(2*$F$6*$F$7*$F$4*
        SQRT(PI*$F$8*$F$3*$K$9/$F$4))*
        EXP(-((D12-$F$3*$K$9/$F$4)^2/
        (4*$F$8*$F$3*$K$9/$F$4)))*EXP(-$F$5*$K$9)
    ```

    This expression is derived from the formula shown in case 1 in the technical note on page 59.

- **Momentary Injection and Two-Dimensional Dispersion (Rows 30-35)**

    The contents of cell D30, for example, are as follows:

    ```
    1000*$K$4/(4*PI*$F$7*$F$6*$F$3*
        SQRT($F$8*$F$9)*$K$9)*
        EXP(-((D$12-$F$3*$K$9/$F$4)^2/
        (4*$F$8*$F$3*$K$9/$F$4)+
        $A15^2/(4*$F$9*$F$3*$K$9/$F$4)+$F$5*$K$9))
    ```

    This expression is derived from the formula shown in case 2 in the technical note on page 60.

- **Continuous Injection and One-Dimensional Dispersion (Row 61)**

    The contents of cell D61, for example, are as follows:

    ```
    D40/2*EXP(D12/2/F8)*(D51*EXP(-D12*D41/2/F8)-
        D59*EXP(D12*D41/2/F8))
    ```

    This expression is based on various interim calculations and is derived from the formula shown in case 3 in the technical note on page 60.

- **Continuous Injection and Two-Dimensional Dispersion (Rows 112-17)**

    The contents of cell D112, for example, are as follows:

$D$40/(4*PI*SQRT($F$8*$F$9))*EXP(D$12/2/$F$8)*D105

This expression is based on various interim calculations and is derived from the formula shown in case 4 in the technical note on page 61.

Every time the time the model is run, it does all of these calculations. Then, on the basis of the data entered in cells K4, K5, and K7, the model decides which of the four sets of results to display in the "scoreboard" at the top of the worksheet.

Operation

Operation of the model is quite straightforward. It suffices to enter the data on the aquifer where the spill has occurred, select which of the scenarios (momentary or continuous, one dimension or two) you want to model, enter the elapsed time you want to simulate, and wait while the model does its job.

Note that the model always goes through a number of iterations before it finishes, even though iteration is required for only one of the four scenarios (i.e., a a continuous injection and a two-dimensional flow). This is because the model calculates dispersion for all four scenarios, even though it displays the results (in the scoreboard at the top) only for the scenario selected. So, if you have selected one of the first three scenarios and are too impatient to wait for the model to finish, you can safely interrupt it![6]

Interpretation and Use

The graphing facilities of the spreadsheet can be used to illustrate how the pollutant is dispersing. Consider first the case of a one-time spill, in which a momentary mass of (say) 100 kg is injected into an aquifer under the conditions depicted in Listing 2.1. Figures 2.1 and 2.2 show the predicted dispersion of the pollutant 100 days after the original spill. Figure 2.1 assumes one-dimensional flow in the groundwater, while Figure 2.2 assumes two-dimensional flow.

---

6.   Ctrl-Break, or Ctrl-C, will usually do the trick.

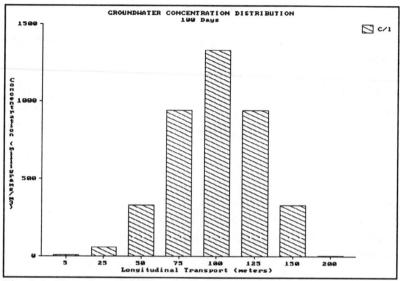

Fig. 2.1. Momentary Injection (100 kg) in One-Dimensional Flow

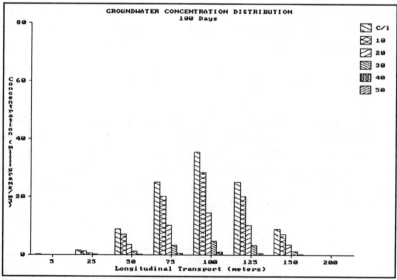

Fig. 2.2. Continuous Injection (1 kg/day) in One-Dimensional Flow

As might be expected, both graphs show the maximum concentration of the pollutant not at the source but about 100 m downstream. This is because the pore velocity of the groundwater flow is given (in Listing 2.1) as 1 m/day. The maximum concentration in the one-dimensional case (about 1.3 g/m³) is higher than the maximum concentration in the two-dimensional case (about 40 mg/m³). This is because, in the one-dimensional case, the pollutant does not dissipate "sideways", whereas it does in the two-dimensional case.

Further examination shows that pollutant concentrations are distributed normally on either side of the peak, along the direction of flow and (in the case of two-dimensional flows) on either side of the flow as well. In fact, it is notable that traces of the pollutant can be found at considerable distances not only behind but also ahead of the point of maximum concentration. After 100 days of one-dimensional flow, for example, concentrations of more than 5 mg/m³ are predicted at 200 m from the source. That is twice the distance of the physical flow of groundwater in that period of time. Even in two-dimensional flows, minute traces (0.3 mg/m³) are predicted at 200 m from the source.

Thus, as time goes by, there are two basic processes are at work in the case of a momentary injection of pollution into the groundwater. One is that the peak concentration gradually moves away from the source, subject to the pore velocity of the groundwater. The other process, according to Kinzelbach (1986, p. 209 and Figure 7.1.1), is that the concentration distribution gradually flattens out, longitudinally and (in the case of two-dimensional flows) laterally as well, until it eventually dissipates.

Now consider the case of leaching, in which (say) 1 kg per day is injected continuously into an aquifer under the conditions depicted in Listing 2.1. Figures 2.3 and 2.4 show the predicted dispersion of the pollutant 100 days after the leacing began. Figure 2.3 assumes one-dimensional flow in the groundwater, while Figure 2.4 assumes two-dimensional flow. In this case, both graphs show the maximum concentration of the pollutant to be at the source. Essentially, this is because the quantity of pollutant is constantly being refreshed at the source. Again, the maximum concentration in the one-dimensional case (about 1.0 g/m³) is higher than the maximum concentration in the two-dimensional case (about 80 mg/m³). As before, this is because, in the one-dimensional case, the pollutant does not dissipate "sideways", whereas it does in the two-dimensional case.

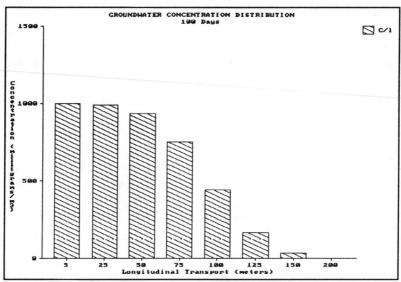

Fig. 2.3. Momentary Injection (100 kg) in Two-Dimensional Flow

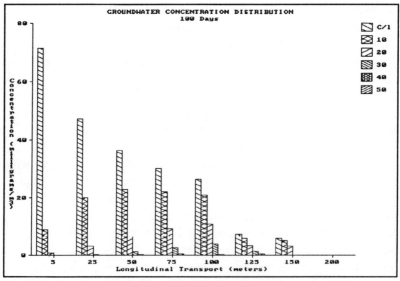

Fig. 2.4. Continuous Injection (1 kg/day) in Two-Dimensional Flow

In both cases, moreover, the leading edge of the pollutant has advanced to the point where you would expect it to be after 100 days, given the pore velocity of groundwater flow specified in Listing 2.1. Once again, traces of the pollutant are predicted to appear up to 100 m in advance of the leading edge in the case of one-dimensional flows, and up to 60 m in advance in the case of two-dimensional flows.

In the long run, continuous leaching should reach a stable concentration distribution. For one-dimensional flows, Kinzelbach suggests (ibid., Figure 7.1.3) that the shoulder-like profile shown in Figure 2.3 gradually expands out from the source, flattening out somewhat until it reaches a stable profile sloping down from the source. For two-dimensional flows like the one shown in Figure 2.4, there is an additional and more complicated flattening-out process taking place in the lateral dimension. This is because the lateral dispersion of the pollutant does not peak at the source, but instead increases with distance from the source (at least up to a certain point). Thus, the concentration distribution in the lateral dimension is quite steep close to the source but flattens out further away from the source. Thus, continuous leaching into a two-dimensionsal flow creates a kind of "cone" of pollution that can be expected to flatten and spread out both longitudinally and laterally, as it moves away from the source, until it reaches a stable profile. According to Kinzelbach,

> For large times, t, the plume reaches an asymptotic shape, which is character-ized by an equilibrium between lateral dispersive transport and longitudinal convective-dispersive transport. . . . In [the] axial direction, the concentrations decline steeply near the source. This is caused by dilution due to lateral dispersion, which is important near the source, where a large concentration gradient prevails. In the further course of the plume, the widening slows down [ibid., pp. 217-18].

In summary, one-dimensional scenarios may seem idealized. However, there are aquifers which are geologically or hydrologically of just such a nature. More-over, one-dimensional modeling may be quite appropriate in certain special cases where (for example) there are multiple injection points at right angles to the direction of groundwater flow, or where injection is from a fairly large source (e.g., a body of surface water) rather than from a single point. In such cases, a one-dimensional flow model may give better results than a two-dimensional model. In any case, it should be noted that the concentration distribution, or profile, of the pollutant is essentially the same for one- and two-dimensional models. The difference is largely a matter of the quantity of pollutant to be found on the centerline.

Evaluation and Extension

Great caution is required in applying models like these to field situations. First, these models assume an aquifer that is uniform and homogeneous. While this may be a reasonable assumption in many cases, the reality is that we often do not know. It is not as easy to build up a picture of an underground aquifer as it is a picture of the topological and atmospheric conditions affecting the dispersion of air pollution, for example. Moreover, even if the assumption of uniformity and homogeneity is reasonable for an aquifer as a whole, it may not be appropriate for the point at which a field sample is obtained.

Second, most groundwater models assume that the pollutant enters the aquifer in the same state as it enters the ground and that it instantly spreads evenly across the vertical dimension of the aquifer. As Kinzelbach notes:

> This is often not the case. A pollutant is carried into the aquifer via the unsaturated zone and is gradually dispersed downward due to vertical dispersion. Only after a certain distance has the vertical mixing proceeded to a point where the distribution can be looked upon as homogeneous in [the vertical or] z-direction [ibid., p. 220].

However, as Kinzelbach wryly acknowledges, it is often difficult to determine how far this "certain distance" actually is![7]

Nevertheless, groundwater models like the one presented here are useful in providing us with a picture of what happens to pollutants when they enter the groundwater. The models also enable us to examine how the various parameters of the aquifer—its pore velocity, its dispersivity, its thickness, etc.—can affect the impact of pollutants that enter into it. As the pressures on our planet increase, groundwater is becoming a more and more precious resource (Das Gupta 1990). It behooves us all to learn as much as we can about this invisible resource.

---

7.   See Sayre 1973 for discussion of how waters "mix" in the vertical dimension.

Technical Note

The model in this chapter provides for four different transport scenarios: one-dimensional dispersion of momentary and continuous injections, and two-dimensional dispersion of momentary and or continuous injections. Each scenario has its own algorithm. A more detailed treatment can be found in Kinzelbach 1986 (chapters 6 & 7) and Hunt 1983 (chapter 6).[8]

1. Momentary Injection and One-Dimensional Dispersion

The concentration of a pollutant, c (in milligrams per cubic meter), at time t (in days) and distance x (in meters downstream from the source of contamination) is calculated as follows:

$$c(x,t) = \frac{MI}{2*M*EN*R*\sqrt{\pi*AL*u*t/R}} *$$

$$\exp(\frac{-(x-u*t/R)^2}{4*AL*u*t/R}) * \exp(-LA*t)$$

where

MI	=	mass of the injection (in grams),
u	=	pore velocity (in meters per day),
R	=	coefficient of retardation due to adsorption, according to a linear adsorption isothermal,
LA	=	decay constant (in liters per day),
M	=	vertical thickness of saturated aquifer (in meters),
EN	=	effective porosity, and
AL	=	longitudinal dispersivity (in meters)

---

8.   There is one discrepancy between the results obtained in the model presented here and those obtained with BASIC models like those in Kinzelbach 1986. This discrepancy arises in the fourth case (continuous injection and two-dimensional flow). I think the different results may stem from the fact that the exponent function in BASIC—EXP(n)—generates an error for any number, n, greater than approximately ±89. In spreadsheets, however, the range of non-error values for n in the exponent function is typically much bigger (e.g. ± 127 in SuperCalc4 and ± 11,355 in Lotus 1-2-3 Release 3.1.) See also the technical note to Chapter 1 for further discussion of exponents.

2. Momentary Injection and Two-Dimensional Dispersion

The concentration of a pollutant, c (in milligrams per cubic meter), at time t (in days) and location x,y (in meters downstream from the source of contamination) is calculated as follows:

$$c(x,y,t) = \frac{MI}{4*\pi*EN*M*u*\sqrt{AL*AQ}*t} *$$

$$\exp(\frac{-(x-u*t/R)^2}{4*AL*u*t/R} + \frac{y^2}{4*AQ*u*t/R} + LA*t)$$

where AQ is the transverse dispersivity (in meters) and all the other variables are the same as in case 1 above.

3. Continuous Injection and One-Dimensional Dispersion

The concentration of pollutant, c (in milligrams per cubic meter), at time t (in days) and distance x (in meters downstream from the source of contamination) is calculated as follows:

$$c(x,t) = \frac{MP}{2*M*EN*u} * \exp(\frac{x}{2*AL}) *$$

$$(F1*\exp(\frac{-x*GA}{2*AL}) - F2*\exp(\frac{x*GA}{2*AL}))$$

where   MP   =   mass of the flow (in grams per day), and
        GA   =   1 + (4 * LA * AL * R/u).

F1 and F2 are the following complementary error functions (between 0 and 2):

$$F1 = \frac{x - u*t*GA/R}{2*\sqrt{AL*u*t/R}}$$

$$F2 = \frac{x + u*t*GA/R}{2*\sqrt{AL*u*t/R}}$$

and are calculated using an approximation function given in Abramowitz and Stegun 1970, p. 1065. The other variables are all the same as in cases 1 and 2 above.

4. Continuous Injection with Two-Dimensional Dispersion

The concentration of pollutant, c (in milligrams per cubic meter), at time t (in days) and horizontal location x,y (in meters downstream from the source of contamination) is calculated as follows:

$$c(x,y,t) = \frac{MP/(M*EN*u)}{4*\pi*(AL*AQ)^2} * \exp(\frac{x}{2*AL}) * W$$

where W is the Hantush function (Hantush 1956) calculated by logarithmic integration (using Simpson's rule) of two variables, A1 and A2, where

$$A_1 = \frac{(x^2+AL*y^2/AQ)*R}{4*AL*u*t}$$

$$A_2 = \frac{\sqrt{x^2+AL*y^2/AQ}*GA}{2*AL}$$

Bibliography and References

Abramowitz, Milton, and Irene A. Stegun, eds. 1970. *Handbook of Mathematical Functions, with Formulas, Graphs, and Mathematical Tables.* New York: Dover.

American Society of Agricultural Engineers (ASAE). 1988. *Modeling Agricultural, Forest, and Rangeland Hydrology.* Proceedings of the International Symposium, Chicago, December 12-13, 1988. St. Joseph, Mich.: ASAE.

Aral, M.M. 1990. *Ground Water Modeling in Multilayer Aquifers: Unsteady Flow.* Chelsea, Mich.: Lewis. Contains PC-compatible program.

Bear, J., and M. Jacobs. 1965. "On the Movement of Waterbodies Injected into Aquifers." *Journal of Hydrology,* 3.1, pp. 37-57.

Bouwer, Herman. 1978. *Groundwater Hydrology.* New York: McGraw-Hill.

Celia, M.A., ed. 1988. *Computational Methods in Water Resources.* Proceedings of the 7th International Conference. 2 vols. Developments in Water Science, nos. 35 and 36. Amsterdam: Elsevier.

Clarke, David K. 1987. *Microcomputer Programs for Groundwater Studies.* Developments in Water Science, no. 30. Amsterdam: Elsevier.

Committee on Ground Water Modeling Assessment. 1990. *Ground Water Models: Scientific and Regulatory Applications*. National Research Council. Washington, D.C.: National Academy Press.

——. 1988. *Groundwater Discharge Tests: Simulation and Analysis*. Developments in Water Science, 37. Amsterdam: Elsevier.

Das Gupta, A. 1990. "Modeling of Groundwater Overdraft and Related Environmental Consequences." Chapter 9, pp. 84-119, in *Environmental Modelling for Developing Countries*. Edited by Asit K. Biswas, T.J. Khoshoo, and Ashok Khosla. London: Tycooly.

Hantush, M.S. 1956. "Analysis of Data for Pumping Tests in Leaky Aquifers." *Transactions of the American Geophysical Union*, 37.6, pp. 702-14. Cited in Kinzelbach 1986.

Hunt, Bruce. 1983. *Mathematical Analysis of Groundwater Resources*. London: Butterworth.

Huyakorn, P.S., and G.F. Pinder. 1983. *Computational Methods in Subsurface Flow*. New York: Academic Press.

Kinzelbach, Wolfgang. 1986. *Groundwater Modeling: An Introduction with Sample Programs in Basic*. Developments in Water Science, no. 25. Amsterdam: Elsevier.

Mariño, M.A., and J.N. Lutkin. 1982. *Seepage and Groundwater*. Developments in Water Science, no. 13. Amsterdam: Elsevier.

Nielsen, D.R., M.Th. van Genuchten, and W.A. Jury. 1989. "Transport Processes from Soil Surfaces to Groundwaters." In *Groundwater Contamination*, edited by Linda M. Abriola, pp. 99-108. International Association of Hydrological Sciences Publication no. 185. Wallingford, Oxford: IAHS Press.

Olsthoorn, T.N. 1985. "The Power of the Electronic Worksheet: Modeling without Special Programs." *Ground Water*, 23.3 (May-June), pp. 381-90.

Sayre, W.W. 1973. "Natural Mixing Processes in Rivers." In *Environmental Impact on Rivers*, ed. H.W. Shen, chapter 6. Fort Collins, Colo.: Murphy.

Whittington, Dale, and Giorgio Guariso. 1983. *Water Management Models in Practice: A Case Study of the Aswan High Dam*. Amsterdam: Elsevier.

Chapter 3

PRESERVING A SPECIES: FINDING THE MINIMUM VIABLE POPULATION

Everything has a smell of its own for those that have noses to smell.
Wahb had been learning smells all his life, and knew the meaning of most
of those in the mountains. It was as though each and every thing had
a voice of its own for him; and yet it was far better than a voice,
for every one knows that a good nose is better than eyes and ears together.

E.T. Seton, The Biography of a Grizzly

In the long march of time since the beginning of the Cambrian period some 550 million years ago, species are thought to have been going extinct at an average rate of about one species per year. Today, however, the rate of species extinction is much higher. Norman Myers, an ecologist, estimates that the rate of extinction reached 100 species per year in 1974, and that it will increase to about 100 species *per day* by the year 2000—due largely to the loss of tropical rain forest, which is estimated to be occurring at a rate of some 47 hectares (ha) per day (Myers 1979).

Extinction can be defined as the failure of a species or population to maintain itself through reproduction (Frankel and Soulé 1981, p. 10). As far as animals and birds are concerned, humans have been responsible directly or indirectly for the extinction of numerous species, including the dodo, the great auk, the passenger pigeon, the Carolina parakeet, the Mongolian wild horse, Burchell's zebra, the Caucasian wisent, and many marine animals. In other cases, extinction has been caused by essentially random, or stochastic, events, such as natural disasters or diseases.

All this has led to efforts to conserve and protect species that seem threatened with extinction, and even to reintroduce them into suitable habitats. Various agencies keep up-to-date lists of endangered species. The International Union for the Conservation of Nature (IUCN) in Switzerland, for example, publishes a *Red Data Book* of endangered species. Such efforts to prevent extinction naturally raise the question of how many is enough—what is the minimum viable population? In other words, for a given species in a given location, what is the smallest population that will suffice to ensure its survival more or less indefinitely?

According to one study (Gilpin and Soulé 1986, p. 19), the concept of minimum viable population implies "some threshold for the number of individuals, or some multivariate set of thresholds and limits, that will ensure (at some acceptable risk) that a population will persist in a viable state for a given time interval." Of course, there can be no exact determination of minimum viable population, since the survival of any species depends to some extent on chance. So the best we can do is to describe the probability of a species surviving for a given period of time. Or, as Daniel Goodman (1988, p. 11) has put it, we can try to "determine the minimum number of individuals that will suffice to confer upon a population an expected time to extinction that is, by some criterion, acceptably remote."

Purpose of the Model

The purpose of the model discussed in this chapter is to simulate the evolution of a population over a relatively long period of time, in order to see whether the population survives. The long-term evolution of any species is subject to two basic factors: births and deaths. Thus, its population at the beginning of any period is equal to the population at the beginning of the previous period plus any births and minus any deaths that occurred in the intervening time. In order to *forecast* how a population is likely to evolve over time, we can divide the population into a number of age-sex cohorts and then compute (on the basis of fertility and mortality data) how each cohort will "age" into the next one. That is, births are computed from application of a specific fertility rate (and, if appropriate, an average litter size) to the number of females in the appropriate age cohorts (subject, of course, to there being a sufficient number of males to make that possible!) Similarly, deaths in each cohort are computed by means of an appropriate mortality rate. In this way, total births and deaths can be estimated for appropriate periods of time, as far into the future as you feel confident of relying on the fertility and mortality rates. Such a model is easy to construct and has been found to give reasonably good forecasts.

However, such a model takes no account of unpredictable or stochastic factors that can affect a population (Kaufman and Mallory 1986). Following the work of Mark Shaffer (Shaffer and Samson 1985), Tony Oliveri (1991) has identified four potential sources of stochastic threat to a population:

- *Demographic stochasticity*, which arises from chance events in the survival and reproductive success of a population, such as imbalances in sex ratios (e.g., too many males), in age structure (more sexually immature than mature individuals), etc.;

- *Environmental stochasticity*, which occurs as a result of temporal variation in habitat parameters, in populations of competitors, predators, or parasites, or in the incidence of diseases;

- *Genetic stochasticity*, which results from changes in gene frequencies due to inbreeding or other factors; and

- *Catastrophic stochasticity*, which can include floods, fires, droughts, and any other natural disasters that occur at random intervals over time.

In an earlier work, Shaffer (1981) describes in graphic detail how these various factors combined and interacted in the early years of this century to cause the eventual extinction of the North American heath hen in 1932.

One way of adding a stochastic dimension to a simulation model is to include random numbers in the model, and then run the model over and over again. Subject to the usual statistical considerations, the results can then be treated as a sample of all possible outcomes and interpreted accordingly.[1] For example, if the model (with its random variables) is run 50 times for, say, 100 years and the species survives 45 times and becomes extinct 5 times, we might conclude that the species has a 90% probability of survival over a period of 100 years. This technique is usually attributed to John von Neumann (1951) and Stanislaw Ulam (Metropolis and Ulam 1949),

---

1.   See also chapters 6 and 10, below. In chapter 6, the Klein model employs a stochastic factor in some of its equations; and, in chapter 10, the capacity model uses a full-scale Monte Carlo simulation similar to the one used here. See also Hammersley and Handscomb 1973; Kleijnen 1974, 1975; Swartzman and Kaluzny 1987, chapter 7; and Gould and Tobochnik 1988, part II.

who called it the "Monte Carlo" method—because it seemed to them to rely on little more than gambling! Since then, the Monte Carlo method has found many different applications in the natural and social sciences.

Conceptual Basis

Numerous analytical and empirical models of population dynamics have been developed, many of them designed to run on a computer. Most models deal with the evolution of a single species, but there are also models showing how species interact with each other, often in a competitive (or even predator-prey or host-parasite) relationship. There have even been models (like the one shown here) written specifically for use on a microcomputer and in a spreadsheet environment.[2] However, the model in this chapter has a more specific purpose than most population models, since it deals with the issue of extinction.

The present model is derived from another study by Mark Shaffer, this one of data on the population of grizzly bears (*Ursus arctos horribilis*) in and around Yellowstone National Park in the United States.[3] The data were originally collected by J.J. Craighead, F.C. Craighead, Jr., and colleagues over a 12-year period (1959-70). According to Shaffer (1983, p. 133), these data provide "the most detailed information and longest continuous record available for any grizzly bear population."

The number of grizzlies alive today is not known. Two hundred years ago, the grizzly bear ranged all over western North America from the Pacific Coast east to the Ontario-Manitoba border, and from the shores of the Beaufort Sea to northern Mexico (Storer and Trevis 1955). According to their diaries, the explorers Lewis and Clark had numerous encounters with grizzlies on their travels in 1803-5 (Russell 1967). Today, however, the grizzly bear has almost disappeared from the United

---

2.   For useful reviews of biological population models generally, see Watt 1968 or Collier et al. 1973. On models of species interaction, see Varley et al. 1973, chapter 4, or Krebs 1978, chapter 13; van Ballenberghe 1985 is particularly interesting. An early micro-computer simulation of the interaction of sharks and fishes on the mythical planet of "Wa-Tor" is given in Dewdney 1984. Spreadsheet models of the nearly extinct great bustard (*Otis tarda*) have been made by William Silvert (1983, 1989) and his colleague, A. Longhurst (1985).

3.   The elk population of Yellowstone Park has also been the subject of a similar study: see Fowler and Barmore 1979.

States (except Alaska and Montana), and in Canada it is confined to British Columbia, western Alberta, and the Yukon and Northwest Territories. As a species, the grizzly is probably not yet in danger of extinction. However, "there is little doubt that [specific] populations and even ecotypes (one or more populations with similar ecological characteristics) continue to be threatened" (Pearson 1975, p. 8). One such population is the Yellowstone grizzly, which (some experts say) may disappear within the next 30 years (McNamee 1984, pp. 66ff.)

The grizzly bear has been studied surprisingly little. After the polar bear, the grizzly is the largest terrestrial carnivore on the North American continent. According to a booklet published by the Canadian Wildlife Service, the grizzly bear,

> like the polar bear . . . has a prominent hump over the shoulders formed by muscles of their massive forelegs. The grizzly's unique features are its somewhat dished face, and its extremely long [10 cm] claws. Its color ranges from nearly white or ivory yellow to black. . . . Although grizzly bears have been known to weigh as much as 550 kg, the average male weighs 250-350 kg and the female about half that.

> Young grizzlies are born in a winter den usually during January or February. They are very small, weighing about 400 g, and measuring less than 225 mm. . . . The young grow very rapidly and when they leave the den with the mother in spring weigh around 8 kg . . . . [and] enter the winter den approaching 45 kg. Usually they remain with the mother until June of their second year [CWS 1984, p. 2].

Grizzlies are solitary animals. Their home range is usually 900-1,800 square kilometers (km^2) for males and 200-600 km^2 for females. Grizzlies mate in May or June, but gestation, which takes 6-10 weeks, does not begin until the female enters her winter den in November. In the meantime, the fertilized eggs (or blastocysts)— usually two in number, but sometimes one, three, or even four—remain dormant, floating about in the womb instead of attaching themselves to the wall of the uterus as they do in other mammals. Grizzlies do not mate until they are 5 or 6 years old. After each litter, females do not mate again until their cubs are fully grown. In most cases, this means every 3 years, with the result that the reproductive rate of grizzlies is one of the lowest known for mammals (McNamee 1984, p. 12). Grizzlies are thought to have a natural life-span of 20-25 years, although some are known to have lived for as long as 40 years.

Contrary to popular belief, grizzlies do not hibernate. They enter a den in November and become lethargic, even immobile, until March (males) or late April (females, especially if with cubs). But grizzlies do not fall into a deep sleep; their temperature and respiration do not fall much below normal, and some grizzlies may remain occasionally active throughout the winter. Animals that do hibernate (like the woodchuck, whose temperature plummets to just above the freezing point, whose heartbeat drops from around 100 beats per minute to about 4, and who breathes only about once every 6 minutes) become nearly insensible. Yet animals that hibernate must also waken to the fully active state every few days, in order to move around, urinate, and in some cases eat, drink, and defecate. If they did not do so, they would die of starvation or uremic poisoning. Grizzlies (as well as black, brown, and probably polar bears) have developed what seems to be a unique ability to reabsorb their own wastes, combining them with products catabolized from their stored fat and recreating living tissue. According to McNamee (1984, pp. 254-55), "This state of simultaneous protein catabolism and anabolism is unknown in any other animal. There is a phoenix inside a midwinter's bear, creating a new self from the ashes of the old."

In short, grizzlies are a relatively long-lived species with relatively low reproductive rates, able to occupy and adapt to stable habitats at a level near to their carrying capacity. Ecologists sometimes refer to such species as "K-selected species" (where K stands for carrying capacity) and, as McNamee goes on to say (ibid., p. 61), "the population trend of any K-selected species is determined [primarily] by the survival and reproductive rate of the females. There are frighteningly few adult female grizzly bears still alive in Yellowstone."

Shaffer's approach to modeling the Yellowstone population was to extract from the Craigheads' data six key variables: total population, number of adults, number of subadults, overall population mortality rate, percentage of adult females reproducing, and average litter size. Next, Shaffer determined that all variables except reproductive rate were normally distributed over time. Nor was there any trend or oscillation apparent in the variables.[4] Accordingly, Shaffer proposed a simulation model in which stochastic factors are accounted for by what are called random-normal adjustments: that is, instead of applying fixed rates of mortality, fertility, and

---

4.   See Shaffer 1983 for details of the statistical methods employed. Note that the mortality rate was normally distributed only after transformation to a natural-log scale.

litter size, he used rates that were adjusted to reflect the pattern observed in the available historical data (assuming they were normally distributed). For each rate, Shaffer computed the mean and standard deviation of the observed rates and assumed that the future pattern of rates would vary within the same parameters. In other words, instead of assuming that rates would vary from the mean in a completely random fashion (which would hardly be realistic), Shaffer assumed that random variations would occur within a normal distribution.[5]

The model presented here uses the same approach. An initial population is created and distributed by age-sex cohorts. Each year, each age-sex cohort is aged in accordance with male and female mortality rates, with each rate being adjusted by an appropriate random-normal factor. Similarly, births are simulated on the basis of the number of adult females multiplied by mean annual reproductive rate and mean litter size, with each rate again being adjusted by an appropriate random-normal factor.

The model can be run in two different modes. In *standard* mode, the model keeps track of only the mean time to extinction for the cycles in which extinction occurs. In *tally* mode, however, the model records the total population for every year of each cycle. The latter is the more interesting mode, because, at the end of the simulation run, the model can provide a graphic picture of how the population has evolved over time. However, tally mode is limited to a relatively small number of cycles, each of fairly limited length. This is because of the need to record all the yearly data somewhere in the worksheet. In standard mode, however, the model can run through as many cycles as you like, with each cycle lasting for as long as you like.

Data Requirements

In order to run the model and determine the minimum viable population for a given species, two sets of data are required. First, we need an initial age-sex distribution of the population: that is, the number of bears of each sex and age-group alive at the beginning of the simulation.

---

5.   Strictly speaking, the random numbers generated by computers are usually only pseudo-random, although the distinction is usually glossed over. For further discussion of this and related issues, see appendix A, below.

Second, we need data on five characteristics of its evolution over time:

- the mean mortality rate for males (and its standard deviation);
- the mean mortality rate for females (and its standard deviation);
- the mean reproductive rate for females in the appropriate age cohorts;
- the mean litter size; and
- the mean percentage of any litter that is female (or male).

Then it is up to the user to specify how many times or *cycles* the model is to run, and how many *years* it is to run each cycle.

Overview

The worksheet consists of 199 rows and 18 columns (A to R). To some extent, the number of rows is flexible, because it depends on the maximum number of years the model is expected to run when it is in tally mode (i.e., when it needs space to record results). Here (see Listing 3.1), the worksheet is designed to run in tally mode for a maximum of 100 years in each cycle. It could just as well be more or less.

The worksheet consists of four parts, arranged one above the other. The top 20 rows (cells A1:R20) provide the area for data entry and recording the principal results. Next is a scratchpad area (cells A24:R48) that serves two purposes. On the left is a lookup table that is used for creating random-normal distributions (see below); on the right is an area for calculating the age-sex distribution of the population in each year, before it is written into the visible part of the worksheet above.

Below all this are two macros that run the model for whatever number of cycles and years per cycle the user may specify. One macro (Alt-S) runs the model in standard mode. As the simulation proceeds, the program records on the screen the number of times the population goes extinct and the mean time to extinction when it does.

Listing 3.1

```
 |A||  B  ||  C  ||   D  |   |  G  ||  H  ||   I  ||  J  ||  K  ||  L  |      |O||  P  ||  Q  ||   R  |
```

MINIMUM VIABLE POPULATION   © 1993 by T.J. Cartwright

Age	Male	Fem	Tot		Species: Grizzly Bear				Age	Male	Fem	Tot
Cub	5	6	11		Mean mortality, male	.17	.08		Cub	4	5	11
1	1	3	4		Mean mortality, fem	.17	.08		1	4	5	4
2	2	2	4		Mean reproduction	.30	.07		2	1	3	4
3	1	1	2		Mean litter size	2.21	.23		3	2	1	2
4	2	1	3		Female offspring	50%			4	1	1	3
5	1	1	2						5	2	1	2
6	2	2	4		No of cycles to run	10			6	1	1	4
7	1	1	2		No of years/cycle	100			7	2	2	2
8	1	1	2						8	1	1	2
9	1	1	2		Current cycle	11			9	1	1	2
10	1	1	2		Current year	1			10	1	1	2
11	1	1	2						11	1	1	2
12	1	1	2		No of times extinct	0			12	1	1	2
13	1	1	2		Mean time to extinct	0			13	1	1	2
14	1	1	2						14	1	1	2
15+	2	2	4		Total mature males	13			15+	1	1	4
Tot	24	26	50		Total mature females	12			Tot	24	26	50

Look-Up Table for Actual
Deviations of Means

Calcs for Next Year

From	To	Dev		Age	Male	Fem	Tot
	.00	-2.81		Cub	4	1	5
.00	.05	-2.81		1	3	4	7
.05	.10	-1.65		2	3	5	8
.10	.15	-1.28		3	1	2	3
.15	.20	-1.04		4	2	1	3
.20	.25	-.84		5	1	1	2
.25	.30	-.67		6	2	1	3
.30	.35	-.52		7	1	1	2
.35	.40	-.39		8	2	2	4
.40	.45	-.25		9	1	1	2
.45	.50	-.13		10	1	1	2
.50	.55	.13		11	1	1	2
.55	.60	.25		12	1	1	2
.60	.65	.39		13	1	1	2
.65	.70	.52		14	1	1	2
.70	.75	.67		15+	1	1	2
.75	.80	.84		Tot	26	25	51
.80	.85	1.04					
.85	.90	1.28					
.90	.95	1.65					
.95	1.00	2.81					

Run-Time Macros (S = Standard Run; T = Tally Run)

```
{\S}  {home}{paneloff}                           home cursor; dialog off
      {let K16,0}{let K17,0}                     zero extinction counters
      {let K13,1}                                initialize cycles
      {let K14,1}                                [C56] initialize years
      /c B4:D20,P4,v                             copy initial pop to current
      !                                          [C58] calc next pop
      {if K14=K11}{let K13,K13+1}{branch C56}if max years, reinitialize
      {if K13>K10}{branch C69}                   if max cycles, quit
```

```
   |A|| B || C || D |   | G || H || I || J || K || L |   |O|| P || Q || R |
61        {let K14,K14+1}                           increment year
62        {if P20<1}{branch C66}                     if no males left....
63        {if Q20<1}{branch C66}                     if no females left....
64        /c P27:R43,P4,v                            copy next pop to current
65        {branch C58}                               loop to next year
66        {let K17,(K17*K16+K14)/(K16+1)}            [C66] adjust mean time
67        {let K16,K16+1}{beep}                      increment times extinct
68        {let K13,K13+1}{branch C56}                reinitialize
69        {beep}{panelon}                            [C69] dialog on
70
71   {\T} {home}{paneloff}                           home cursor; dialog off
72        {if K10>10}{let K10,10}                    limit max cycles
73        {if K11>100}{let K11,100}                  limit max years
74        /b G100:P199 ~                             blank records
75        {let K16,0}{let K17,0}{let K13,1}          initialize counters
76        {let K14,1}                                [C76] initialize years
77        /c B4:D20,P4,v                             copy initial pop to current
78        !                                          [C78] calc next pop
79        /c R20,[B101+K13;B103+K14],v               update records
80        {if K14=K11}{let K13,K13+1}{branch C76}    if max years, reinitialize
81        {if K13>K10}{branch C91}                   if max cycles, quit
82        {let K14,K14+1}                            increment year
83        {if P20<1}{branch C87}                     if no males left....
84        {if Q20<1}{branch C87}                     if no females left....
85        /c P27:R43,P4,v                            copy next pop to current
86        {branch C78}                               loop to next year
87        {let K17,(K17*K16+K14)/(K16+1)}            [C87] adjust mean time
88        {let K16,K16+1}{beep}                      increment times extinct
89        {if K14<K11}{let [B101+K13;B103+K14+1],0}    add zero pop to records
90        {let K13,K13+1}{branch C76}                reinitialize
91        /b [B101+K13;B103+K14] ~                   [C91] blank the last cycle
92        {beep}{panelon}                            dialog on
```

Source: Adapted from a model in Mark Shaffer, "Determining Minimum Population Sizes for the Grizzly Bear," International Conference on Grizzly
Bear Research and Management, 5, pp. 133-39.

Tally Sheet

	Year	Cycle Population									
		1	2	3	4	5	6	7	8	9	10
Col offset	1	50	50	50	50	50	50	>	>	50	50
6	2	51	51	54	53	54	53	>	>	56	55
Row offset	3	53	52	57	55	56	54	>	>	53	55
99	4	57	54	52	61	62	57	>	>	56	53
	5	61	54	56	65	61	55	>	>	55	61
	6	62	61	57	68	62	56	>	>	59	65
	7	66	61	63	70	62	63	>	>	63	64
	8	69	59	63	73	64	63	>	>	62	66
	9	70	66	63	70	63	67	>	>	63	71
	10	74	70	71	63	64	69	>	>	66	68
	11	74	68	72	71	66	71	>	>	64	71
	12	76	70	75	75	73	70	>	>	65	67
	93	139	95	136	90	109	110	>	>	>>	127
	94	133	89	133	88	105	110	>	>	>>	125
	95	124	89	125	80	111	106	>	>	>>	115
	96	126	89	120	77	113	116	>	>	>>	115
	97	125	87	125	74	116	113	>	>	>>	122
	98	127	100	125	82	110	117	>	>	>>	121
	99	124	107	115	85	104	109	>	>	>>	122
	100	116	104	109	90	98	112	>	>	>>	126

The other macro (Alt-T) runs the simulation in tally mode. That is, it does exactly the same, except that a detailed record, or tally, is kept of the total population in every year of every cycle. Furthermore, in tally mode the run is limited to a maximum of 10 cycles and 100 years per cycle.[6] Finally, the last part of the worksheet (cells A98:R199) is the area used for recording the population in each year (up to 100) of each cycle (up to 10) when the model is run in tally mode. This area also provides the data for a graph of the population dynamics, year by year, for each cycle, at the end of the tally-mode run.

Operation

The worksheet is surprisingly simple. Apart from a few totals, there are only two formulas in the entire model: one to age the population cohorts each year, and one to compute new births. The rest of the model is composed of a lookup table and two macros. In the top ("visible") part of the worksheet, the initial population totals are calculated with the following formulas:

    B20   SUM(B4:B19)
    C20   SUM(C4:C19)
    D20   SUM(D4:D19)

Calculation of the total number of mature males and females in cells K19 and K20 is a bit more complicated. Shaffer (1983, p. 135) proposes that both sexes reach sexual maturity at 4½ years of age. Since most grizzlies are born in January or February, that could be interpreted to mean that they become mature during the breeding season (May or June) when they are 4 years old. However, we have taken the more conservative view that grizzlies do not become sexually active until age 5.[7] Thus, the totals are calculated as follows:

---

6.   The maximum number of cycles (10) reflects the maximum number of variables that can be graphed by the spreadsheet rather than memory requirements. If you want to run the model for more cycles, by all means alter the macro to increase the number of cycles and use the columns to the right of P100:P199.

7.   Even this may be optimistic in some cases. Some studies have found that female maturity may not occur until age six or even seven, especially when the climate is less benign or food is less plentiful than is the case in an environment like Yellowstone Park (e.g. Pearson 1975).

```
K19          SUM(P9:P19)
K20          SUM(Q9:Q19)
```

The formula for aging the population is quite straightforward. The number of grizzlies in any age-sex cohort in a given year is equal to the number of bears in the *preceding* cohort in the *previous* year, adjusted by the survival rate (or one minus the mortality rate). The mortality rate is given in cells K4 (for males) and K5 (for females). However, the calculation is not quite as simple as this, because we do not want to use a fixed mortality rate. We want mortality to vary slightly from cohort to cohort and from year to year, in order to reflect the stochastic dimension of population change, as discussed above. We achieve this effect by taking the mean mortality rate and adjusting it, each time it is used, by a certain number of standard deviations. The precise number of standard deviations that we add to or subtract from the mean rate in each case is determined by selecting a value, at random, from a table of standard deviations that is skewed in the shape of a normal distribution.

The table required to generate this random-normal distribution is shown in cells B27:D48. There are no formulas in this table: the values are just copied from a standard textbook on statistics (e.g., Edwards 1967, Table 3).[8] What happens is this: in order to age each age-sex cohort, the model generates a random number (between 0 and 1) and "looks up" that number in column C of the table, until it finds that number or the closest *lower* number. When it finds that number in column C, the model returns the corresponding number from column D. Then, this number is used to determine how many standard deviations are added to or subtracted from the mean.[9] For example, suppose that, for a particular age-sex cohort, the computer generates a random number, say .3585. From the random-normal table, this is equivalent to a weighting of -.52. (See cells B35, C35, and D35 in Listing 3.1). Then, if the mean mortality rate is .17 and the standard deviation is .08 (which is the case depicted in Listing 3.1), the mortality rate for this particular cohort would be .17 plus (-.52)*.08, or about .13. Similarly, if the next age-sex cohort generates a random number of, say, .8056, then its weighting is 1.04 (from cell D45); so the mortality rate for that cohort would be .17 + (1.04)*.08, or about .25. And so on.

---

8.   In fact, there are ways of computing a random-normal distribution directly, without the need for a lookup table: see the technical note beginning on page 81, below.

9.   Column B, incidentally, is unnecessary as far as the computer is concerned; it is included only to clarify the operation of the lookup table for the user.

Once the lookup table is organized, the remaining formulas in the worksheet are quite simple. Consider the formula in cell P28, which is designed to calculate the number of 1-year old male grizzlies:

```
ROUND(P4*(1-(K$4+L$4(VLU(RAN,C$28:D$48,1))))))
```

The result is determined by multiplying the number of cubs in the current year (cell P4) times one minus the mortality rate (cell K4), adjusted by the standard deviation (cell L4) times a random-normal factor found in the vertical lookup table (VLU) located in cells C28:D48—with the expression rounded to the nearest whole number, since you cannot have a fraction of a grizzly bear running around! The formula for 1-year old female grizzlies in cell Q28 is exactly the same (*mutatis mutandis*):

```
ROUND(Q4*(1-(K$5+L$5(VLU(RAN,C$28:D$48,1))))))
```

Finally, cell R28 is defined simply as the sum of P28 and Q28.

Thanks to the absolute cell-references (dollar signs) in the above three formulas, they can all be copied from row 28 down into rows 29 to 42. There is one slightly awkward matter, and that is how to age the last cohort (15 years and older). We cannot just assume they all die off at the end of a year in the cohort; otherwise, there would be never be any grizzlies 16 years old or older. Equally, we can hardly assume that the mortality rate remains close to the mean in extreme old age. Accordingly, we have arbitrarily assumed that the population in the last cohort (15 years and older) is computed as the sum of the preceding cohort (14 year-olds) in the previous period, aged by the appropriate survival rate, plus an entirely random proportion of the last cohort (15 years and older) in the previous period. Thus, the formulas for males in the last cohort (cell P42) and females in the last cohort (cell Q42) in the last cohorts are respectively:

```
ROUND(P18*(1-(K$4+L$4(VLU(RAN,C$28:D$48,1))))
       +RAN*(P18*(1-(K$4+L$4(VLU(RAN,C$28:D$48,1)))))))

ROUND(Q18*(1-(K$5+L$5(VLU(RAN,C$28:D$48,1))))
       +RAN*(Q18*(1(K$5+L$5(VLU(RAN,C$28:D$48,1)))))))
```

That leaves only the question of calculating the number of new births in cells P27 and Q27. In this case, the number of cubs of each sex is found by multiplying the number of (sexually) mature females (in cell K20) by the rate of reproduction (in cell K6), by the mean litter size (in cell K7), and by the appropriate sex ratio (in, or derived from, cell K8). Once again, however, we want to adjust two of the fixed parameters (the reproductive rate and the mean litter size) by a random-normal value in order to reflect the influence of stochastic factors. Exactly the same procedure is used as with the mortality rate: that is, the lookup table is used to determine how much influence the specified standard deviations will have on each mean. Thus, the formulas for new male births (cell P27) and new female births (cell Q27) are respectively:

```
ROUND(K20*(K6+L6*VLU(RAN,C28:D48,1))
      *(K7+L7*VLU(RAN,C28:D48,1))*(1-K8))*(K19>0)

ROUND(K20*(K6+L6*VLU(RAN,C28:D48,1))
      *(K7+L7*VLU(RAN,C28:D48,1))*K8)*(K19>0)
```

The last expression in both formulas (K19 > 0) is a simple way of zeroing the result, if there are no mature males around.[10]

Finally, there are the two macros, which are in fact very similar to each other. They work like this:

1. The screen "window" is moved to the upper left-hand corner of the worksheet, so that we can see what is going on; and the various counters (K13, K14, K16, and K17) that record the results are initialized. If the second macro has been activated and the simulation is running in tally mode, the control counters (K10 and K11) are also checked to see that they do not exceed the allowable maxima (10 and 100 respectively); if they do exceed the maxima, they are reset to the maxima.

2. The starting population distribution, which the user has entered on the left side of the screen, is written into the current-population area on the right side.

3. The population for the following year is calculated in the scratch pad area to the right of what is visible on the screen.

---

10.   If cell K19 is greater than zero, the expression (K19 > 0) is true; so it returns a one, and the result of the calculation is unaffected. If cell K19 is zero, the expression (K19 > 0) is false; so it returns a zero and the result of the whole calculation becomes zero and there are no cubs born that year.

4. The model checks to see (a) if the maximum number of years and/or cycles has been reached and (b) if there are still mature males and females in the population capable of reproducing. If the maximum years and/or cycles have been reached or if there are no longer any mature males and females, appropriate action is taken.

5. If everything is satisfactory, the results for the next year are copied from the scratch pad area where they were calculated up into the area on the right side of the screen.

6. The macro then loops back and calculates the population for the next year, does its checks, and (if everything is satisfactory) writes the new results to the visible part of the model. The model keeps on looping back like this, until each cycle of the prescribed number is completed.

At the end of the run, the summary results appear in cells K16:K17, which are in the middle of the screen near the bottom. The results generated by the model are (a) the number of times the population became extinct and (b) the mean time it took to do so when it did.

If the model has been run in tally mode, a detailed record of how these results came about will be available lower down in the model (from cell A98 onwards). After the run is completed, this record can be reviewed directly, or it can be translated into a line graph. See Figure 3.1 for a graph of the evolution of what is apparently (at least after ten runs) a stable and secure population, as generated from the data shown in Listing 3.1.

Interpretation and Use

Naturally, models like the one presented here have to be used with a good deal of caution. It must always be remembered that the results are only as good as the parameters provided. In this particular case, data on grizzly-bear mortality rates, reproductive rate, and so on are based on observations of a single population over only 12 years. Clearly, that is not a very strong basis for generalization.

In addition, it should be noted that the model presented here differs in several minor respects from Shaffer's original version (1983):

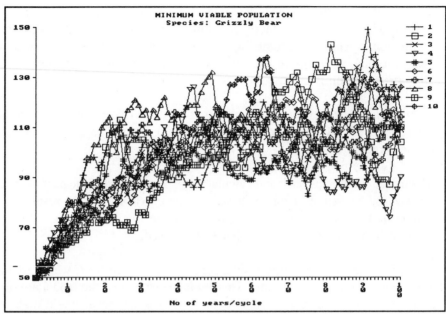

Figure 3.1. Population Runs Based on Data in Listing 3.1

1. The model here assumes that the fertility rate is normally distributed over time, whereas Shaffer (ibid., p. 133) found that, for his grizzly bear population at least, fertility was not normally distributed.

2. The model here assumes that the mortality rate is normally distributed over time, whereas Shaffer (ibid., pp. 133, 135) found that, for his grizzly bear population, the mortality rate had to be transformed to a natural-log scale in order for it to appear normally distributed.

3. Shaffer tested two criteria for extinction—the death of all animals, and the death of all mature females—and found that it made little difference which criterion he used (ibid., p. 137). The method used here is a bit different. The number of mature members of both sexes is factored directly into the birth calculations in cells P27 and Q27 (see above); so only one criterion of extinction is used—the death of either all the males or all the females.

Finally, the time-scale of the model—a few centuries—is scarcely enough to allow for genetic and other long-run kinds of stochasticity. In fact, if the model shown here is run for long periods of time (e.g., 500 or 1,000 years), populations that appear secure over shorter time spans can "suddenly" run into a spot of statistical bad luck (so to speak) and go extinct. Perhaps this just represents the futility of looking for certainty in an uncertain world, or perhaps no small population can ever be totally secure from extinction.

In any case, Shaffer's model of the grizzly-bear population in Yellowstone Park has had two important effects on the debate about minimum viable population. First, the immediate effect of his study was to cause estimates of the minimum viable population in Yellowstone Park to be revised upwards from 35 to 50 bears. Shaffer ran 18 different tests of sensitivity with his model, most of them using an initial population of 35 bears and running over 50 cycles of 100 years each. He found that a fairly minor (e.g., 10%) increase in mean mortality coupled with a fairly minor (e.g., 10%) decrease in mean litter size had the effect of doubling the minimum viable population. By contrast, making fairly drastic changes in the minimum age of fertility in females (e.g., raising it from 4½ to 6½ years) and in the proportion of female to male cubs (e.g., reducing it to 41%) had relatively little effect: the minimum viable population increased by only 10 bears. Thus, he concludes:

> Of the four modifications which had a substantial impact on the initial estimate of minimum viable population size, the possible underestimate of mean mortality rate caused the greatest concern. Because of this, [my] initial estimate of minimum viable population size of 35 was raised to 50 for the Yellowstone population [ibid., pp. 137-38].

These results were subsequently confirmed in other models (e.g., Goodman 1988 and Belovsky 1988; see also Oliveri 1991). Shaffer expects that refinement of his model will, if anything, lead to further upward revision of the number. Indeed, some ecologists (e.g., Seal 1989) maintain that population viability analysis (or PVA) may be the key to environmental conservation.

Perhaps more important than this, however, was Shaffer's discovery that even highly endangered populations can sometimes persist for long periods of time. As Shaffer puts it,

The most remarkable and sobering feature [of my study] was the relatively high survival times for populations, even when their probability of surviving was quite low. For example . . . a population of 10 bears had only a 16% chance of surviving the 100-year test period; yet the average survival time for the 42 populations that became extinct [out of the 50 simulated] was 31 years [1983, p. 138].

This conclusion is echoed by Thomas McNamee:

What is most frightening about Shaffer's model is how long that last doomed population hangs on: without sophisticated analysis, even an experienced wildlife expert would not necessarily come intuitively to the belief that the population was in trouble until long after it was too late. So, rough and ragged and confusing as present computer simulations may be, they are indispensable for peering into the future, and worthy of continuing refinement; they have already made it indisputably clear that the reproductive capacity of the grizzly bear cannot possibly compensate for the Yellowstone [Park] population's present level of mortality [1984, pp. 65-66].

Evaluation and Extension

Among possible extensions of the model shown here are the following:

* Replacement of the single mortality rate and standard deviation for each sex with a table of rates and deviations with one set for each age cohort.

Shaffer reports (1983, p. 138) that he tried this, using the data collected by the Craigheads and their colleagues, but that it did not make much difference to his results. He cites another study of black bears, which seems to show that mortality rates in bears tend to vary only in subadult age-cohorts.

* Making the fertility rate and/or litter size inversely dependent on the population size (i.e., density dependent), on the assumption that fertility tends to decline with crowding.

Shaffer (1983, p. 134) analyzed the data collected by the Craigheads and their colleagues, and found negative correlations of only .10 (for fertility rate) and .05 (for litter size). In other words, density dependence did not seem to be a significant factor among grizzlies. Shaffer also notes that lagging the population data by a single year did not seem to make much difference.

- Constructing models of species interaction, specifically where one is predator and the other is prey.

For example, Fowler and Barmore (1979) have studied dynamics of an elk herd (*Cervus canadensis*) in Yellowstone Park and the effect of reintroducing wolves into the park.

In summary, the model shown here is a reasonably robust and pragmatic approach to determining minimum viable population. While this model does not and cannot reflect reality in every detail, it is sufficiently compelling to have had an effect on prevailing thinking about minimum viable population levels for endangered populations such as the grizzly bear in Yellowstone Park.

Technical Note

If desired, a random-normal distribution can be computed directly, without the need for a lookup table. One method of doing this is to use either of the following formulas (Abramowitz and Stegun 1965, p. 953):

$$\sqrt{-2 \ln R1} * \sin(2 * \pi * R2)$$
$$\sqrt{-2 \ln R1} * \cos(2 * \pi * R2)$$

where R1 and R2 are two random deviates from a uniform random distribution on the interval (0,1). This technique is known as the Box and Muller method, after its inventors. In a spreadsheet, the formulas would be written:

```
SQRT(-2*LN(RAN))*SIN(2*PI*RAN)
```

and

```
SQRT(-2*LN(RAN))*COS(2*PI*RAN)
```

Repeated use of either formula generates a random distribution about a mean of zero and with a standard deviation of one.

The simulation model described in this chapter can be altered to make use of the Box and Muller technique by the simple expedient of substituting one of the above formulas for

VLU(RAN,C28:D48,1)

in all the cells where it appears (namely, cells P27:Q42). However, while this method may seem more elegant than the lookup table, the latter is much faster, at least in a spreadsheet. Presumably, this is because each calculation using the Box and Muller method requires the computer to generate a pair of random numbers, whereas each calculation with the lookup table requires only one. In any case, my computer runs the full "tally" mode (10 cycles of 100 years each) in about 6 minutes with the lookup table, but needs nearly 9 minutes to do it without the lookup table. The latest versions of some spreadsheets (e.g. Excel 4.0) have a built-in ability to generate not only random-normal as well as random numbers.

Incidentally, a spreadsheet can be used to test the efficacy of the Box and Muller method. In a blank worksheet, enter either of the above formulas in (say) cell A4. Copy the formula into a suitable number of additional cells (e.g., A5:A2003). This will give you a column of 2,000 numbers all calculated with the same formula. (How many calculations you will be able to do on your system will depend on the capacity of its memory and/or the limitations of the software you are using.) In some other suitable cells, enter a formula for the mean of these 2,000 vallues—i.e., AV(A4:A2003)—and its standard deviation—i.e., STD(A4:A2003). They should be close to zero and one, respectively.

What you do next depends on how sophisticated a spreadsheet you are using. The simplest step is to sort the column in ascending order (using /Arrange in SuperCalc) and then graph the column as a bar chart or line graph. This will yield a nice, smooth S-curve that is indicative of a normal distribution. In order to get the bell-shaped curve that most people recognize as the sign of a normal distribution, an additional step is required: namely, to make a frequency distribution. That is, you need to allocate the 2,000 results into a number of classes and record how many results there are in each class.

The first step is this process is to define the boundaries of the classes you intend to use and list them in a nearby column (say, column C). The next step depends on your spreadsheet.

- If your spreadsheet has a command specifically for allocating a range of data to a set of classes, or "bins," you should use it (since it is much faster than the alternative). In SuperCalc5, for example, the command is

    // D, A, D, datarange, binrange

where *datarange* is the range of data to be allocated (including a "label" line at the top) and *binrange* is the range of classes to which the data are to be allocated. The results are then written into the column of cells immediately to the right of the *binrange*.

• If your spreadsheet lacks such a command, it should have a function that will search a data range and count the number of cells in that range that meet a particular criterion. This function is often named DCOUNT() or similar. In SuperCalc, for example, the syntax is as follows:

DCOUNT(datarange, offset, criterion)

where *datarange* is as defined above, *offset* is the offset (from the left-most column of the *datarange*) of the column that is to be searched, and *criterion* is the range where the search criterion is defined.

Listing 3.2. illustrates how the second method might be implemented, with a total of 30 classes defined in columns E and F. The key formulas are those in cells A4:A2003 (or as far as you can go before your spreadsheet runs out of memory!) Consider the formula in cell A4:

```
SQRT(-2*LN(RAN))*COS(2*PI*RAN)
```

This formula is then copied into the cells directly below it (e.g., A5:A2003). The other key formulas are:

```
G4    DCOUNT(A$3:A$2003,0,J4:J5)
G5    DCOUNT(A$3:A$2003,0,K4:K5)
J5    AND(A$4>F3,A$4<=F4)
K5    AND(A$4>F4,A4$<=F5)
```

Cells J4 and K4 contain just the text "Sample"—which has to be the same as the label at the top of the "database" in cell A3. This block of six cells (G4:K5) can then be copied down into cells G6, G8, G10, etc. as far as cell G32.[11] Note that SuperCalc requires the data range to be formally "input" before the DCOUNT function will take effect. Use the command, //D,I,datarange, and be sure to include the label in the topmost cell as part of *datarange*. Other spreadsheets may have other peculiarities involving the use of their database functions.

---

11. The awkwardness of this side-by-side arrangement is dictated by the fact that most of the data management functions in SuperCalc are based on the assumption that databases will be arranged vertically. Thus, "labels" must appear above, not beside, the data they describe.

Listing 3.2

	A	B	C	D	E	F	G	H	J	K
1	RANDOM-NORMAL DISTRIBUTION Using the Box and Muller Method									
2	---									
3	Sample		Classes	From:	To:		Frequency		DCOUNT	labels
4	.10723			$-\alpha$	-10		0		Sample	Sample
5	1.6672			-10	-4		0		0	0
6	1.8311		Mean	-4	-3		5		Sample	Sample
7	.39761		.02922	-3	-2.75		3		0	0
8	1.3100			-2.75	-2.5		5		Sample	Sample
9	2.3023		StDev	-2.5	-2.25		6		0	0
10	1.6692		.99112	-2.25	-2		22		Sample	Sample
11	.32182			-2	-1.75		28		0	0
12	-2.788			-1.75	-1.5		51		Sample	Sample
13	.12450			-1.5	-1.25		85		0	0
14	-.7062			-1.25	-1		104		Sample	Sample
15	.93116			-1	-.75		123		0	0
16	-.2340			-.75	-.5		153		Sample	Sample
17	-.1346			-.5	-.25		180		0	0
18	-.4864			-.25	0		202		Sample	Sample
19	-.1739			0	.25		183		0	1
20	1.3356			.25	.5		215		Sample	Sample
21	-.8134			.5	.75		170		0	0
22	.72159			.75	1		147		Sample	Sample
23	.17850			1	1.25		108		0	0
24	.87228			1.25	1.5		78		Sample	Sample
25	1.0229			1.5	1.75		50		0	0
26	2.0627			1.75	2		40		Sample	Sample
27	.33735			2	2.25		17		0	0
28	-.3513			2.25	2.5		12		Sample	Sample
29	.49589			2.5	2.75		8		0	0
30	1.8773			2.75	3		1		Sample	Sample
31	-1.386			3	4		3		0	0
32	-.3550			4	10		1		Sample	Sample
33	-.2238			10	α		0		0	0
34	.60009									
35	.49978									
36	.56521									
37	-1.327									
38	.04463									
39	-1.491									
40	-1.706									
41	-1.175									
42	.91877									
43	.20547									
·	·	·	·	·	·					
2002	-.8788									
2003	1.0634									

Finally, graph the results (which will be in column G, next to the column of class boundaries) as a bar chart. The result will be the familiar bell-shaped curve of the normal distribution, thus substantiating the Box and Muller method.

Bibliography and References

Abramowitz, Milton, and Irene A. Stegun, eds. 1965. *Handbook of Mathematical Functions, with Formulas, Graphs, and Mathematical Tables*. New York: Dover. First published as the National Bureau of Standards Applied Mathematics Series 55 (June 1964).

Ballenberghe, Victor van. 1985. "Wolf Predation on Caribou: The Nelchina Case History." *Journal of Wildife Management*, 49.3, pp. 711-20.

Belovsky, Gary E. 1988. "Extinction Models and Mammalian Persistence." In Soulé 1988, pp. 35-57.

Canadian Wildlife Service (CWS). 1984. *Grizzly*. Hinterland Who's Who Series. Ottawa: Supply and Services Canada.

Collier, B.D., G.W. Cox, A.W. Johnson, and P.C. Miller. 1973. *Dynamic Ecology*. Englewood Cliffs, N.J.: Prentice-Hall.

Dewdney, A.K. 1984. "Computer Recreations." *Scientific American*, 251.6 (December), pp. 14-22. Contains notes on the "Wa-Tor" simulation.

Edwards, Allen L. 1967. *Statistical Methods*. 2nd ed. New York: Holt, Rinehart, and Winston.

Fowler, Charles W., and William J. Barmore. 1979. "A Population Model of the Northern Yellowstone Elk Herd." In *Proceedings of the Conference of Scientific Research in National Parks*, 5, pp. 427-34. National Park Service, Department of the Interior.

Frankel, O.H., and Michael E. Soulé. 1981. *Conservation and Evolution*. Cambridge: Cambridge University Press.

Gilpin, M.E., and Michael E. Soulé. 1986. "Minimum Viable Populations: Processes of Species Extinction." In *Conservation Biology: The Science of Scarcity and Diversity*, edited by M.E. Soulé, pp. 19-34. Sunderland, U.K.: Sinauer Associates.

Goodman, Daniel. 1988. "The Demography of Chance Extinction." In Soulé 1988, pp. 11-34.

Gould, Harvey, and Jan Tobochnik. 1988. *An Introduction to Computer Simulation Methods: Applications to Physical Systems*. 2 vols. Reading, Mass.: Addison-Wesley.

Hammersley, J.M., and D.C. Handscomb. 1973. *Monte Carlo Methods*. New York: Wiley.

Kaufman, Les, and Kenneth Mallory, eds. 1986. *The Last Extinction*. Cambridge, Mass.: MIT Press.

Kleijnen, Jack P.C. 1974, 1975. *Statistical Techniques in Simulation*. 2 vols. New York: Marcel Dekker.

Krebs, C.J. 1978. *Ecology: The Experimental Analysis of Distribution and Abundance*. New York: Harper and Row.

Longhurst, A., and William Silvert. 1985. "A Management Model for the Great Bustard in Iberia." *Bustard Studies*, 2, pp. 57-72.

McNamee, Thomas. 1984. *The Grizzly Bear*. New York: Knopf.

Metropolis, Nicholas, and Stanislaw Ulam. 1949. "The Monte Carlo Method." *Journal of the American Statistical Association*, 44, pp. 335-41. Reprinted in *Stanislaw Ulam: Sets, Numbers, and Universes; Selected Works*, edited by W.A. Beyer, J. Mycielski, and G.-C. Rota. Paper no. 38, pp. 319-25. Cambridge, Mass.: MIT Press, 1974.

Mundy, K.R.D., and D.R. Flook. 1973. *Background for Managing Grizzly Bears in the National Parks of Canada*. Canadian Wildlife Report Series, no. 22. Ottawa: Information Canada.

Myers, Norman. 1979. *The Sinking Ark: A New Look at the Problem of Disappearing Species*. Oxford: Pergamon.

Neumann, John von. 1951. "Various Techniques Used in Connection with Random Digits." Reprinted in *John von Neumann Collected Works*, edited by A.H. Taub, vol. 5, no. 23. New York: Pergamon, 1961.

Oliveri, Tony. 1991. "Modeling Extinction in a Spreadsheet." In "Environmental Models in a Spreadsheet." Discussion Paper. Faculty of Environmental Studies, York University. Toronto: photocopy.

Pearson, Arthur M. 1975. *The Northern Interior Grizzly Bear Ursus arctos L.* Canadian Wildlife Series, Report no. 34. Ottawa: Information Canada.

Russell, Andy. 1967. *Grizzly Country*. New York: Knopf.

Seal, Ulysses S., ed. 1989. *Conservation Biology and the Black-Footed Ferret.* New Haven: Yale University Press.

Seton, Ernest Thompson. 1900. *The Biography of a Grizzly.* Toronto: Copp Clark.

Shaffer, Mark L. 1981. "Minimum Population Sizes for Species Conservation." *Bioscience*, 31.2, pp. 131-34.

———. 1983. "Determining Minimum Viable Population Sizes for the Grizzly Bear." *International Conference on Bear Research and Management*, 5, pp. 133-39.

———, and F.B. Samson. 1985. "Population Size and Extinction: A Note on Determining Critical Population Sizes." *The American Naturalist*, 125.1, pp. 144-52.

Silvert, William. 1983. "Ecological Modelling with Electronic Spreadsheets." *Canadian Mathematical Society Applied Mathematics Notes*, 8, pp. 1-8.

———. 1989. "Modelling for Managers." *Ecological Modelling*, 47, pp. 53-64.

Soulé, Michael E., ed. 1988. *Viable Populations for Conservation.* Cambridge University Press.

Storer, T.I., and L.P. Trevis. 1955. *California Grizzly.* Berkeley and Los Angeles: University of California Press.

Swartzman, Gordon L., and Stephen P. Kaluzny. 1987. *Ecological Simulation Primer.* New York: Macmillan.

Varley, G.C., G.R. Gradwell, and M.P. Hassell. 1973. *Insect Population Ecology: An Analytical Approach.* Berkeley and Los Angeles: University of California Press.

Watt, Kenneth E.F. 1968. *Ecology and Resource Management: A Quantitative Approach.* New York: McGraw-Hill.

Chapter 4

SUSTAINABLE YIELD: MANAGING THE FOREST FOR THE TREES

What is this life if, full of care,
We have no time to stand and stare.

No time to stand beneath the boughs
And stare as long as sheep or cows.

No time to see, when woods we pass,
Where squirrels hide their nuts in grass.

W.H. Davies, Leisure

As rising levels of carbon dioxide become more and more threatening and ecological diversity comes more and more under attack from urbanization and industrialization, forests become ever more precious resources. For centuries, we have been accustomed to cutting down trees indiscriminately, whether to clear the land for agriculture, to make way for houses and subdivisions, or just to meet the endless demand for wood. The effect is the same: we are cutting down trees faster than we are replacing them. Nor can this continue, because trees are not just aesthetically pleasing. Trees also play a vital role in combating the "greenhouse effect" by recycling carbon dioxide into oxygen. Trees provide an indispensable habitat for many of the world's flora and fauna. And trees represent an important source of genetic wealth. In short, we need trees.

On the other hand, this does not mean we have to deny ourselves a vital resource. Trees can be successfully "farmed," meaning that the cultivation of trees can make good economic as well as ecological sense. Agroforestry is a small but nevertheless growing component of agriculture in both industrialized and developing countries. Critical to any development of this sector, however, is an understanding of the long-term dynamics of a forest—how does a forest evolve in terms of the age and quality of its trees? and what is the effect of outside interventions, including natural disasters and human activities?

The purpose of this chapter is to present a model that simulates the growth and development of a forest over time, subject to the effect of forest fires on the one hand and various harvesting strategies on the other.[1] To program this kind of model in a spreadsheet is quite challenging. Human population models work quite well on a microcomputer, for they usually involve ages up to 70 or 80 years old, only two different categories (male and female), and time-horizons of perhaps 20 or 50 years. Forests, on the other hand, require age-categories up to 200 years old, several different yield categories, and time-horizons measured in centuries! Even using 5-year age-groups, only 3 different yield categories, and 5-year cycles, the present model must cope with a matrix of 45 cells by 60 cells by 3 cells, with many of the cells containing formulas running to lengths of 100 or more characters. In addition, the model requires some complex (and rather inelegant) macros in order to do the simulation. Partly because of these macros, the model presented here runs rather slowly. On a 386-class computer with a mathematics coprocessor, a single run of the model over 300 years can take 6-7 minutes; on a similar 486 machine, it takes about 1 minute. However, while the model may be a bit slow, it does work.

Purpose of the Model

Essentially, the model does two things. It defines a particular forest (in terms of the age and quality of its trees) and then simulates its development over time, subject to the effects of forest fires and (optionally) harvesting. Second, the model computes the merchantable volume (MV) and the extracted volume (EV) of timber in the forest as a result of these impacts. Thus, the model has three main purposes:

● To show how different kinds of forests evolve over time. For example, you can model a *natural* forest (where the trees vary widely in age), a *replanted* forest (where a lot of trees are of the same age), or a *stable* forest (in which the age distribution is more or less uniform).

---

1. There is also a growing number of habitat evaluation models being developed in spreadsheets for use in wildlife management (e.g., Gray and Keith 1988).

- To illustrate the impact of forest fires. The model allows the user to make various adjustments to the projected effect of fires, in order to illustrate best-case and worst-case scenarios.

- To illustrate the effect of different kinds of human interventions. In other words, you can see the effect of leaving the forest completely alone and compare that with the effect of various forms of controlled cutting.

Conceptual Basis

The model presented here is derived from a model (FIRFOR) developed by R.M. Newnham, of the Petawawa National Forestry Institute, Chalk River, Ontario, Canada (Newnham 1987). FIRFOR is programmed in Fortran-77 and runs on a DEC VAX/785 minicomputer. FIRFOR is designed to simulate a forest about 100,000 hectares (ha) in size.[2] In Canada, at least, this is a fairly typical management unit, capable of sustaining a local manufacturing facility. Larger and smaller forests can be input into the model; however, some of its assumptions may not be realistic for forests of widely divergent sizes. The forest is assumed to be dominated by a single species of tree. There is provision in the model for dividing the forest into three different sites, on the basis of the rate of growth and thus the timber yield of trees per unit area of land. This could reflect either varying growing conditions (seed, sun, soil, etc.) or the relative density of the species in question (including in-filling by other species). But fundamentally the model is concerned with only one kind of tree. The minimum unit of measurement in the model is not the individual tree but the unit area of land on which the trees grow (here, the hectare).

In the forest, trees are seeded (naturally or by reforestation), they grow, and they die. At the beginning of any run, the trees in the forest can be distributed either uniformly or randomly by age. That is to say, the number of trees in each age-group can be the same for all age-groups, or it can be varied randomly.[3] The user also

---

2. The model is presented here in metric units (hectares and cubic meters) but it could as well be designed in Imperial units (acres and cubic yards).

3. Actually, the model specifies not the number of trees but the quantity of land "under" the trees. Thus, in a uniformly distributed model of a forest of 50,000 hectares, in which the trees live to a maximum of 150 years, there would be 50,000/150 or 333 ha of trees in each age-class.

specifies the maximum age of the trees in each site. The model assumes that trees are replaced as soon as they are cut down, burned up in a forest fire, or reach the age of 225 years. The effect of the model's equations (as discussed below) is to ensure that the merchantable volume of any particular tree declines to zero by the time its age reaches about 200 years; moreover, forest fires have a disproportionate effect on older trees. In any case, the model assumes that any trees still in existence after 225 years are replaced by new trees. As presented here, the model can simulate runs of up to 300 years at a time. For simulating evern longer periods, the model can simply be run again without reinitializing the forest.

From time to time, there are fires in the forest and these are incorporated into the model. The model assumes that, in most cases, fires cause relatively little damage (say, 1-2% of the forest is lost). Occasionally, however, fires may devastate much larger tracts—say, 25% or even more—and have a major impact on the long-term yield of the forest. To this end, the model uses a "tuneable" Weibull function for simulating the incidence of forest fires. Its precise operation is discussed further in a technical note at the end of this chapter. Of course, fire-damaged trees can sometimes be salvaged and sold; so the model allows the user to decide whether this is the case and, if so, what proportion of the burned trees are assumed to be salvaged.

The model also allows for deliberate cutting of trees in the forest to obtain timber. Once again, trees of different ages normally produce different quantities of merchantable timber. Up to a certain age, most trees yield an ever-increasing quantity of lumber as they grow older; after that, the potential yield steadily declines. Such data can be found in yield tables (such as those in Plonski 1974 for species found in Ontario) and incorporated directly into the model as a lookup table. Alternatively, the tables can be converted into polynomial regression equations, as is done by Newnham 1987 (pp. 3ff.), and used in the model in that form. In either case, yield data are normally specific to a single species of tree. The equation used here is reported to give accurate results (to within 2 m^3/ha of the tabular data), although the equation may in fact be more complex than it need be.[4] (See also the technical note beginning on page 113, below, for further discussion of this point.)

---

4.   Newnham (1987, p. 3) states: "In practice, it is anticipated that less complex equations than [the one shown] would be used in the model and that, within a region, these equations would have a common form regardless of species or stand type."

Finally, the model can simulate three distinct forest management options, each with or without the option of salvaging any proportion of the fire-damaged trees:

- no cutting at all;
- cutting of overmature trees (i.e., trees that have reached a particular age); or
- cutting of overmature trees plus a so-called "rotating" cut at a lower age.[5]

Naturally, provision is made for the user to specify the ages to be used in either of the cutting options.

Data Requirements

The model requires data about the forest, about the incidence of forest fires, and about the management strategy to be followed. For the forest, the user must enter the total size of each yield area, or site, in the forest. The model can accommodate up to three different sites; just enter zero for the area of any site you do not need. Note that the model assumes the sites are identified in order of yield quality, with site 1 being the best, site 2 the next best, and site 3 the poorest; see the technical note at the end of the chapter. Then you must enter the maximum age of the trees in each site and decide whether you want trees to be distributed evenly or randomly among the different age-classes.

For forest fires, the user has to decide whether to assume that fires will consume a constant proportion of the forest each year, or whether the incidence of fire will be simulated by the Weibull distribution. In the latter event, there is provision to "tune" the Weibull function with two parameters; these are discussed further in the technical note below. In either case, the model distributes the effect of fires according to the age and yield of each class of trees.

---

5. The result of the third strategy is to create two "cut-off" ages instead of just one, as in the second strategy. Suppose you wanted to harvest trees as they reached 100 years of age. With the second strategy, you would get a relatively big harvest at the beginning, as all the trees over 100 were cut; then there would be an immediate drop-off, as only the trees that were 99 years old would become available for harvesting the following year. With the third strategy, however, you might set the age of maturity higher than 100 years (say, at 120) and make a rotating cut at age 100. That way, instead of one bumper harvest in year 1, you would get a slightly better harvest for the first 20 years, as the age-group between 100 and 120 years emptied out. After 20 years, the effect of the two strategies would be the same.

As far as harvesting the trees is concerned, the user can choose a particular management strategy. This can range from no harvesting at all to strict annual cutting. In addition, the user has to decide what proportion (if any) of any fire-damaged trees are to be salvaged as timber. Finally, the user has to decide how many years to run the model for.

Overview

The model consists of 15 columns and 245 rows. It can be divided into five main parts (see Listing 4.1). At the top (rows 3-7) is an area for entering the parameters of the model: the size, age, and distribution of age-groups in the forest, the incidence of forest fires, and the harvesting strategy to be followed. Below this (in rows 9-74) are two areas for recording the development of the forest. The area on the right is used for displaying a distribution of the results of each five-year cycle, whereas the area on the left contains a one-line summary of the results of every five-year run. Tucked in at the bottom of the right-hand table are some labels for the graphs.

Below this (in rows 77-130) are three working tables showing a distribution by age-group of the average annual forest area, forest-fire damage, and salvaged yield (if appropriate). Below these tables (in rows 133-86) are three more working tables showing a distribution by age-group of the average annual harvest, the merchantable volume of the forest, and the extracted volume. Finally, at the bottom of the model (rows 189-240) are two macros to run the model through up to 60 five-year cycles and space for a scratch pad.

The model begins with an initial distribution of trees in the forest. This is located in cells D86:F130. Of course, the user must first provide the parameters for the forest, including the total size and maximum age of its various yield sites. Then the precise distribution of trees by age-group in each site is then accomplished by invoking the "initialization" macro (Alt-I), which is one of the two macros found at the bottom of the model. Essentially, the macro proceeds by first zeroing the top three cells (cells D86:F86) and then copying these zeroes into the rest of the table below (cells D87:F130). Then a "recalc" instruction is issued to clear the rest of the model of any data left over from previous runs.

Listir g 4.1

	A	B	C	D	E	F	H	I	J	K	L	M	N	O
1	FOREST MANAGEMENT MODEL (BLACK SPRUCE)									© 1993 by T.J. Cartwright				

Forest Site	Site1	Site2	Site3	Fire Damage		Harvest Option	2
Area (ha)	25000	50000	25000	Fixed	.00%	0 = none	
Maximum age	150	160	170	Weibull b	.001	1=over age of:	160
Random distrib (Y=1;N=0):			0	Weibull c	.350	2=over age of &	
Years to run (max 300):			300	% salvaged	25%	rotating cut:	130

Sim Year: 300   Burn 1.0%

To Yr:	5-Yr Burn (ha)	5-Yr Salv (ha)	5-Yr Cut (ha)	MV (m3*1000)	EV
5	381	95	38639	10114	7485
10	601	150	27097	9234	5286
15	308	77	27084	8781	5278
20	2589	647	31088	8325	6424
25	59	15	31096	7621	6382
30	465	116	31080	6966	6384
35	576	144	15547	6329	3228
40	2454	614	15493	6335	3239
45	115	29	15492	6326	3212
50	387	97	15482	6338	3213
55	801	200	15462	6363	3213
60	332	83	15450	6402	3206
65	2258	565	15394	6487	3217
70	141	35	15384	6580	3190
75	195	49	15370	6720	3188
80	130	33	15355	6911	3184
85	662	166	15326	7138	3186
90	874	219	15291	7387	3182
95	311	78	15267	7641	3169
100	434	108	15240	7910	3166
105	164	41	15220	8180	3157
110	2343	586	15148	8446	3179
115	287	72	15129	8673	3141
120	260	65	15113	8917	3137
125	3	1	15106	9150	3131
130	73	18	15104	9370	3132
135	63	16	37659	9570	7217
140	2498	625	26660	8910	5238
145	347	87	26387	8588	5144
150	393	98	32553	8281	6744
155	265	66	30114	7622	6245
160	95	24	30521	7039	6325
165	248	62	15618	6421	3237
170	344	86	17433	6434	3606
175	2282	570	15113	6372	3158
180	1541	385	15346	6384	3197
185	749	187	15732	6406	3266
190	1501	375	15249	6435	3177
195	1509	377	17069	6513	3546
200	1050	262	14989	6542	3120
205	539	135	15019	6683	3120
210	919	230	14926	6877	3106
215	17	4	15426	7104	3196
220	864	216	15594	7347	3242
225	46	12	15032	7579	3116
230	8	2	15133	7852	3136

5-YEAR SUMMARY, ALL SITE By Age Groups

Age Group	Area	5-Yr Burn (ha)	5-Yr Salv (ha)	5-Yr Cut (ha)	MV (m3*1000)	EV
Totals>> >>>>>		283	71	17005	6542	3516
0 4	3050	6	2	0	0	0
5 9	6278	17	4	0	0	0
10 14	5839	19	5	0	0	0
15 19	6298	22	6	0	0	0
20 24	5746	21	5	0	0	0
25 29	5572	21	5	0	0	0
30 34	7139	27	7	0	27	0
35 39	2875	11	3	0	27	0
40 44	2911	11	3	0	42	0
45 49	3014	10	3	0	58	0
50 54	3679	12	3	0	91	0
55 59	3413	11	3	0	129	0
60 64	2916	8	2	0	149	0
65 69	2941	8	2	0	188	0
70 74	2927	7	2	0	231	0
75 79	3195	8	2	0	300	0
80 84	2995	7	2	0	326	0
85 89	3074	6	2	0	378	0
90 94	3017	6	2	0	412	0
95 99	3109	6	1	0	465	0
100 104	3598	7	2	0	579	0
105 109	3238	6	2	0	559	0
110 114	3179	6	1	0	582	0
115 119	3253	6	2	0	625	0
120 124	3342	7	2	0	669	0
125 129	3402	7	2	17005	703	3513
130 134	0	0	0	0	0	0
135 139	0	0	0	0	0	0
140 144	0	0	0	0	0	0
145 149	0	0	0	0	0	0
150 154	0	0	0	0	0	0
155 159	0	0	0	0	0	0
160 164	0	0	0	0	0	0
165 169	0	0	0	0	0	0
170 174	0	0	0	0	0	0
175 179	0	0	0	0	0	0
180 184	0	0	0	0	0	0
185 189	0	0	0	0	0	0
190 194	0	0	0	0	0	0
195 199	0	0	0	0	0	0
200 204	0	0	0	0	0	0
205 209	0	0	0	0	0	0
210 214	0	0	0	0	0	0
215 219	0	0	0	0	0	0
220 224	0	0	0	0	0	0

	A	B	C	D	E	F	H	I	J	K	L	M	N	O
60	235	130	32	14857	8125	3082								
61	240	573	143	16924	8403	3510	Graph Subtitles (1-3)							
62	245	3923	981	14819	8580	3138								
63	250	574	144	14761	8782	3070	Current Age Distribution of Trees (ha)							
64	255	299	75	14496	9023	3011	Timber Loss due to Fires and Cuts (ha)							
65	260	38	10	14560	9264	3019	Merchantable & Extracted Volume (m3*1000)							
66	265	7	2	36193	9487	6939								
67	270	124	31	28065	8886	5492	Variable Labels			Simulation Year				
68	275	3305	826	25594	8544	5043								
69	280	95	24	31549	8232	6532								
70	285	212	53	29086	7624	6032	Fire		Cuts					
71	290	2169	542	29225	7090	6086	Timber							
72	295	93	23	15159	6506	3140	Timber							
73	300	283	71	17005	6542	3516	Age-Class (Years)							
74	Mean	746	187	19988	7664	4087								

Detailed Annual Distribution by Yield Sites and Age Group
(Site 1 = High; Site 2 = Medium; Site 3 = Low)

		Total Forest Area (ha)			Mean Annual Burn (ha)			Mean Annual Salvage (ha)		
	Site: > >	1	2	3	1	2	3	1	2	3
Totals > >		25000	50000	25000	13	28	15	3	7	4
Age Groups										
0	4	809	1522	719	0	1	0	0	0	0
5	9	1660	3133	1486	1	2	1	0	0	0
10	14	1555	2914	1370	1	2	1	0	0	0
15	19	1668	3142	1488	1	2	1	0	1	0
20	24	954	3248	1543	1	2	1	0	1	0
25	29	913	3162	1496	1	2	1	0	1	0
30	34	1547	2895	2697	1	2	2	0	1	1
35	39	767	1434	673	1	1	1	0	0	0
40	44	776	1452	682	1	1	1	0	0	0
45	49	801	1504	709	1	1	1	0	0	0
50	54	963	1837	880	1	1	1	0	0	0
55	59	898	1704	811	1	1	1	0	0	0
60	64	777	1455	683	0	1	0	0	0	0
65	69	784	1468	690	0	1	0	0	0	0
70	74	780	1461	686	0	1	0	0	0	0
75	79	845	1594	755	0	1	0	0	0	0
80	84	797	1495	704	0	1	0	0	0	0
85	89	816	1534	724	0	1	0	0	0	0
90	94	802	1505	709	0	1	0	0	0	0
95	99	824	1552	733	0	1	0	0	0	0
100	104	940	1796	861	0	1	0	0	0	0
105	109	854	1616	767	0	1	0	0	0	0
110	114	840	1587	752	0	1	0	0	0	0
115	119	857	1624	772	0	1	0	0	0	0
120	124	878	1668	795	0	1	0	0	0	0
125	129	893	1698	812	0	1	0	0	0	0
130	134	0	0	0	0	0	0	0	0	0
135	139	0	0	0	0	0	0	0	0	0
140	144	0	0	0	0	0	0	0	0	0
145	149	0	0	0	0	0	0	0	0	0
150	154	0	0	0	0	0	0	0	0	0
155	159	0	0	0	0	0	0	0	0	0
160	164	0	0	0	0	0	0	0	0	0
165	169	0	0	0	0	0	0	0	0	0
170	174	0	0	0	0	0	0	0	0	0
175	179	0	0	0	0	0	0	0	0	0
180	184	0	0	0	0	0	0	0	0	0
185	189	0	0	0	0	0	0	0	0	0

	A	B	C	D	E	F	H	I	J	K	L	M	N	O
124	190	194		0	0	0		0	0	0		0	0	0
125	195	199		0	0	0		0	0	0		0	0	0
126	200	204		0	0	0		0	0	0		0	0	0
127	205	209		0	0	0		0	0	0		0	0	0
128	210	214		0	0	0		0	0	0		0	0	0
129	215	219		0	0	0		0	0	0		0	0	0
130	220	224		0	0	0		0	0	0		0	0	0

Detailed Annual Distribution by Yield Sites and Age Group
(Site 1 = High; Site 2 = Medium; Site 3 = Low)

		Mean Annual Cut (ha)			Merchant Volume (m3*1000)			Extracted Volume (m3*1000)		
	Site: > >	1	2	3	1	2	3	1	2	3
140	Totals > >	892	1698	811	>>>	2931	674	259	344	101
141	Age Groups									
142	0 4	0	0	0	0	0	0	0	0	0
143	5 9	0	0	0	0	0	0	0	0	0
144	10 14	0	0	0	0	0	0	0	0	0
145	15 19	0	0	0	0	0	0	0	0	0
146	20 24	0	0	0	0	0	0	0	0	0
147	25 29	0	0	0	0	0	0	0	0	0
148	30 34	0	0	0	27	0	0	0	0	0
149	35 39	0	0	0	27	0	0	0	0	0
150	40 44	0	0	0	42	0	0	0	0	0
151	45 49	0	0	0	58	0	0	0	0	0
152	50 54	0	0	0	87	4	0	0	0	0
153	55 59	0	0	0	97	32	0	0	0	0
154	60 64	0	0	0	98	51	0	0	0	0
155	65 69	0	0	0	112	76	0	0	0	0
156	70 74	0	0	0	125	99	7	0	0	0
157	75 79	0	0	0	149	133	17	0	0	0
158	80 84	0	0	0	153	148	25	0	0	0
159	85 89	0	0	0	169	175	34	0	0	0
160	90 94	0	0	0	177	193	42	0	0	0
161	95 99	0	0	0	193	220	52	0	0	0
162	100 104	0	0	0	232	277	70	0	0	0
163	105 109	0	0	0	220	269	70	0	0	0
164	110 114	0	0	0	225	281	76	0	0	0
165	115 119	0	0	0	237	303	85	0	0	0
166	120 124	0	0	0	249	326	94	0	0	0
167	125 129	892	1698	811	258	344	101	258	343	101
168	130 134	0	0	0	0	0	0	0	0	0
169	135 139	0	0	0	0	0	0	0	0	0
170	140 144	0	0	0	0	0	0	0	0	0
171	145 149	0	0	0	0	0	0	0	0	0
172	150 154	0	0	0	0	0	0	0	0	0
173	155 159	0	0	0	0	0	0	0	0	0
174	160 164	0	0	0	0	0	0	0	0	0
175	165 169	0	0	0	0	0	0	0	0	0
176	170 174	0	0	0	0	0	0	0	0	0
177	175 179	0	0	0	0	0	0	0	0	0
178	180 184	0	0	0	0	0	0	0	0	0
179	185 189	0	0	0	0	0	0	0	0	0
180	190 194	0	0	0	0	0	0	0	0	0
181	195 199	0	0	0	0	0	0	0	0	0
182	200 204	0	0	0	0	0	0	0	0	0
183	205 209	0	0	0	0	0	0	0	0	0
184	210 214	0	0	0	0	0	0	0	0	0
185	215 219	0	0	0	0	0	0	0	0	0
186	220 224	0	0	0	0	0	0	0	0	0

	A	B	C	D	E	F	H	I	J	K	L	M	N	O
												**	SCRATCH	PAD **

Row	Macro	Comment	M	N	O
189	{\S}{comment} SIMULATION MACRO				
190	{paneloff}	dialog off			
191	{home} =A20 ~	move to see	809	1522	719
192	{let C9,0}	set year	1660	3133	1486
193	{if F7=C9} {branch B206}	exit if finished	1555	2914	1370
194	{branch B202}	skip growth	1668	3142	1488
195	/c D86:F130,M192,v	[B195] copy to pad	954	3248	1543
196	/c I86:K130,M192,-	less annual burn	913	3162	1496
197	/c D142:F186,M192,-	less annual cut	1547	2895	2697
198	/c M236:O236,M191,v	copy dead into new	767	1434	673
199	/c I84:K84,M191,+	add total burn	776	1452	682
200	/c D140:F140,M191,+	add total harvest	801	1504	709
201	/c M191:O235,D86,v	copy from scratch	963	1837	880
202	{let C9,C9+5}	[B202] increment yr	898	1704	811
203	! {down}	recalc; move down	777	1455	683
204	/c K13:O13,[2;13+C9/5],v	copy to summary	784	1468	690
205	{if C9<F7} {branch B195}	repeat	780	1461	686
206	{panelon} {home} {beep}	[B206] dialog on	845	1594	755
207			797	1495	704
208	{\I}{comment} INITIALIZATION MACRO		816	1534	724
209	{paneloff} {home}	dialog off	802	1505	709
210	= A80 ~	move to see	824	1552	733
211	{let D86,0}	zero site 1	940	1796	861
212	{let E86,0}	zero site 2	854	1616	767
213	{let F86,0}	zero site 3	840	1587	752
214	/c D86:F86,D87:D130,v	copy to all	857	1624	772
215	!	recalc	878	1668	795
216	{let D86,5*D4/D5}	set site 1	893	1698	812
217	{let E86,5*E4/E5}	set site 2	0	0	0
218	{let F86,5*F4/F5}	set site 3	0	0	0
219	/c D86,D87:[4;85+D5/5],v	copy	0	0	0
220	/c E86,E87:[5;85+E5/5],v	copy	0	0	0
221	/c F86,F87:[6;85+F5/5],v	copy	0	0	0
222	{if F6<>1} {branch B236}	skip randomizing	0	0	0
223	/c D86:F86,M240,v	store exact values	0	0	0
224	=A80~ =D86	setup	0	0	0
225	M240*(RAN+.5)	[B225] randomize	0	0	0
226	{down}	down and check okay	0	0	0
227	{if crow<=(85+D5/5)} {branch B225}		0	0	0
228	=A80~ =E86~	setup	0	0	0
229	N240*(RAN+.5)	[B229] randomize	0	0	0
230	{down}	down and check okay	0	0	0
231	{if crow<=(85+E5/5)} {branch B229}		0	0	0
232	=A80~ =F86~	setup	0	0	0
233	O240*(RAN+.5)	[B233] randomize	0	0	0
234	{down}	down and check okay	0	0	0
235	{if crow<=(85+F5/5)} {branch B233}		0	0	0
236	/b B14:F73 ~	[B236] blank last	0	0	0
237	{let C9,0}	zero year			
238	{panelon}	dialog on			
239	{home}	ready for options			
240	{beep}	signal	833	1563	735
241					
242					

243   Source: Adapted from a Fortran model in R.M. Newnham, FIRFOR -- A Simple
244   Forest Management Model, Canadian Forestry Service (Ottawa: Supply and
245   Services Canada, 1987).

The next step is to compute the mean area of each age-group of trees in each site (based on the total area of each site and maximum age of the trees in it) and to enter those figures into the top three cells (cells D86:F86). Then these mean values are copied down into the cells below, as far down as is warranted by the maximum age specified in cells D5, E5, and F5 respectively. The result is an equal distribution of trees among the different age-classes.

If (in cell F6) the user has opted for a random distribution of trees by age-group, then a randomizing subroutine is executed from within the macro: see cell B222. This subroutine stores the mean values for each site in three cells at the bottom of the worksheet (cells M240:O240). Then the subroutine goes through the distribution of trees throughout the forest (cells D86:F130), replacing the mean value in each cell with a value adjusted by a random factor in the range .5 to 1.5. In other words, the mean area of trees in each age-group in each site is adjusted at random by up to ±50% of its original value.

Finally, when the forest is filled with trees, the macro blanks out the data from any previous run, zeros the current year (cell C9), and homes the cursor, ready for the user to set the parameters for the next simulation.

The next step is to set the options for forest fires, salvaging, harvesting, and the duration of the simulation run. For fire damage, the user has the option of entering a fixed percentage in cell K4. For example, it might be felt reasonable to assume that, in an average year, half of one percent of the forest is destroyed by fire. In that case, you would enter .005 in cell K4 and ignore cells K5 and K6. If, on the other hand, you prefer to rely on the Weibull distribution to determine the incidence of forest fires, then enter a zero in cell K4 and examine cells K5 and K6 to see if they need adjusting.

The effect on the distribution of cells K5 and K6 can be seen in cell F9, which contains the following formula:

```
5*K4+(K4=0)*(K5*(-LN(1-RAN))^(1/K6)+
    K5*(-LN(1-RAN))^(1/K6)+K5*(-LN(1-RAN))^(1/K6)+
    K5*(-LN(1-RAN))^(1/K6)+K5*(-LN(1-RAN))^(1/K6))
```

The formula works like this. If a fixed annual burn is desired, then cell K4 will have a positive value. Thus, the expression K4=0 will evaluate as false, which is treated as 0, and the entire cell will evaluate as simply 5*K4. (The number 5 is there because we want a value for a five-year period). If, instead of a fixed annual burn, the Weibull distribution is to be used, then cell K4 will have a value of zero, the expression K4=0 will evaluate as true, which is treated as 1, and the entire cell will evaluate as zero (5*K4) plus the sum of five random draws from the Weibull distribution.[6]

Regarding the salvaging of timber damaged by fire, the user has the option of specifying in cell K7 how much of the timber that has been damaged by forest fires will be salvaged. For example, the user might find it reasonable to assume that 25% of the burned timber will be salvaged; in that case, the figure .25 would be entered in cell K7. (The model is programmed to display the value in percentage terms, even though it is entered as a decimal.)

Regarding forest management strategies, the user enters his or her desired harvesting strategy in cell O3. As described above, there are three options, which are chosen by entering the corresponding digit:

0   no cutting or management at all;
1   cutting of overmature trees only; or
2   cutting of overmature trees plus a rotating cut.

Cell O5 allows the user to specify the age at which trees are deemed to become mature and, therefore, subject to cutting in options 1 and 2; similarly, cell O7 contains the specific age for the rotating cut (if there is one). Finally, the user specifies in cell F7 how long the model is to be run. The maximum allowed run is 300 years; however, longer runs can be simulated by running the model again without initializing the forest with the Alt-I macro, as described above. This can be repeated as many times as you like.

Now we are ready to begin the simulation proper. This is accomplished by invoking the second macro in the model (Alt-S). The core of this macro is a series of routines, beginning with cell B195, that "grow" the forest through one cycle. First, the current distribution of trees in the forest (cells D86:F130) is written into

---

6.   Five draws are used, rather than one draw multiplied five times, because we want five distinct draws, each with a different random number. This is because the model assumes (rightly or wrongly) that the incidence of fire each year is wholly independent of its incidence in previous years.

a "scratch pad" (in cells M191:O236). Each age-group is then adjusted by subtracting any trees that were burned or cut in the last cycle. Any trees at the bottom of the scratch pad in row 236 are trees that will be more than 225 years old in the next cycle; so they are copied up into the first age-class (cells D86:E86). To them are added all the trees that have to be replaced due to burning or cutting in the current cycle (i.e. the totals in cells I84:K84 and D140:F140). The whole scratch pad is then copied back into the original cells (cells D86:F130), and the year is incremented.

The other important function of the simulation macro (Alt-S) is to issue a "recalc" command to have the model calculate the results for the next cycle, and then copy the results of the last cycle into the summary area at the left of the worksheet in rows 14-73. Just before the summary table is updated, the cursor is moved one row down, so that data are recorded against a new cycle (see cells B203:B204). The rest of the macro just involves checking that the run has not exceeded the duration specified in cell F7.

Most of the rest of the model consists of five other formulas, applied to each of the age-classes in each of the three yield sites. These five formulas, some of them quite complicated, are as follows:

● **Mean Annual Burn (Cells I86:K130)**

For example, the formula in cell I86 is as follows:

```
$F$9/5*D86*5*(.0284+.5047/10^11*$I14^5-
     .9224/10^4*$I14^2+.0007492*LN($I14)-
     .3353/10^8*$I14^4+.003540*$I14-
     .4139/10^8*$I14^3*I$82+.8554/10^6*$I14^3+
     .7035/10^4*$I14*I$82)
```

Thus, the mean annual burn in a given age/yield category is computed by multiplying the total area of trees in that category (cell D86) by the mean incidence of fire for the year (cell $F$9/5) times 5 (because the result is desired for a five-year period) times the proportion of the category estimated to burn (which is calculated by means of the regression function inside the outer parentheses).

- **Mean Annual Salvage (Cells M86:O130)**

For example, the formula in cell M86 is as follows:

```
I86*$K$7
```

In other words, the quantity salvaged is simply a fixed proportion (cell $K$7) of the quantity burned (I86).

- **Annual Cut (Cells D142:F186)**

For example, the formula in cell D142 is as follows:

```
($O$3>0)*(OR($I14>=$O$5,AND($O$3>1,
       AND($I14>$O$7-5,$I14<$O$7+5)))) *(D86-I86)
```

The mean annual cut in a given age/yield category depends on the harvesting option entered in cell $O$3. If cell $O$3 is zero, then the first part of the formula evaluates as zero, causing the whole cell to be evaluated as zero. On the other hand, if either of two other conditions is true, all the unburned trees in the category (D86-I86) will be harvested. The two conditions are: (a) that the trees are overmature (i.e., $I14> =$O$5), or (b) that option 2 has been selected (in cell $O$3) and the trees are within 5 years of the rotation age entered in cell $O$7 (i.e., AND($I14>$O$7-5,$I14<$O$7+5)).

- **Merchantable Volume (Cells I142:K186)**

For example, the formula in cell I142 is as follows:

```
D86/10^3*MAX(0,-265.8+3.381*$I14+
      .7876*$I14*D$82-.3519/10^6*$I14^4+
      156.4*D$82-319.4*LN(D$82)-
      .4719*$I14*D$82^2+.001176*$I14^2*D$82^2)
```

Thus, the merchantable volume of timber in any age/site category is computed by multiplying the total area of trees in that category (divided by 1,000 to change the scale to thousands of cubic meters) times the greater of zero and the result obtained from applying the yield function. The MAX(0,...) function is included in order to suppress any negative yields that might arise in areas of extremely old trees.

• **Extracted Volume (Cells M142:O186)**

For example, the formula in cell M142 is as follows:

```
IF(D86=0,0,(D142+M86)/D86*I142)
```

Thus, the extracted volume is computed by adding the area burned to the area cut (D142+M86), expressing that result as a proportion of the total area (cell D86), and multiplying that by the total merchantable volume (cell I142). The IF() function merely forces a zero in cases where there are no trees of the particular age/yield category (D86=0); otherwise, an error message would appear in cell M142.[7]

The totals at the tops of each of the above tables (e.g. I84:K84) are just sums of the values below them. For example, the formula in cell I84 is:

```
SUM(I86:I130)
```

The remaining part of the model is found in cells J14:O58 near the top of the worksheet. Here the formulas are quite straightforward, as this part is a consolidation (by age-group) of the site-specific data from the six tables below it. Thus, each cell in this part is the sum of the three site-specific cells in the corresponding table below. For example, the formulas in the cells in row 14 are as follows:

```
J14    SUM(D86:F86)
K14    5*SUM(I86:K86)
L14    5*SUM(M86:O86)
M14    5*SUM(D142:F142)
N14    SUM(I142:K142)
O14    5*SUM(M142:O142)
```

As can be seen, four of the totals—the mean annual burn (K14), the mean annual salvage (L14), the annual cut (M14), and the extracted volume (O14)—are each multiplied by 5, because they are annual "flows" and we require the total for 5 years. The other two—forest area (J14) and merchantable volume (N14)—are not multiplied by 5, because they are end-of-period assets or "stocks," rather than annual flows.

---

7.   The same effect could be achieved either with the ISERROR() function or with a formula using a Boolean expression: e.g., (D86=0)*(D142+M86)/D86*I142.

Once again, the totals at the tops of these columns (cells J13:O13) are all just sums of the values below them. For example, the formula in cell J13 is:

```
SUM(J14:J58)
```

Finally, there are five similar formulas in cells B74:F74, which are simple averages of the data for each five-year period entered above them. For example, the formula in cell B74 is as follows:

```
IF(ISERROR(AV(B14:B73))," ",AV(B14:B73))
```

The IF() function suppresses any error messages that might arise from blank cells within the range to be averaged.

Operation

The model is very simple to run. You just enter the description of the forest and other parameters in rows 3-7 and then use the two macros. Alt-I initializes the forest with either a uniform or random distribution of trees by age-class. (You can enter the details of the forest by hand, if you wish, but it is tedious to have to do so.) Then Alt-S runs the model for the number of years (in cycles of five years) specified in cell F7. To run the model for multiples of that overall time-period, just invoke the Alt-S macro again *without* reinitializing the forest with the Alt-I macro.

As the model proceeds, you will notice that the screen moves down the worksheet row by row, so you can see what is happening. You may want to use the window command of your spreadsheet so you can keep the titles and other display information visible at the top of the screen.[8]

---

8.   If you decide to use a window split, you will probably find you do not need the contents of cell B191 in the Alt-S macro. Since you cannot have an empty cell inside a macro (because the macro stops when it finds an empty cell), you cannot just blank out cell B191. So the easiest thing to do is to replace the contents of cell B191 with a dummy instruction (such as "{comment}").

Interpretation and Use

The model shown here can be used either to simulate the growth and development of forests that are left alone to follow nature's course, and to see the effect of different management strategies on forests and their yields. Perhaps the clearest lesson to emerge from running this model is how much a forest can change and how sharply yields can vary in what might be thought to be a fairly stable system!

Essentially, the pattern that emerges from use of the model is this:

- If the forest is fairly homogeneous in terms of the age of its trees, then there will be sustained periods in which relatively high yields are possible; but change, when it comes, can be quite dramatic.

- If the forest is fairly heterogeneous in terms of age of its trees, then the long-term yield may be quite consistent while at the same time varying dramatically from one five-year period to the next.

This is not always the case, of course, but it is a reasonably good generalization—and it underlines the importance, both for maintaining and harvesting the forest, of (a) promoting variety and (b) adopting a prudent management strategy.

To illustrate this, let us look at three different cases, using the parameters shown in Listing 4.1 above:

- a *stable forest*, in which trees are distributed uniformly over a wide range of ages;

- a *replanted forest*, in which trees are concentrated uniformly in a fairly small number of age-classes; and

- a *natural forest*, in which there is a random distribution of trees over a wide range of age-classes.

To facilitate comparison, we will use the Weibull function to simulate the impact of forest fires in each case; and we will assume that harvesting consists of both a cut of mature trees (over the age of 160) and a rotating cut (at age 130), and that 25% of fire-damaged timber is salvaged.

Figure 4.1 depicts a stable forest under these conditions. This forest begins with a uniform distribution of trees by age-class. After a run of 300 years, however, the originally uniform distribution is affected (see Figure 4.1a) by regeneration following the large initial harvest allowed by the management system adopted for this run. This large harvest is due to the fact that, for the first 30 years, there is a double cut in effect: there is the rotating cut of all trees as they reach a particular age (in this case, 130 years); and there is the high, initial cut of all overmature trees (in this case, that meant all trees over the age of 150 years) with an annual cut for the next 20 years of other trees that missed the first rotating cut (because they were over the age of 130), as they reach the age of maturity. The result is that, twenty years after the first harvest, there are no longer any trees older than the age of the rotating cut (in this case, 130 years old).

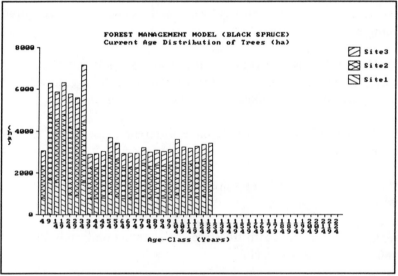

Fig. 4.1a. Age Distribution, Stable Forest, Subject to Fires and Cutting

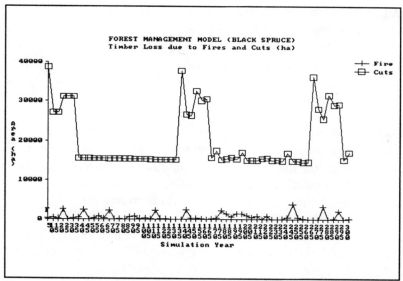

Fig. 4.1b. Timber Losses in a Stable Forest due to Fires and Cutting

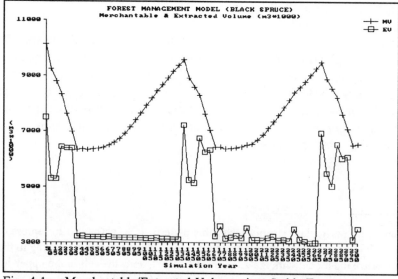

Fig. 4.1c. Merchantable/Extracted Volumes in a Stable Forest

Over time, the "bulge" caused by regeneration after the relatively large initial cuts matures in its own right, until eventually it reaches the age of the rotating cut and the harvest suddenly rises once again (in year 135 in Figure 4.1b). When these trees are in turn regenerated, another bulge appears (in year 270); and so the process goes on. In the long term, this pattern will gradually be attenuated by the "noise" of regeneration after forest fires.

Figure 4.1c shows the total extracted timber (timber deliberately cut as well as timber salvaged after forest fires) compared to total merchantable timber. From this graph, it can be seen that the current harvesting strategy appears to be sustainable at all times in this particular forest, in the sense that the amount of timber extracted does not exceed the estimated merchantable volume at any point.

Now consider the case of a replanted forest, with a relatively small number of different age-classes. Such a forest could take many forms, of course, but the one we have chosen to model here is one in which each of the three sites was planted in a staggered sequence over three different 25-year periods. Thus, site 1 was planted with 5,000 ha in each of the five age-classes from 0-24 years, site 2 with 10,000 ha in each of the five age-classes from 25-49 years, and site 3 with 5,000 ha in each of the five age-classes from 75-99 years.[9]

Figure 4.2 shows the state of this replanted forest after 300 years. In Figure 4.2a, the original planting patterns are still visible. Clearly, it will take millenia for this forest to evolve into a stable state. In the meantime, the effect of the current harvesting strategy is to induce wild fluctuations in the quantity of merchantable timber (see Figure 4.2b). Indeed, there are times when the rate of extraction exceeds the estimated supply of merchantable timber. This implies that logging is occurring at a rate that the forest cannot sustain.

---

9.   There is no particular significance to this pattern: it is just meant to represent a forest planted according to some entirely exogenous criteria. To enter such a planting in the model, you first initialize the model with the Alt-I macro. Next, zero the cell in the upper-left corner of the forest table (cell D86) and copy it into the rest of the table (i.e. into cells D86:F130). Then you are ready to enter into specific cells the area that you want planted with each age-class in each site. In this case, that would mean 5,000 in cells D86:D90, 10,000 in cells E91:E95, and 5,000 in cells F101:105.

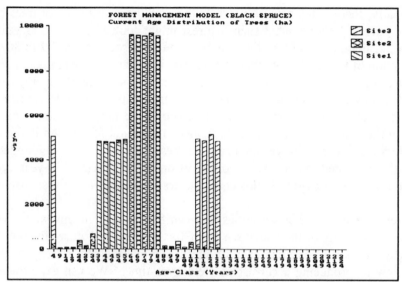

Fig. 4.2a. Age Distribution, Replanted Forest, Subject to Fires and Cutting

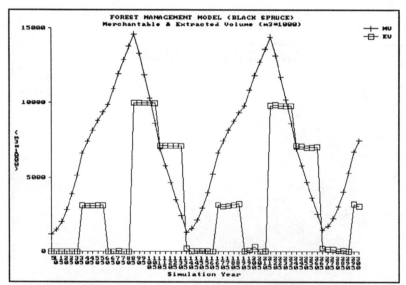

Figure 4.2b: Merchantable/Extracted Timber in a Replanted Forest

Finally, there is the natural forest, with a random distribution of trees by age-class.   Figure 4.3 shows what such a forest might look like after 300 years.   In Figure 4.3a, there is again the slight bulge in the age-groups from 5-9 to 30-34 that is a reflection of two cycles of regeneration after the heavy initial cutting allowed by the harvesting practices in effect.   This is revealed more clearly in Figure 4.3b, which also shows that there was a succession of moderately severe forest fires at the beginning of the third century.   Figure 4.3c shows the fairly smooth, cyclical rise and fall in the quantity of merchantable timber in a natural forest.   As in the case of the stable forest, current harvesting practices mean that the extracted volume comes close to the estimated merchantable volume at one point in each 130-year cycle but does not exceed it.   Logging in this case appears, therefore, to be sustainable.

There are many other variations that can be modeled in this way.   To start with, we can run any of the above scenarios further into the future (for 600 or 900 years, or even more).   We can explore alternative management strategies, perhaps with a view to producing a consistent yield over time.   We can experiment with greater variation in the maximum age and make-up of the three sites in the forest, as well as with the age-distribution of tress within each site.   There are also mechanisms

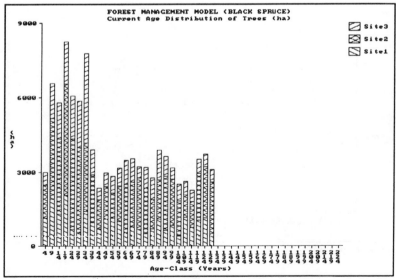

Fig. 4.3a.   Age Distribution, Natural Forest, Subject to Fires and Cutting

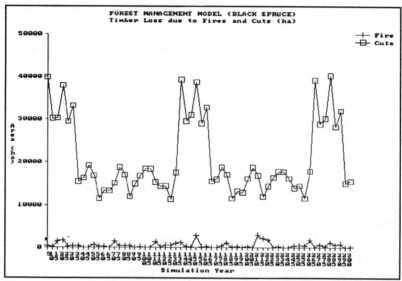

Figure 4.3b: Timber Losses in a Natural Forest, Subject to Fires and Cutting

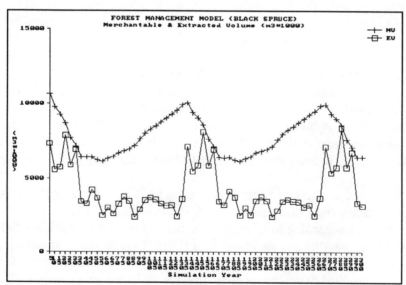

Fig. 4.3c. Merchantable/Extracted Volume in a Natural Forest

for adjusting the incidence of forest fires, the effect of which is not just to destroy parts of the forest but also through regeneration to provide for attenuating some of the more artificial effects of fixed harvesting policies.

Evaluation and Extension

There are several ways in which the model presented here could be extended. First, we could build in other management options besides the three shown here. The original version of this model (Newnham 1987) includes two other options:

- The **Highest Yield Option**, which allows cutting of areas with the highest yield (i.e. merchantable volume per hectare), up to a fixed annual limit on total allowable volume.

- The **Area of Approved Cutting (AAC) Option**, which allows cutting of areas with the lowest year-to-year growth, up to a fixed annual limit on the total area of approved cutting.

In addition, we might also introduce another factor (either user-specified or perhaps a random factor) to simulate the extent to which the management option, whichever one is selected, is actually complied with.

Second, the model could be adapted to different species. The model presented here uses a yield function for the black spruce tree (*Picea mariana*). Newnham has developed a similar function for the red pine (*Pinus resinosa*), which has a considerably higher yield than black spruce (ibid., pp. 3 and 65). It would a simple matter (see the technical note at the end of the chapter) to program this formula into cell I142 and then copy it down through the rest of the yield table (cells I142:K186).

Third, there are several other assumptions built in to the model that could be made into parameters. For example, the model is constructed so that the impact of forest fires varies according to age-class and site. Some users might like to build in the option of suppressing this variation, so that fires have an equal impact on all age-classes. Moreover, while the model allows for forest fires, it makes no provision for other kinds of natural disasters, such as diseases or invasions of insects. To some extent, we can "tune" the forest-fire function to reflect the impact of not only fires but also other stochastic events. However, the Weibull function used in the model has a particular character which is appropriate for forest fires but may not be appropriate for other kinds of disasters.

Fourth, the model assumes that any trees harvested for timber or destroyed by fire are immediately regenerated. Some users might like to add the option of specifying a number of years by which regeneration is lagged, or to arrange for losses to accumulate until the area affected reaches a certain minimum size before regeneration occurs.

In summary, this model provides many opportunities for further development. The model is also a good example of how a computer can be used not so much to do complicated mathematical operations as to do many little calculations. This model contains about 12,300 discrete calculations; so if you run the model for the full 300 years (i.e. through 60 five-year cycles), the computer may well do more arithmetic in a few minutes than you and I will do in our entire lifetimes! None of the calculations is particularly difficult: e.g., finding the natural logarithm of a number is about the most complex operation in the model. But none of us would have the patience to do so much tedious work by hand even once, let alone do it repeatedly with minor variations in the parameters. Moreover, when the computer has finished, it can (thanks to the capabilities of the spreadsheet) present you with a graph of the results.

Saving the forests is going to take a lot of commitment as well as knowledge. But it may be that better understanding of the link between sustainability and yield, such as is provided by a model like this one, can contribute in some small way to more enlightened action.

Technical Note

Newnham's model and the model shown here depend on two complex formulas, one governing the incidence of forest fires and their impact on the different sites and age-classes in the forest, and the other governing the yield of merchantable timber from the forest.

1. Forest Fires

The impact of forest fires is modeled in two discrete steps: the first step is to simulate the proportion of forest that is burned in any given year; the second step is to estimate the impact of such a fire on the various age-classes and sites in the forest.

Unless the user selects a fixed rate of damage, the incidence of forest fires is derived
from a specific statistical function: namely, the Weibull distribution, which has been found
to give realistic results for forests in North America (Van Wagner 1983; Reed and Errico
1985; Newnham 1987, p. 45). The Weibull function takes the form

$$F(x) = 1 - \exp\left(-\frac{x}{b}^{c}\right)$$

where x is the proportion of the forest that is burned in any year, and b and c (b > 0 and
c > 0) are parameters governing the character of the distribution—b being the parameter that
affects the "scale" of the curve and c the parameter that affects its shape. Values of .0001
and .35, respectively, are used here.[10] In order to determine whether there will be a forest
fire in any given year and how extensive it will be, a proportion is drawn at random from a
Weibull distribution. This is accomplished by means of the following formula:

$$p = b * (-\ln(1 - RND))^{1/c}$$

where p is the proportion of the area of the forest that is burned in a given year, b and c are
the tuneable parameters described above, and RND is a random number between the values
of zero and one. The result, as described by Newnham, is that

> on the average, approximately 0.5% of the area was burnt each year but, on one
> occasion during the 300-year period, 25% of the area was burnt. Fires of 25,000 ha
> do occur from time to time, so that this "disaster" is not entirely unexpected. The
> values of b and c would vary with the area of the forest unit: larger areas would have
> a smaller range in the proportions of the area burnt each year while smaller areas
> would have a greater range. (In an extreme case, the range could be from 0 to 100%;
> i.e., either there was no fire or the whole forest was burnt.) The values of b and c
> should be chosen with care to take into account the size of the forest and also the fire
> history in the region [1987, p. 4].

Thus, the b and c parameters can be used to tune the function in such a way as to reflect the
historical incidence of fire in the forest that is to be simulated.

Once the incidence of forest fires is determined, the model proceeds to estimate the
impact of fire on the forest. As Newnham notes (1987, p. 6), "forest stands are more
susceptible to fire at certain stages of their development (e.g., at the onset of crown closure

---

10.   Newnham (1987, pp. 5-6 et passim) provides some graphic illustrations of the effects
of selecting different parameters. You can do the same with your spreadsheet.

and at stand break-up)." Nor do fires affect trees of different quality in the same way: up to maturity, fast-growing, high-quality trees tend to be more resistant to fire than slow-growing, lower-quality trees; thereafter, the relationship is reversed and better trees become more susceptible to fire (ibid.) The result is that the proportion of the forest that burns is not uniformly distributed across all age-classes or across all sites. In other words, the effect of a given fire on any particular unit area of forest varies according to the age and quality of the trees in that area:

$$P = P(A, Q_s)$$

where P is the proportion of any unit area that is burned in any year, A is the age in years of the trees, and Q_s is the yield quality of the site where the burn occurs.

In this model, sites are assigned three quality ratings—class 1 (the highest) is assigned a rating of 1, class 2 a rating of 2, and class 3 a rating of 3. On this basis, the following formula is used to allocate the impact of any fire:[11]

$$P = .0284 + \frac{.5047 * A^5}{10^{11}} - \frac{.9224 * A^2}{10^4} +$$
$$.0007492 * \ln(A) - \frac{.3353 * A^4}{10^8} +$$
$$.003540 * A - \frac{.4139 * A^3 * Q_s}{10^8} +$$
$$\frac{.8554 * A^3}{10^6} + \frac{.7035 * A * Q_s}{10^4}$$

where P, A, and Q_s are all as specified above. This is the formula that is programmed into cells I86:K130 in the model. Figure 4.4 on the next page shows the relative impact of a fire on the different age-classes of trees in the different categories of sites.

2.   Yield Functions

The volume of merchantable timber in the forest is calculated by means of an equation derived by Newnham (ibid.) from yield tables in Plonski 1974. The example used here is the same as the one used in the original FIRFOR model: namely, the black spruce tree

---

11.   There appears to be a typographical error in the formula as given on p. 4 of Newnham 1987; the version in the listing on p. 59 is correct.

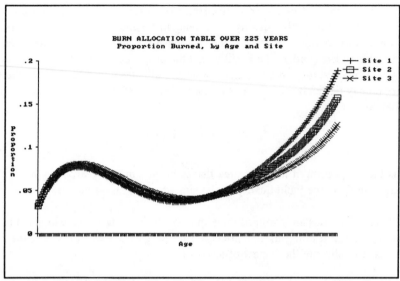

Fig. 4.4. Impact of Forest Fires on Trees of Varying Age and Quality

(*Picea mariana*). For this species, the following yield equation is used:

$$MV = -265.8 + 3.381*A + .7876*A*Q_s - \frac{.3519*A^4}{10^6} + 156.4*Q_s - 319.4*\ln(Q_s) - .4719*A*Q_s^2 + .001176*A^2*Q_s^2$$

where MV is the merchantable volume in cubic meters per hectare (m³/ha), A is the age of the tree in years, and Q_s is the yield quality of the site on a scale of 1 to 3 (1 being the highest quality).[12]   This equation is translated into the formulas that appear in cells I142:K186; for example, the formula in cell I42 (as noted on p. 102, above) is as follows:

---

12.   There is a leading minus sign missing from the formula on p. 3 in Newnham 1987; the version in the listing on p. 65 appears to be correct. There are also errors in the next formula as it appears on p. 6 of ibid.; the version on p. 67 appears to be correct.

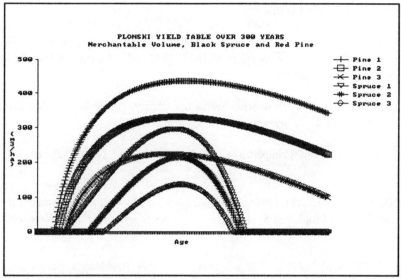

Figure 4.5: Plonski Yield Functions for Black Spruce and Red Pine

```
D86/10^3*MAX(0,-265.8+3.381*$I14+
    .7876*$I14*D$82-.3519/10^6*$I14^4+
    156.4*D$82-319.4*LN(D$82)-
    .4719*$I14*D$82^2+.001176*$I14^2*D$82^2)
```

For comparative purposes, Newnham (1987, p. 65, routine 30) presents a similar yield function for the red pine (*Pinus resinosa*): namely,

$$MV = -920.3 + 359.0*\ln(A) - 22.88*\ln(A)*Q_s - 2.183*A + 2.900*\ln(A)*\ln(Q_s) + \frac{.1559*A^2*Q_s}{10^2} - \frac{.3473*A^2*Q_s^2}{10^3}$$

The age-related yield curves for the two species, each in three different site qualities, are shoown in Figure 4.5, above. In volume terms, it is clear that the red pine is more productive (in cubic meters per hectare) than black spruce. The pine becomes productive sooner, grows more quickly, and sustains its yield for longer than the spruce.

Bibliography and References

Armstrong, G.W., et al. 1984. "Relaxing Even-Flow Constraints to Avoid Infeasibility with the Timber Resources Allocation Method (RAM)." *Canadian Journal of Forest Research*, 14, pp. 860-63.

Arney, J.D. 1985. "A Modelling Strategy for the Growth Projection of Managed Stands." *Canadian Journal of Forest Research*, 15, pp. 511-18.

Bella, I.E. 1971. "A New Competition Model for Individual Trees." *Forest Science*, 17, pp. 364-72.

Brodie, J.D., and R.G. Haight. 1985. "Optimization of Silvicultural Investment for Several Types of Stand Projection Systems." *Canadian Journal of Forest Research*, 15, pp. 188-91.

Chappelle, D.E., et al. 1976. "Evaluation of Timber RAM as a Forest Management Planning Model." *Journal of Forestry*, 74, pp. 288-93.

Ek, A.R., and R.A. Monserud. 1974. "Trials with the Program FOREST: Growth and Reproduction Simulation of Mixed Species Forest Stands." In Fries 1974, pp. 56-73.

Fries, J., ed. 1974. *Growth Models for Tree and Stand Simulation*. Research Note no. 30. Department of Forestry Yield Research. Stockholm: Royal College of Forestry.

Gray, P.A., and J. Keith. 1988. "The Application of a Spreadsheet Formatting Procedure in the Development and Implementation of a Habitat Evaluation Procedure for the Northwest Territories". In *Perspectives on Land Management: Workshop Proceedings*, edited by R. Gélinas, D. Bond, and B. Smit. Proceedings of a Workshop, November 17-20, 1986. Montreal: Polyscience.

Hegyi, F. 1974. "A Simulation Model for Managing Jack Pine Stands." In Fries 1974, pp. 74-90.

Johnsen, K.N., et al. 1977. "Rotation Flexibility and Allowable Cut." *Journal of Forestry*, 75, pp. 699-700.

Marshall, P.L. 1986. "A Decision Context for Timber Supply Modelling." *Forestry Chronicles*, 62, pp. 533-36.

Newnham, R.M. 1975. "LOGPLAN—A Model for Planning Logging Operations." Forest Management Institute Information Report no. FMR-X-77. Ottawa: Department of the Environment.

———. 1987. *FIRFOR—A Simple Forest Management Model*. Petawawa National Forestry Institute Information Report PI-X-72. Canadian Forestry Service. Ottawa: Supply and Services Canada.

Payandeh, B. 1974. "Nonlinear Site Index Equations for Several Major Canadian Timber Species." *Forestry Chronicles*, 50, pp. 194-96.

Plonski, W.L. 1974. *Normal Yield Tables (Metric) for Major Forest Species of Ontario*. Toronto: Ministry of Natural Resources.

Reed, W.J., and D. Errico. 1985. "Assessing the Long-Run Yield of a Forest Stand subject to the Risk of Fire." *Canadian Journal of Forest Research*, 15, pp. 680-87.

Tait, D.E. 1986. "A Dynamic Programming Solution of Financial Rotation Ages for Coppicing Tree Species." *Canadian Journal of Forest Research*, 16, pp. 799-801.

Van Wagner, C.E. 1978. "Age-Class Distribution and the Forest Cycle." *Canadian Journal of Forest Research*, 8, pp. 220-27.

——— 1983. "Simulating the Effect of Forest Fire on Long-Term Annual Timber Supply." *Canadian Journal of Forest Research*, 13, pp. 451-57.

Chapter 5

HERE COMES THE SUN: SOLAR ENERGY FROM A FLAT-PLATE COLLECTOR

*The amount of solar radiation reaching the surface of the earth
totals some 3.9 million exajoules a year. (An exajoule is
one billion joules of energy, approximately equivalent to the amount of
heat released during the combustion of 22 million tons of oil.)*

Israel Dostrovsky, "Chemical Fuels from the Sun"

As energy prices begin to edge upwards again—and as the environmental costs of burning fossil fuels become daily more apparent—we can expect to see a renewed interest in the sun as a source of energy. For one thing, technology has improved a good deal since the energy crisis of the 1970s; so the cost-effectiveness of solar energy is now better than it was, particularly in regard to photovoltaic systems (Hunt 1982). For another thing, public attitudes and policies have generally become more favorable to the kind of decentralized approaches to energy collection and distribution that tend to favor solar energy (Lovins 1977, 1979; Elder 1984). Third, the inexhaustible supply of sunlight and the lack of any major environmental hazards associated with its collection make the sun uniquely attractive as a long-term source of energy.[1]

The sun shines everywhere, of course, but the fundamental question is always, does the sun shine enough at a given location for it to be a viable source of energy? The mathematics of calculating the incidence of solar radiation is well known, but it

---

1. There are some indirect hazards associated with some kinds of solar-energy production (see Fowler 1984, pp. 446-49; Hunt 1982, and Lovins 1975, pp. 81-82); but solar energy usually gets good marks from environmentalists.

is complicated. To calculate the solar flux at a particular time on a particular day at a particular location on the surface of the earth takes approximately 70 arithmetic operations. To get—even for a single location—a matrix of, say, eleven hourly forecasts for an "average day" of each month of the year thus requires about 10,000 calculations. Change the location or even the orientation of the collector, and you have to do the calculations all over again. Clearly, this kind of modeling cries out for computerization. Indeed, the calculations involved are quite within the capabilities of even the most basic microcomputer running standard, off-the-shelf software.

Computer simulations of solar radiation are not at all unusual; some have even been written for microcomputers (e.g., Mobus 1981). The purpose of this chapter is to present a spreadsheet version of a simulation model for calculating the amount of solar energy that can be derived from a flat-plate collector located at any point on the surface of the earth. We will also discuss how such a model can be used for evaluating the feasibility of solar-energy production, for analyzing its possible uses, and for designing an optimum installation.

Purpose of the Model

There are various mechanical devices in use for collecting solar radiation and converting it to energy. According to one expert,

> The solar-energy collector most commonly used involves a thin plate of metal (usually copper, stainless steel, or, sometimes, aluminum) sealed behind a glass panel. A working medium (water, air, or antifreeze) passes behind it to carry away the heat. The plate is coated with a black, light-absorbing substance such as flat black engine paint. The flat-plate collectors are connected in parallel by some appropriate ducting or tubing, mounted facing, generally, in a southerly direction. The fluid is circulated through the collectors and then to a storage device, usually a bed of pebbles for air systems or a tank of water for water-based systems. The heat contained in the storage devices is removed and circulated to the point of use by a thermostatically controlled recovery system [Mobus 1981, p. 48; see also Fowler 1984, p. 449].

The simulation model presented here calculates the amount of solar radiation reaching the earth, adjusts it for the distance it must travel through the earth's atmosphere, calculates how much of that radiation will be absorbed by the collector, and adjusts the results for both the estimated mean cloud cover at the site and the thermal efficiency of the collector. The model is designed to answer three basic kinds of questions in relation to solar energy:

- How much solar radiation (and ultimately energy) is available at a given location on the earth, and how does this quantity vary over the hours of the day and the months of the year?

- What are the effects of different strategies for orientation of the collector—that is, for adjusting its azimuth relative to south and its tilt relative to the horizontal.

- How well suited is solar energy for various purposes at different places on the earth: e.g., solar energy for space heating means maximum output is required during the coldest months of the year; solar energy for water heating may mean maximum output is required in the early morning or late afternoon; and solar energy for refrigeration and cooling may mean that a *minimum* level of energy must be maintained throughout the day.

As Mobus points out:

> Since the design process involves trade-offs, it would be helpful to be able to predict the gains and losses of alternative strategies, particularly when it comes to the placement of the solar collector array.
>
> The governing factor in orienting collectors is that the actual light that can be absorbed falls off as the cosine of the angle of incidence increases between the light beam and the collector surface. When the beam of light is exactly normal (perpendicular) to this surface, as it would be when the collectors face the sun directly, the cosine of the angle (zero degrees) is 1.0 and all of the light is available for conversion into heat. At angles of more than zero degrees, the available light falls off, slowly at first, then more rapidly as the angle of incidence increases.
>
> It can be seen from this that the proper orientation of the collectors (i.e., the tilt angle with respect to the horizontal and the azimuth angle with respect to due south) is extremely important [op. cit., p. 48].

Conceptual Basis

The mechanics of a simulating the collection of solar energy can be taken from a standard textbook on solar engineering (such as Kreith and Kreider 1978, chapter 2). The version presented here is derived from a BASIC model written by George Mobus (1981). Essentially, such a model entails the following five steps:[2]

1. The average amount or incidence of solar radiation reaching the earth in a given unit of time is assumed to be a constant: namely, 116.4 calories per square centimeter per hour (cal/cm²-hr). This equivalent to 429 British Thermal Units (BTUs) per square foot per hour or 1.353 kilowatts per square meter).[3] However, because the earth follows an elliptical (annual) orbit about the sun, the amount of solar radiation reaching the earth (I_o) varies throughout the year. Thus, the daily solar flux is given by the equation:

$$I_o = 116.4 * (1 + .034 * \cos(\frac{360 * N}{365}))$$

   where N is the sequential day of the calendar year (e.g., February 1st is day 32, and so on). To get results in BTUs per square foot per hour, replace the number 116.4 in the above equation with 429; to get results in kilowatts per square meter, use 1.353 instead of 116.4.

---

2. The critical factor at almost every stage of the model is the angle formed by the sun's rays and the surface of the collector. Thus, the standard trigonometric ratios (sine and cosine) inevitably play a key part in the model. Non-mathematical readers may wish to be reminded that the sine and cosine of an angle are computed by drawing a line from one side, or "arm", of the angle to the other, so that a right angle is formed at one of the sides. Then the sine of the original angle is defined as the ratio of that line to the hypotenuse of the right triangle created by the line, and varies from zero to one as the angle varies from 0° to 90°. The cosine of the angle is the ratio of the other, adjacent side of the right triangle to its hypotenuse, and varies from one to zero as the angle varies from 0° to 90°).

3. In fact, the solar constant is not constant. According to Foukal (1990, p. 37), "satellite instruments reveal that [the solar constant] varies by as much as 0.2% over time scales of weeks. This relatively short-term variation is caused by the passage of dark sun-spots and bright faculae across the solar disk as the sun executes its approximately monthly rotation." Others (e.g. Kreith and Kreider 1978, chapter 3) put the effect of sunspots as high as 4%.

2. Because the axis of the earth's (daily) rotation is not at right angles to the plane of its orbit about the sun—i.e., because the earth's axis is tilted slightly with respect to the sun's radiation—the angle at which the sun's radiation strikes a given point on the earth at a given time of the day will vary throughout the year. This is called the angle of solar declination (D) and is given by the equation

$$D = 23.45 * \sin(360 * \frac{284+N}{365}) * N$$

where D is measured in radians and N is again the sequential day of the calendar year.

3. To reach any point on the surface of the earth, the sun's radiation must pass through the earth's atmosphere. Because the earth is constantly rotating about its own axis, and because this axis is tilted with respect to the sun, the mass of air through which the radiation must pass to reach a given point on the earth's surface is constantly changing—from a relatively large air mass on winter mornings and evenings to a relatively small air mass in the middle of a summer day (Norton 1992, chapter 2). The mass of air (M) through which the sun's radiation must pass (relative to the air mass it would have to pass through if the sun was directly overhead) is given by the equation

$$M = \sqrt{1229 + (614 * \sin E)^2} - 614 * \sin E$$

where

$$\sin E = \sin L * \sin D + \cos L * \cos D * \cos H$$

and E is the angle of solar elevation in radians above the horizon, L is the latitude in degrees above (positive) or below (negative) the equator, D is the angle of solar declination in radians as defined above, and H is the solar hour of the day.[4]

4. Thus, the amount of solar radiation actually striking the earth at sea-level at a given point and at a given time of day (I_s) is defined as the solar incidence or

---

4. The solar hour is an angle of arc, where 1 hour is equivalent to 360°/24 or 15°.

solar flux (I_o) attenuated by the air mass (M). According to Bouger's law, the effective solar radiation (I_s) is calculated as follows:[5]

$$I_s = I_o * .56 * (\exp(-.65*M) + \exp(-.095*M))$$

5.   Finally, the amount of solar energy actually reaching a fixed, flat surface or plate located at a given point at sea level on the surface of the earth at a given time and on a given day (Ip) is defined by the equation:

$$I_p = I_s * \cos i$$

where

$$\cos i = \sin D * (\sin L * \cos T - \cos L * \sin T * \cos A) +$$
$$\cos D * \cos H * (\cos L * \cos T +$$
$$\sin L * \sin T * \cos A) +$$
$$\cos D * \sin T * \sin A * \sin H$$

and i is the angle of incidence in radians at which the radiation strikes the plate, D is the angle of solar declination in radians, L is the latitude in degrees relative to the equator, H is the solar hour expressed in degrees of arc, A is the azimuth or orientation of the plate in degrees east (positive) or west (negative) of true south, and T is the angle of tilt in degrees with respect to the horizontal.

All these calculations pertain to what is called direct, or beam, radiation on a perfectly clear day in a standard atmosphere. In practice, the presence of water vapor (clouds, fog, haze, etc.) and atmospheric particulate matter (such as pollution) can affect solar radiation in at least two important ways: they can reduce the amount of direct radiation reaching the surface of the earth, and they can increase the amount of diffuse radiation. Furthermore, the model described here assumes a perfectly efficient collector; in other words, no allowance is made for heat loss from the collector. Naturally, these factors can have a significant and sometimes rapidly changing effect on the amount of available solar energy at any site (Hottel 1976). Consequently, some refinements will have to be added to the model in order to allow users to adjust for the effect of these factors.

---

5.   Exponentiation is discussed in a technical note to chapter 1: see page 37, above.

Data Requirements

In order to run the model, you need the following data:

- **Location:** You must enter the latitude (and optionally the name) of the site where the solar collector is located. Use degrees and decimal fractions of degrees (not minutes and seconds), and enter a positive value for north latitude and a negative value for south latitude. Thus, values may range from -90° to +90°.

- **Typical Day:** You must choose a particular day of the month (e.g., the 21st) to be used as a typical day for each set of monthly calculations; any convenient day will do.

- **Cloud Cover:** Enter the mean daily cloud cover (from zero to one) for each month at the selected location.

- **Efficiency:** Enter the collector's thermal efficiency (from zero to one), which is the rate at which the collector converts available solar radiation into usable solar energy. Typically, this index varies from .5 to close to 1 for different types of collectors and also varies for a given type according to the spread between the operating temperature inside the collector and the ambient temperature; in some jurisdictions, manufacturers are required to declare the index of thermal efficiency of their products according to standardized testing procedures.

- **Azimuth:** Enter the orientation of the collector in degrees east (positive) or west (negative) of true south; thus, azimuth may range from -180° to +180° (both of which would in fact indicate true north).

- **Tilt or slope:** Enter the angle of the collector in degrees above the horizontal; thus, tilt may range from 0° to 90°.

Using these inputs, the model provides an estimate of the amount of solar energy in calories per square centimeter per hour that can be collected each hour from 7:00 a.m to 5:00 p.m. on a particular day of each month at the selected location.

Overview

The model is contained in 17 columns and 110 rows (see Listing 5.1). The top 20 rows are designed both for entering the model's parameters and for displaying its results, while the rest of the spreadsheet consists of tables of calculations. With both input and output visible on a single screen, it is easy for users to see exactly how changes in the first affect the second.

Data input is done at several points across the top and down the left-hand side of the first twenty rows:

● **Location**: the name of the site is entered in cell O4 and its latitude at cell M2; use negative values to indicate southern latitudes.

● **Typical Day**: the desired typical day for each set of monthly calculations is entered at cell A7.

● **Cloud cover**: the mean daily cloud cover for each month is entered in the range B8:B19.

● **Efficiency**: this value is entered in cell A4.

● **Azimuth**: the orientation of the collector is entered in cell M3; use negative values for orientations west of true south.

● **Tilt**: the angle of tilt of the collector is entered in cell M4.

The results are displayed in the range D8:N19 and summarized in row 20 and in columns P and Q.

Below this "visible" part of the worksheet (rows 1 to 20) are six principal work areas. First there are some rows (rows 27-31) containing labels used in the graphs shown in the various figures used in this chapter. The next part of the worksheet (rows 34-36) converts data in degrees to radians. The model's calculations

Listing 5.1

```
     | A || B|  |D||  E  ||  F  ||  G  ||  H  ||  I  ||  J  || K || L  ||  M  ||  N  |  |  P  ||  Q  |
1    DAILY SOLAR RADIATION                                        © 1993 by T.J. Cartwright
2    Thermal      Latitude (degrees north of the equator)       43
3    Efficiency   Azimuth (degrees east of true south)           0        Location:
4     .7          Tilt (degrees above the horizontal)           30        Toronto
5
6    Day Mean| 7am  8am  9am 10am 11am Noon  1pm  2pm  3pm  4pm  5pm | Total  Var
7     21  DCC|      (Calories per cm2 per hour per day)                        .40
8    Jan  .5|  0    4   10   16   19   21   19   16   10    4    0 |  119   .69
9    Feb  .5|  1    7   14   20   24   25   24   20   14    7    1 |  157   .60
10   Mar  .4|  4   13   21   28   33   34   33   28   21   13    4 |  233   .51
11   Apr  .4|  7   15   24   31   35   37   35   31   24   15    7 |  262   .45
12   May  .3| 10   20   29   37   41   43   41   37   29   20   10 |  315   .41
13   Jun  .2| 12   23   33   41   47   49   47   41   33   23   12 |  361   .40
14   Jul  .1| 13   25   37   47   53   55   53   47   37   25   13 |  403   .41
15   Aug  .1| 10   23   36   46   52   55   52   46   36   23   10 |  389   .44
16   Sep  .2|  6   17   28   37   43   46   43   37   28   17    6 |  307   .51
17   Oct  .3|  1   10   19   27   33   34   33   27   19   10    1 |  215   .60
18   Nov  .4|  0    5   12   19   23   24   23   19   12    5    0 |  141   .70
19   Dec  .5|  0    2    9   14   17   19   17   14    9    2    0 |  104   .74
20   Mean .32|  5   14   23   30   35   37   35   30   23   14    5 |  251   .54
21
22
23
24
25
26
27   Graph Labels
28   Main         DAILY SOLAR RADIATION ON A FLAT-PLATE COLLECTOR
29   Sub          [See O4]
30   X-axis       Months                          Time of Day
31   Y-axis       Calories/cm2-hour               Calories/cm2-hour
32
33
34   Convert degrees        Latitude   .750
35   to radians:            Azimuth      0
36                          Elevation  .524
37
38
39   Compute D (angle of solar declination for each selected day of the year):
40
41   Jan   21  -.4
42   Feb   52  -.2
43   Mar   80  >>>
44   Apr  111  .20
45   May  141  .35
46   Jun  172  .41
47   Jul  202  .36
48   Aug  233  .21
49   Sep  264  >>>
50   Oct  294  -.2
51   Nov  325  -.4
52   Dec  355  -.4
53
54
```

| | A | B | | D | | E | | F | | G | | H | | I | | J | | K | | L | | M | | N | | P | | Q | |

55 Compute E (hourly angle of solar elevation above horizon for each day):
56
57 Solar Hou 75   60   45   30   15   0 -15 -30 -45 -60 -75
58 in radian 1.3 1.05 .785 .524 .262   0 -.26 -.52 -.79 -1.0 -1.3
59 Seg. Day

Seg. Day	75	60	45	30	15	0	-15	-30	-45	-60	-75
21	-.1	.109	.251	.360	.428	.452	.428	.360	.251	.109	-.06
52	.05	.226	.374	.488	.560	.585	.560	.488	.374	.226	.053
80	.18	.361	.512	.629	.702	.727	.702	.629	.512	.361	.184
111	.32	.495	.644	.757	.829	.853	.829	.757	.644	.495	.322
141	.41	.578	.720	.829	.898	.921	.898	.829	.720	.578	.413
172	.45	.607	.746	.852	.919	.942	.919	.852	.746	.607	.445
202	.42	.581	.723	.832	.900	.923	.900	.832	.723	.581	.416
233	.32	.497	.645	.759	.831	.855	.831	.759	.645	.497	.324
264	.19	.363	.515	.631	.704	.729	.704	.631	.515	.363	.187
294	.05	.219	.367	.481	.553	.577	.553	.481	.367	.219	.046
325	-.1	.104	.246	.355	.424	.447	.424	.355	.246	.104	-.06
355	-.1	.064	.203	.310	.377	.400	.377	.310	.203	.064	-.10

74 Compute M (hourly air mass between sun and earth's surface for each day):
75 Seg Day

Seg Day	75	60	45	30	15	0	-15	-30	-45	-60	-75
21	85.	8.66	3.94	2.76	2.33	2.21	2.33	2.76	3.94	8.66	84.6
52	15.	4.36	2.66	2.04	1.78	1.71	1.78	2.04	2.66	4.36	15.3
80	5.3	2.76	1.95	1.59	1.42	1.38	1.42	1.59	1.95	2.76	5.30
111	3.1	2.01	1.55	1.32	1.21	1.17	1.21	1.32	1.55	2.01	3.08
141	2.4	1.73	1.39	1.21	1.11	1.09	1.11	1.21	1.39	1.73	2.41
172	2.2	1.65	1.34	1.17	1.09	1.06	1.09	1.17	1.34	1.65	2.24
202	2.4	1.72	1.38	1.20	1.11	1.08	1.11	1.20	1.38	1.72	2.40
233	3.1	2.01	1.55	1.32	1.20	1.17	1.20	1.32	1.55	2.01	3.06
264	5.2	2.74	1.94	1.58	1.42	1.37	1.42	1.58	1.94	2.74	5.24
294	17.	4.49	2.71	2.07	1.81	1.73	1.81	2.07	2.71	4.49	16.7
325	89.	8.96	4.01	2.80	2.35	2.23	2.35	2.80	4.01	8.96	88.6
355	130	13.4	4.84	3.20	2.64	2.49	2.64	3.20	4.84	13.4	130.

90 Compute I (angle of incidence at which radiation strikes the plate):
91
92 Solar hou 75   60   45   30   15   0 -15 -30 -45 -60 -75
93 in radian 1.3 1.05 .785 .524 .262   0 -.26 -.52 -.79 -1.0 -1.3
94 Seg. Day

Seg. Day	75	60	45	30	15	0	-15	-30	-45	-60	-75
21	.16	.380	.569	.715	.806	.837	.806	.715	.569	.380	.159
52	.20	.434	.632	.784	.879	.912	.879	.784	.632	.434	.204
80	.25	.486	.687	.842	.940	.973	.940	.842	.687	.486	.251
111	.29	.522	.720	.872	.967	1.00	.967	.872	.720	.522	.292
141	.31	.535	.724	.870	.961	.992	.961	.870	.724	.535	.314
172	.32	.536	.722	.864	.953	.983	.953	.864	.722	.536	.321
202	.31	.535	.724	.869	.960	.992	.960	.869	.724	.535	.315
233	.29	.523	.720	.872	.967	1.00	.967	.872	.720	.523	.293
264	.25	.486	.688	.843	.940	.974	.940	.843	.688	.486	.251
294	.20	.431	.629	.780	.876	.908	.876	.780	.629	.431	.201
325	.16	.378	.567	.712	.803	.834	.803	.712	.567	.378	.158
355	.14	.357	.543	.685	.774	.804	.774	.685	.543	.357	.142

---

109 Source: Adapted from a BASIC model in George Mobus, "Harvesting the Sun's
110   Energy," BYTE, 6.7 (July 1981), pp. 48-58.

are all in radians.[6] However, most of us think in terms of degrees rather than radians; so provision is made to convert user-provided data on latitude, azimuth, and tilt into radians. In rows 39-52, the model computes the sequential-day value of the selected day for each month, and the corresponding angle of solar declination. Next, in rows 55-71, the model computes the angle of solar elevation above the horizon for each solar hour of each selected day. Then, in rows 74-87, the model calculates the the air mass between the sun and the earth's surface for each solar hour of each selected day. Finally, at the bottom of the worksheet (rows 90-106), the model computes the angle of incidence between the solar radiation and the surface of the collector. The model works essentially by feeding the results of calculations in the lower parts of the spreadsheet up into the display area at the top.

Another important aspect of this model is its ability to make use of spreadsheet graphics. Users can easily arrange for the program to draw a bar graph showing the total daily (or even hourly) solar energy available in each month of the year. For example, Figure 5.1 shows the daily solar radiation in Toronto (at Latitude 43° North) on a collector facing true south and tilted at 30°. In Figure 5.2, an hourly breakdown has been added to the daily totals, and the angle of tilt of the collector has been increased to 60°. This illustrates the effect of trying to get more out of the winter sun, which is closer to the horizon at that time of year at the latitude shown. Note that the distribution has become slightly bimodal (i.e., it has two peaks instead of one).

In Figure 5.3, solar radiation is plotted on an hourly basis, and the collector has been oriented towards the west (i.e., the azimuth is -45°). This illustrates the effect of trying to catch more of the afternoon sun. Finally, Figure 5.4 is a pie chart showing how daily solar radiation is distributed over the hours of a day in May, based on the same assumptions as the previous figure. Of course, none of these graphs is adjusted for daylight saving time!

---

6. A radian is the angle subtended by the arc of a circle equal in length to its radius. In any circle, the radius is related to the circumference in the exact proportion of $1:2\pi$ ($C = 2\pi r$). From this, it follows that the number of degrees subtended by an arc equal in length to the radius must bear the same proportion ($1:2\pi$) to the angle subtended by the entire circumference (namely, 360°). Thus, 1 radian = $360°/2\pi$, or approximately 57.3°.

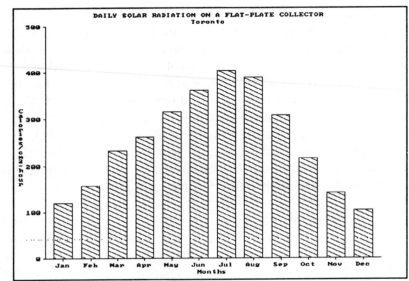

Fig. 5.1. Daily Solar Radiation (Latitude 43°, Azimuth 0°, Tilt 30°)

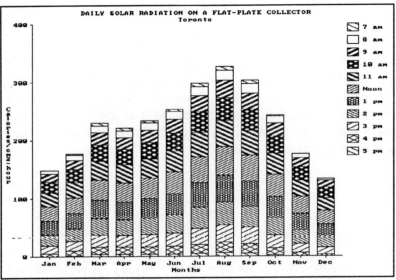

Fig. 5.2. Daily Radiation per Hour (Latitude 43°, Azimuth 0°, Tilt 60°)

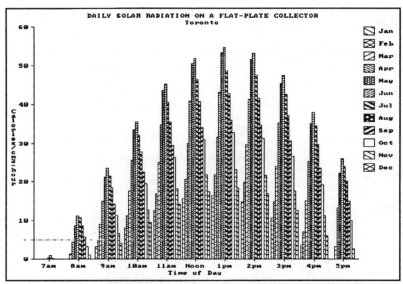

Fig. 5.3. Hourly Radiation (Latitude 43°, Azimuth -45°, Tilt 30°)

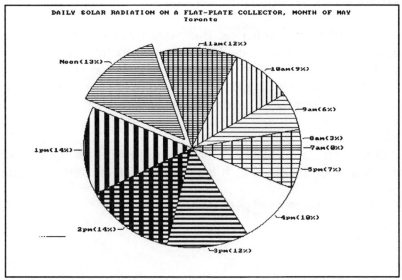

Fig. 5.4: Hourly Radiation in the Month of May from Fig. 5.3

Operation

The key to the operation of this model is its use of the two basic trigonometric functions, SIN() and COS(). In most spreadsheets, these functions work with angles measured in radians rather than degrees; so, as just noted, all data in degrees must be converted to radians, as they are in cells I34:I36 and rows 58 and 93.

The display area in the top 20 rows contains the principal equations in the model. For example, the cell in the upper left-hand corner (cell D8), which refers to 7:00 a.m. on January 21, is as follows:

```
MAX(0,$A$4*(1-$B8)*116.4*
     (1+.034*COS(360*$A60/365*PI/180))*D95*.56*
     (IF(ISERROR(EXP(-.65*D76)+EXP(-.095*D76))=0,
     EXP(-.65*D76)+EXP(-.095*D76))))
```

Although at first sight this formula may seem intimidating, it is just a concise way of rendering the calculations discussed above. The outer MAX() function is to ensure that, if the calculations produce a negative value, the result is set to zero. Similarly, the last part of the formula, IF(ISERROR. . .), ensures that the exponent functions return a value of zero rather than an error.[7]

The rest of the formula is derived from steps 1 and 4 of the model (see pages 124-25, above). If we examine it, the formula given here (for cell D8) refers to five other cells: namely, A60, D76, D95, A4, and B8. The first is quite straightforward; cell A60 converts the calendar day to the sequential day of the year. (For example, January 21 is day 21, February 21 is day 52, etc.). The other cells need a bit more detailed discussion.

Cell D76 contains a calculation of the air mass between the sun and the earth's surface at 7:00 a.m. on January 21. It is calculated with the following formula:

```
SQRT(1229+(614*D60)^2)-(614*D60)
```

---

7.   Different spreadsheets have different tolerances. For example, EXP(n) generates an error for $n > \pm 127$ in SuperCalc4 and $n > \pm 11,355$ in Lotus 1-2-3 Release 3.1.

This formula implements the discussion in step 3, above. This formula refers in turn to cell D60, which is the angle of solar elevation above the horizon (again, at 7:00 a.m. on January 21). Again following step 3, the angle of solar elevation is calculated as follows:

```
SIN($I$34)*SIN($D41)+COS($I$34)*COS($D41)*COS(D$58)
```

This calculation in turn refers to three other cells. Two of these (cells I34 and D58) are already familiar. The first is latitude converted to radians and the second is the solar hour (also converted to radians). The other reference in the above formula, cell D41, is the angle of solar declination (again, at 7:00 a.m. on January 21) and is derived from the sequential day of the year (cell B41) by the following formula (as deescribed in step 2, above):

```
23.45*SIN(360*(284+B41)/365*PI/180)*PI/180
```

Cell D95 calculates the angle of incidence at which solar radiation strikes a collector (once again at 7:00 a.m. on January 21). The formula is as follows:

```
SIN($D41)*(SIN($I$34)*COS($I$36)-
    COS($I$34)*COS($I$35)*SIN($I$36))+
    COS($D41)*COS(D$93)*(COS($I$34)*COS($I$36)+
    SIN($I$34)*COS($I$35)*SIN($I$36))+
    COS($D41)*SIN($I$35)*SIN($I$36)*SIN(D$93)
```

This formula follows the discussion in step 5 (see page 126, above) and refers in turn to various other cells containing data on the position of the sun (cells D41 and D93) and the location and orientation of the collector (cells I34:I36).

Finally, the references to cells A4 and B8 provide a way of adjusting all these calculations of *potential* solar energy available at 7:00 a.m. on January 21 for the thermal efficiency of the collector (in cell A4) and for the mean daily cloud cover in January (in cell B8).

When all the hourly calculations are complete, the model computes the total monthly radiation data in column P using the SUM() function and the mean hourly radiation data in row 20 using the AVERAGE() function. Column Q contains some calculations of variance. The top figure (cell Q7) is the coefficient of variation of the *monthly* radiation. Subject to checking for errors in both calculation, the coefficient is calculated by dividing the standard deviation of the 12 monthly totals by their mean:

```
IF(AND(ISERROR(STD(P8:P19))=0,AV(P8:P19)>=1),
     STD(P8:P19)/AV(P8:P19)
```

The 12 cells beneath cell Q7 (i.e., cells Q8:Q19) are coefficients of variation for *hourly* radiation and are computed, subject to various error checks, by dividing the standard deviation of the hourly radiation by its mean. For example, the contents of cell Q8 are as follows:

```
IF(AND(ISERROR(STD(D8:N8))=0,AV(D8:N8)>=1),
     STD(D8:N8)/AV(D8:N8))
```

The last cell in the column (Q20) is the mean of all the hourly coefficients of variation:

```
AV(Q8:Q19)
```

The purpose and use of these and other parts of the model are the subject of the next section.

Interpretation and Use

Perhaps the most obvious use of the model is to determine how much solar energy is available at a given location. Since actual results will depend on assumptions about the thermal efficiency of the collector and the mean daily cloud cover at each site, we shall ignore these here and assume a perfectly efficient collector on a perfectly clear day. Let us begin by examining how the performance of such a collector can vary in relation to latitude on the earth's surface. Consider, therefore, a collector lying flat on the ground (i.e., with azimuth and tilt set to zero). Running the model for each 10° of latitude in the northern hemisphere gives the results shown in Table 5.1.

Obviously, other things being equal, the amount of solar energy declines as the northerly latitude of the site increases. What also emerges from running the model at different latitudes is that the monthly distribution of solar energy is bimodal for some angles of tile in some latitudes. For example, for 0° of tilt, the distribution

Table 5.1:   Daily Solar Radiation on a Perfectly Efficient, Horizontal, Flat-Plate
Collector at Selected Latitudes above and below the Equator

Latitude	Mean Daily Radiation (Cal/cm²-hr)	Monthly Variation	Mean Hourly Variation
North Pole	76	1.33	--
80°N	127	1.15	--
70°N	190	.98	--
60°N	264	.79	.65
50°N	348	.59	.57
40°N	430	.43	.53
30°N	501	.30	.52
20°N	557	.18	.50
10°N	593	.09	.49
Equator	606	.05	.49
10°S	596	.13	.49
20°S	563	.22	.50
30°S	509	.34	.52
40°S	439	.46	.53
50°S	359	.62	.57
60°S	275	.80	.65
70°S	199	.98	--
80°S	134	1.15	--
South Pole	79	1.35	--

Note: Mean daily radiation is the mean of the total radiation computed for the
21st day of each month; monthly variation is the variation among the 12
observations on the 21st day of each month; and mean hourly variation is
the mean of the variations among the 11 hourly observations computed for
the 21st day of each month. The last calculation becomes inappropriate at
extreme latitudes.

is bimodal between the Equator and the Tropic of Cancer (which is latitude 23° N)
and unimodal elsewhere. In other words, there are two periods of maximum radiation in some parts of the world but only one in others.

However, total radiation is only a crude measure of the availability and
usefulness of solar energy: we may also want to know how consistent it is from
month to month over the year. One measure of such variation is the coefficient of
variation, which is found by dividing the standard deviation of the typical daily totals
for every month by their mean. This gives an index of how much the daily radiation
for each month varies about the mean; the lower the coefficient, the less the variation
and (thus) the more consistent the daily radiation is from month to month over the
year. From Table 5.1, it is apparent that seasonal variation is a function of latitude:

the more northerly or southerly the latitude, the greater the difference that can be expected between solar radiation in the summer and solar radiation in the winter.

In the same way, we might want to know by how much the radiation varies from hour to hour each day. Again, one possible measure of such variation is the coefficient of variation, which is the standard deviation of the hourly totals for each day divided by their mean. This gives an index of the extent to which hourly radiation over the course of a day varies about the mean for that day; the lower the coefficient, the less the variation and (thus) the more consistent the radiation over the day. From Table 5.1, it can be seen that the mean of such variations over the year is also a function of latitude, but not to the same extent as in the case of seasonal variation.

Let us look now at the angle of tilt of the collector. Increasing the angle of tilt should lead to more solar radiation, particularly at higher latitudes, where the angle of solar elevation is often quite acute. For the same reason, it should also be possible to reduce the seasonal variation (assuming that is considered desirable) by tilting the collector more towards the location of the sun during "low" season. In both cases, the objective is to bring the plane of the collector more nearly perpendicular to the radiation from the sun. Consider the case of a collector located somewhere in the world at a latitude of 40°N—which is the approximate latitude of, e.g., Reno, Denver, Columbus, Philadelphia, Madrid, Ankara, Yerevan, Beijing, and Northern Honshu—where there is a fairly significant coefficient of variation (about .43). Running the model for each 10° of tilt, from flat on the ground (0°) to vertical (90°), gives the results shown in Table 5.2.

As expected, solar radiation increases with the angle of tilt, reaching a maximum at about 35° of tilt, after which solar radiation begins to fall off again. In order to minimize seasonal variation, however, a still higher angle of tilt is required. The most consistent annual level of energy is produced at an angle of around 70° above the horizontal. But there is a price to pay for this consistency: the mean amount of solar energy drops by about 17% (from 528 cal/cm^2-hr to 438).[8] On the other hand, the mean hourly variation is not much affected by changing the angle of tilt of the collector.

---

8.   Conventional wisdom has it that you should add 15° to latitude to get the best angle of tilt for that latitude (Mobus 1981, p. 58). But this would appear to be a compromise between the angle that maximizes total radiation (i.e., 35° of tilt at latitude 40°N) and the angle that minimizes its variation (i.e., 70° of tilt at latitude 40°N).

Table 5.2:   Daily Solar Radiation on a Flat-Plate Collector Facing True South at 40° North Latitude, for Selected Angles of Tilt

Angle of Tilt	Mean Daily Radiation (Cal/cm^2-hr)	Monthly Variation	Mean Hourly Variation
0°	430	.43	.53
10°	477	.34	.53
20°	510	.27	.53
30°	527	.21	.53
40°	528	.15	.54
50°	513	.11	.54
60°	483	.09	.56
70°	438	.13	.57
80°	381	.22	.59
90°	314	.35	.63

Note: See Note to Table 5.1.

The third main factor affecting the performance of a solar collector is its azimuth, or orientation in relation to true south. When the collector faces due south, the maximum hourly radiation occurs at noon. Moreover, the collector gets exactly as much radiation before noon as after noon; in fact the hourly distribution is identical, with the data for 7:00 a.m. matching those for 5:00 p.m., the data for 8:00 a.m. matching those for 4:00 p.m., and so on. As illustrated in Figure 5.3, we can alter this distribution by changing the azimuth of the collector. Naturally, an easterly azimuth increases the amount of solar energy in the mornings and a westerly azimuth increases it in the afternoons.

Let us consider the case of a collector at latitude 40°N again, tilted at its optimum angle for that latitude (say, 35°). If we run the model for each 10° of azimuth from 0° (due south) to 120° (which is 30° north of east), we get the results shown in Table 5.3.[9] These results show that solar radiation can indeed be concentrated in the early or late part of the day by changing the azimuth of the collector. At extreme easterly orientations, for example, the maximum level of solar radiation occurs at 10 a.m. instead of noon, and the level or radiation during the first hour of

---

9.   Negative azimuths reflect orientation west of south and generate similar results to the positive azimuths, except that the data for 7:00 a.m. and 5:00 p.m. are reversed.

Table 5.3:   Daily Solar Radiation on a Flat-Plate Collector at 40° North Latitude and an Angle of Tilt of 35°, at Selected Azimuths

Azimuth (in Degrees East of True South	Mean Daily Radiation (Cal/cm²-hr)	Mean Hourly Radiation 7 am	5 pm	Mean 7 a.m. Radiation as Percent of Mean Daily
0°	530	9	9	1.7
10°	527	13	6	2.5
20°	519	16	3	3.1
30°	506	19	0	3.8
40°	491	22	0	4.5
50°	472	24	0	5.1
60°	450	26	0	5.8
70°	426	27	0	6.3
80°	400	28	0	7.0
90°	373	28	0	7.5
100°	345	28	0	8.1
110°	317	27	0	8.8
120°	289	26	0	9.0

Note: See Note to Table 5.1. Mean hourly radiation is the mean of the hourly observations shown on the 21st day of each month; in the last column, mean hourly radiation at 7:00 a.m. is expressed as a percentage of mean daily radiation.

the day (from 7:00 to 8:00 a.m.) can be as much as 44% of this maximum. But all these changes come at a price: namely, a significant decline in total daily radiation.

Evaluation and Extension

The model presented here is for flat-plate collectors, which are the most common kind of solar collectors. However, there is growing interest in cylindrical, or tubular, collectors because (a) they can be encased in a vacuum to reduce energy losses due to convection, and (b) the lower part of the tube can be coated with reflective material to concentrate the energy from the sun. Apparently, such enhancements can add 50-100% to the efficiency of tubular collectors relative to similar flat-plate versions (Kreith and Kreider 1978, pp. 234ff.).

In fact, the model presented here can be adapted relatively easily to tubular collectors rather than flat plates (ibid., p. 59). If the collector is mounted at right angles to the path of the sun through the sky (so that the azimuth of the collector can be disregarded), we can simulate its collection of energy simply by substituting the following formula for cos(i) in the equation presented in step 5 on p. 126, above:

$$\cos(i) =$$

$$\sqrt{1-((\sin(T\text{-}L)*\cos(D)*\cos(H)+\cos(T\text{-}L)*\sin(D))^2}$$

where (as before) D is the angle of solar declination in radians, L is the latitude in degrees relative to the equator, H is the solar hour expressed in degrees of arc, and T is the angle of tilt with respect to the horizontal. To adapt the spreadsheet model to tubular collectors, you would need to amend the formula for cell D95 (see page 135, above) to the following:

```
SIN($D41)*SQRT(1-(SIN($I$36-$I$34)*COS($D41)*
    COS(D$93)+COS($I$36-$I$34)*SIN($D41))^2)
```

and then copy it into cells D95:N106.

In the end, the optimum design and location of a solar collector depends a lot on the use you want to make of its energy. If you want to guarantee hot showers in the morning, then you may want to try and maximize mean solar radiation at 7:00 a.m. throughout the year. If you want to heat a swimming pool, then you may want to maximize early-afternoon radiation during the late spring and early fall months of the year. If, on the other hand, you want to use the energy for conventional space heating, you may want to maximize total daily radiation during the colder months of the year. If you want to run a refrigeration system or purify drinking water, that might mean maximizing total daily radiation during the hotter months of the year. What is best depends on what you want to achieve.

The same principle may also affect matters of more general concern that are raised by viewing the sun as a source of energy, including legal issues such as solar sight lines and rights of light, fiscal matters such as property taxation, and questions of environmental policy such as energy pricing and conservation incentives (Nadler 1977). As noted by Weinberg and Williams,

Solar energy technologies are advancing rapidly, and the prospect for further gains is auspicious. . . . The rules of the present energy economy . . . . . tend to be biased against solar energy. . . . One set of biases involves the taxes and subsidies that encourage the exploitation of fossil fuels and favor operating costs over long-term investments. . . . Another set of biases arises because present energy prices usually do not reflect many of the external social costs of energy production and use, including the dangers from air pollution and nuclear risks and the economic, ecological and human health costs of global climate change. Such costs tend to be lower for solar sources. Policies that take them into account would render solar energy more competitive [1990, p. 154].

With a simulation model like this one, you can optimize your solar collection site for its intended function. Each run involves just over 10,000 arithmetic and trigonometric calculations. But the computer does not mind doing another 10,000 calculations, just to see if another degree of azimuth or another degree of tilt will help you get more out of the sun.

Bibliography and References

Beghi, G., ed. 1983. *Performance of Solar Energy Converters: Thermal Collectors and Photovoltaic Cells.* Lectures at the Joint Research Centre, Ispra, Italy. Boston: D. Reidel Publishing.

Dostrovsky, Israel. 1991. "Chemical Fuels from the Sun." *Scientific American,* 265.6 (December), pp. 102-7.

Elder, P.S. 1984. *Soft Is Hard: Barriers and Incentives in Canadian Energy Policy.* Calgary: Detselig Enterprises.

Foukal, Peter V. 1990. "The Variable Sun." *Scientific American,* 262.2 (February), pp. 34-41.

Fowler, John M. 1984. *Energy and the Environment.* 2nd ed. New York: McGraw-Hill.

Friedman, Herbert. 1986. *Sun and Earth.* New York: Scientific American Library.

Hottel, H.C. 1976. "A Simple Model for Estimating the Transmittance of Direct Solar Radiation through Clear Atmospheres." *Solar Energy*, 18, pp. 129-34.

Hunt, V. Daniel. 1982. *Handbook of Conservation and Solar Energy: Trends and Perspectives*. New York: Van Nostrand Reinhold.

Kreider, Jan F., and Frank Kreith, eds. 1981. *Solar Energy Handbook*. New York: McGraw-Hill.

Kreith, Frank, and Jan F. Kreider. 1978. *Principles of Solar Engineering*. Washington, D.C.: Hemisphere. New York: McGraw-Hill.

Lovins, Amory. 1975. *World Energy Strategies: Facts, Issues, Options*. San Francisco: Friends of the Earth.

———. 1977. *Soft Energy Paths: Toward a Durable Peace*. New York: Friends of the Earth.

———. 1979. "Re-Examining the Nature of the ECE Energy Problem." *Energy Policy*, 7.3 (September).

Meador, Roy. 1978. *Future Energy Alternatives: Long-Range Energy Forecasts for America and the World*. Ann Arbor: Science Publishers.

Mobus, George E. 1981. "Harvesting the Sun's Energy." *BYTE*, 6.7 (July), pp. 48-58.

Nadler, Arnold. 1977. "Planning Aspects of Direct Solar Energy Generation." *Journal of the American Institute of Planners*, 43.4 (October), pp. 339-51.

Norton, Brian. 1992. *Solar Energy Thermal Technology*. London: Springer-Verlag.

Noyes, R.W. 1982. *The Sun, Our Star*. Cambridge, Mass.: MIT Press.

Ontario Research Foundation. 1978. *Solar Collector Technology Study*. Final Report. Ministry of Industry and Tourism. Toronto.

Swartman, R.K., ed. 1975. *The Potential of Solar Energy for Canada*. Presented by the Solar Energy Society of Canada in cooperation with the National Research Council and the Central Mortgage and Housing Corporation. Ottawa: National Research Council.

Weinberg, C.J., and R.H. Williams. 1990. "Energy from the Sun." *Scientific American*, 263.3 (September), pp. 147-55.

Williams, J. Richard. 1975. *Solar Energy Technology and Applications*. Ann Arbor: Science Publications.

———. 1983. *Passive Solar Heating*. Ann Arbor: Ann Arbor Science.

Zirin, Harold. 1988. *Astrophysics of the Sun*. Cambridge: Cambridge University Press.

Part Two

MODELING SOCIAL SYSTEMS

MACROECONOMIC POLICY: ECONOMETRICS AND THE KLEIN MODEL

*I think that Capitalism, wisely managed,
can probably be made more efficient
for attaining economic ends than
any alternative system yet in sight
but that, in itself, it is in many ways
extremely objectionable.*

J.M. Keynes, The End of Laisser-Faire

Nearly every industrial or near-industrial country today has a model of its economy. Indeed, in some cases, the ministry of finance, the central bank, the various commercial banks, university researchers, and even private consultants may all have their own models. The idea behind these models is that they can simulate the performance of an economy under various conditions: what happens if interest rates go up? if the value of the currency goes down? if unemployment gets worse? if productivity gets better? if R&D spending is reduced? if taxes are increased? and so on. Macroeconomic modeling is used to simulate the effect over time of changes like these.

Most of these macroeconomic, or large-scale economic, models consist of numerous equations linking the behavior of certain variables in a given time period with that of similar variables in the same or earlier time periods. For example, the model may include an equation that links changes in interest rates to changes in housing starts, or changes in housing starts in one time period to changes in employment in another time period, and so on. The Dutch economist Jan Tinbergen is generally regarded as the father of macroeconomic modeling. In 1936, he created a model of the Dutch economy made up of 24 equations.[1] By 1939, he had a model of the U.S. economy (50 equations) and, by 1940, a model of the British economy (39 equations).

After world war II, one of the most influential macroeconomic models was developed in the 1950s by Professor Lawrence Klein at the University of Pennsylvania. Klein began work on what he described as "quantifying the Keynesian system" in late 1944, when he joined the Cowles Commission for Research in Economics. His first efforts (Klein 1950) were somewhat disappointing. According to Epstein 1987 (p. 115), "The gross failure [of Klein's models] to predict the post-War surge in inflation and consumer demand compelled model revisions that were so major as to make efficient parameter estimation a matter of distinctly secondary importance." Klein left the Cowles Commission in 1947 but continued work on his models, along with his assistant, Arthur Goldberger. Their combined efforts led to the Klein-Goldberger model, which (ibid., p. 117) "made an auspicious debut with its forecasts for 1954. The 15-equation system was noticeably more accurate than the naive model and correctly predicted a turning point in GNP." This version of the model (published in Klein and Goldberger 1955) led to a host of others, some of which are still in development and use today.[2]

The Klein model and its successors consist essentially of a number of simultaneous equations expressed in terms of a number of variables. Most readers will

---

1.   His first model was published in 1936 under the rather engaging and certainly time-less title (in Dutch), *Is a Recovery in the Domestic Economic Situation of this Country Possible, with or without Action on the Part of Government, Even without an Improvement in our Export Position? and What Can Be Learned about This Problem from the Experience of Other Countries?* It was eventually published in English as "An Economic Policy for 1936" in Klaassen et al. 1959). The first English publication of his ideas (*An Econometric Approach to Business Cycle Problems*) appeared in 1937, when Tinbergen was working at the League of Nations in Geneva, and (according to Klein et al. 1961, p. 1, and Barten 1988, p. 55), it drew a scathing critique from Lord Keynes.

2.   Challen and Haggar 1983 identify five different types of macroeconomic models and name them after the economists who first designed and built them: Keynes-Klein models, Muth-Sargent models, Phillips-Bergstrom models, Walras-Johansen models, and Walras-Leontief (input-output) models. Khayum 1991, p. 8, fn. 2, complains that this classification fails to include disequilibrium models.

recall from their high-school algebra that there is a simple, analytical procedure for solving sets of simultaneous equations—as long as the number of variables (or unknowns) does not exceed the number of equations. In this way, systems of three, four, five, and even six equations can readily be solved by hand. We can program microcomputers—even in BASIC (e.g., McGuire 1983)—to manipulate still larger systems of hundreds and even thousands of equations. But when the number of variables exceeds the number of equations, the analytical procedure does not work, with or without a computer. A different approach is required.

One such technique for solving simultaneous equations is called the Gauss-Seidel method of iteration. Essentially, this method starts not with a "clean sheet" but with a possible answer; then it proceeds by repeated calculation to try to find a better and better solution. You can stipulate that calculations be done a specified number of times, or until a specified value is obtained for one or more of the variables, or until the incremental change in one or more of the variables falls below a specified value. Such a method has been described as a "strategy of successive approximations" (Braybrooke and Lindblom 1963). Of course, this method may not produce an exact answer or an optimum solution; however, a solution that is "close enough" or second best may, in practice, be a lot better than none at all.

Since many spreadsheet programs have a built-in iteration function, they provide a convenient environment for building models that rely on iterative techniques. This chapter presents a simple spreadsheet version of the original Klein model.

Purpose of the Model

Macroeconomic models are designed to simulate the behavior of an economy over time. The Klein-Goldberger model consisted of 16 stochastic equations (i.e., equations that reflect the empirical performance of the economy) and 4 identities (i.e., equations that are true by definition). The model had 38 variables, 20 of them endogenous (i.e., determined by the operations of the model) and 18 of them exogenous (i.e., determined by factors outside the model.[3]

---

3. Details are found in Klein and Goldberger 1955, pp. 34-36 and 51-53. An "updated and extended" version was published in Klein 1965, chapter 8. Klein 1974, chapter 6, provides more discussion of the model, along with three different estimates of suitable

The purpose of the model was both to provide forecasts and to simulate the effect of various policy options. In due course, the Klein model was merged with an input-output model to become the Wharton Econometric Forecasting model. This model consisted of 47 equations and 30 identities, and used 120 variables (77 of them endogenous). Moreover, the Wharton model used quarterly instead of annual data (Evans and Klein 1968, chapter II).

Since its original formulation, the Wharton model has evolved through at least four different versions. The Wharton model led in turn to the Brookings model, the "condensed version" of which contained 68 equations and used 326 variables (Schink 1976). Today's models may consist of many submodels and may contain hundreds of equations and thousands of different variables.

Macroeconomic modeling is also used widely outside the United States (Uebe et al. 1986). Thanks to the influence of Tinbergen and Klein, applications of macroeconomic models began as early as the 1940s and 1950s in the Netherlands (Barten 1988), Canada (Brown 1951), India (Narasimham 1957), and the UK (Kline et al. 1961). In the last three decades, macroeconomic models have been developed for most industrial and many developing countries.[4]

More recently, such models have begun to appear on microcomputers. Ray Fair of Yale University has had a Fortran model of the U.S. economy (FAIR-MODEL) available for microcomputers since 1983 (Fair 1984). The current version consists of 30 stochastic equations and 98 identities. The Economic Council of Canada has a full-scale, macroeconomic model of the Canadian economy written in BASIC, and used it to simulate the effects of a value-added tax (Damus 1986).[5] A similar macroeconomic model of Cyprus has recently been published, complete with BASIC source code (Hudson and Dymiotou-Jensen 1989). No doubt there are others lurking in university Economics departments and consultants' offices in many other countries, too.

---

coefficients.

4.   For a survey of models in Asian countries, for example, as well as a discussion of possible links among them and with models of their major industrial trading-partners, see Ichimura and Ezaki 1985. For an African example (Nigeria), see Oshikoya 1990.

5.   The tax was eventually implemented as the Goods and Services Tax of 1991.

Conceptual Basis

The basic idea of macroeconomic modeling is quite simple. You specify a set of equations which you think describes the way an economy works; then you calibrate these equations to fit known historical data. For example, you might decide that consumption in any time period should be equal to the sum of the wages and salaries people receive in the same time period plus the profits earned by private companies in the previous time period. Such a relationship might be expressed thus:

$$C_t = W_t + P_{t-1}$$

where C, W, and P represent consumption, wages, and profits respectively; and t represents any time period (and t-1 the preceding time period).

When we apply this equation to a set of historical data on consumption, wages, and profits over, say, 20 consecutive time periods, we may find that the equation does not fit the data very well. However, with the help of a regression technique (usually a least-squares method of some kind), we can extract from the historical data a set of coefficients (or, less respectfully, "fudge factors") that yield a better fit between the equation and the data. For example, we might find that the equation

$$C_t = 1.5 * W_t + .8 * P_{t-1}$$

fits the historical data a lot better than the original equation (where the two coefficients are in fact unity). It may also be helpful to add a stochastic factor to the equation to account for unpredictable disturbances. In any case, the accuracy of the model is only as good as its "goodness of fit" with data from the calibration period.[6]

The model presented here is the original version (Klein 1950) and is usually referred to as Klein's "Model I". This version is taken from the presentation in Johansson 1985, which is based on Gujarati 1988 (see example 17.6 on pp. 561-62). Model I is simpler than the Klein-Goldberger version. Instead of 20 equations, Model I has only 6; instead of 38 variables, Model I has only 9. But Model I is similar in principle to the later version and all their successors.

---

6.   Further details can be found in introductory textbooks on econometrics, such as Klein 1974 or Gujarati 1988. The Cyprus case study referred to above (Hudson and Dymiotou-Jensen 1989) also provides a good introduction to the subject of this kind of modeling.

The nine variables in Klein's Model I consist of six endogenous variables and three exogenous variables. All variables are expressed in terms of constant dollars (Klein used 1934 dollars), in order to remove the effects of inflation. The six endogenous variables are:

C   consumption expenditure
I   investment expenditure
W   private-sector wage bill
T   taxes
Y   income after tax
K   capital stock

The three exogenous variables are:

G   government expenditure
W'   government wage bill
P   private-sector profits

Model I treats the exogenous variables in one of two ways: either as having fixed values, or as changing at fixed rates over time.

The six equations in Klein's Model I include three stochastic equations and three identities. The stochastic equations consist of a consumption function, an investment function, and a demand function for labor. These are:

- **Consumption**: consumption is defined as a function of three factors—current profits, current wages, and profits in the previous time period. More precisely,

$$C_t = B0 + B1 * P_t + B2 * (W_t + W'_t) + B3 * P_{t-1} + u1$$

  where B0, B1, B2, and B3 are regression coefficients derived from historical analysis of the economy and u1 is a suitable stochastic factor intended to allow for disturbances of an unpredictable nature.

- **Investment**: investment is a function of current profits, profits in the previous time period, and capital stock in the previous time period. More precisely,

$$I_t = B4 + B5 * P_t + B6 * P_{t-1} + B7 * K_{t-1} + u2$$

where (as before) B4, B5, B6, and B7 are regression coefficients derived from historical analysis, and u2 is a stochastic factor.

- **Demand for labor**: the demand for labor (which is reflected in the private-sector wage bill) is a function of both current and past levels of gross income (less what is paid out in wages for government workers) and of the number of years the model has been running; more precisely,

$$W_t = B8 + B9 * (Y_t + T_t - W'_t) + B10 * (Y_{t-1} + T_{t-1} - W'_{t-1}) + B11 * t + u3$$

where B8, B9, B10, and B11 are regression coefficients derived from historical analysis, t is the time (in years since 1931) and u3 is a stochastic factor.

The other three equations define the levels of taxation, net income (i.e., income after tax), and capital stock.

- **Taxation** in any time period is defined as the difference between production (that is, the sum of expenditure on consumption, investment, and government services) and net income; in other words,

$$T_t = C_t + I_t + G_t - Y_t$$

- **Net income** in any time period is defined as the sum of government wages, private-sector wages, and profits: that is,

$$Y_t = W'_t + W_t + P_t$$

- **Capital stock** consists of capital stock from the previous time period plus any current investment: that is,

$$K_t = K_{t-1} + I_t$$

More details on the equations and their derivation can be found in the original source (Klein 1950) and elsewhere (e.g., Maddala 1977; or Gujarati 1988).

Looked at as a whole, this system of six equations has three notable features: (a) the equations are circular; (b) they are also recursive; and (c) there are more variables than there are equations.

The equations are circular because some of the variables are defined in terms of each other. For example, investment is defined in terms of capital stock and capital stock is defined in terms of investment; similarly, wages are defined in terms of income and income is defined in terms of wages. But, as Johansson explains,

> There is more: not only is this set of equations circular; it is also recursive. That is, the solution to the set of equations for a particular time period depends on the solution of the same equations for preceding time periods. In this particular model, only one preceding time period is needed [period t-1 in relation to period t]. . . . However, it is quite common that one has to go back in time more than one time period to define the model [1985, p. 400].

Clearly, this is a complex system of equations.

A third significant feature of Klein's Model I is that there are more variables (nine) than there are equations (six). That means we are not going to be able to solve the equations in the analytical way we learned in high school. So we must use a method like the Gauss-Seidel iteration method. To quote Johansson again,

> A common method used to solve complex systems of linear equations is the Gauss-Seidel iterative method, which goes roughly like this: take an arbitrary value (often 0) and use it to initialize one or several of the unknowns in order to break all circularities in the system of equations. Now compute the rest of the unknowns through simple substitution. Obviously, what we have now is not a solution because at least one unknown has been given an arbitrary value. Because of the circular nature of the equations, however, we can now compute this variable from the other variables we just computed. In so doing, we will get a better initial value to use in a second round with the other unknowns. Now repeat the process a number of times. During each iteration the values tend to change less and less. After a certain number of iterations, they do not change any more. The process has converged, and the values are stable. We have reached the solution [ibid., p. 402].

Of course, this solution is only an approximation. How close an approximation depends on what we mean when we say the values "do not change any more." This is determined by what is called the "delta value," which is the value below which any change in the value of a variable from one iteration to the next is regarded as being insignificant. At that point, we say that the process has converged to a solution.

Naturally enough, the efficiency of this method (for a given delta value) depends on how close to the correct answer is the initial "arbitrary value" assigned to the variables. Make a good guess and the method may take fewer than ten iterations to reach a solution. Make a bad guess (such as starting at zero or some truly arbitrary value) and the method may require more than 100 iterations. As it happens, spreadsheets are particularly well suited for this kind of iteration. This is because

> spreadsheets like Lotus 1-2-3 always store the results of the last computation (the last solution) with the formulas. When using a spreadsheet [therefore], we will always start the iterations from an approximation that is probably close to the solution. The only exception to this rule is the first time we enter the formulas [ibid.].

After that—and particularly if we make relatively small changes to the parameters of the model—iteration in a spreadsheet is relatively efficient, even if spreadsheets generally are slow compared to more traditional programming environments.

Data Requirements

Two kinds of data are required to run Klein's Model I. One is historical data on the performance of the economy to be modeled, from which to derive the coefficients that will be used for calibrating the model. The other is the set of initial values for the nine variables with (optionally) fixed growth rates for the three exogenous variables. For the first, Klein used data on the performance of the U.S. economy from 1920 to 1941 (Klein 1950; Maddala 1977, p. 242; or Gujarati 1988, p. 611). To get the desired coefficients, Klein used various techniques, including the regression method of ordinary least squares (OLS); however, others now recommend a more sophisticated method called the two-stage, least-squares (2SLS) method.[7]

---

7.   Gujarati 1988 (p. 562) explains that "because of the inter-dependence among the endogenous variables, in general they are not independent of the stochastic disturbance terms, which therefore makes it inappropriate to apply the method of OLS to an individual equation in the system. For . . . the estimators thus obtained are inconsistent; they do not converge to their true population values even when the sample size is very large." However, Klein himself kept to the OLS method right up to his work on the 1979 Wharton model (Epstein 1987, chapter 4). Smith 1973 (chapter 5) has used Monte Carlo simulation to evaluate different ways of estimating parameters for both Model I and the Klein-Goldberger variation. See chapter 10, below for another example of Monte Carlo simulation.

For the second, Klein just relied on appropriate published data. As noted, fixed growth rates can be specified for the exogenous) variables, if that seems appropriate. Naturally, the time period (year) for the nine initial values should be consistent.

In order for us to be able to apply this model to circumstances other than the United States in the 1940s, we will need the appropriate data. On the one hand, this means having historical data with which to recalibrate the model with the appropriate coefficients. On the other hand, it means having data on the current performance of the economy, so that we can enter the initial values required. In most national jurisdictions now, such data are usually available in the national accounts. For subnational regions, however, getting the data may prove more difficult.

Overview

The worksheet consists of 9 columns and 43 rows: see Listing 6.1 below. The 9 columns are sufficient to display a 6-year simulation. The top part of the worksheet shows the structure and parameters of the model, while the lower part presents the results of the simulation for 6 years. However, the size and structure of the model is quite flexible. Longer-term simulations could be accommodated simply by copying the entire last column (column I) into additional columns to the right, as far as the capacity of the computer will allow—and the credibility of the user will accept.

The top part of the worksheet contains a complete description of Klein's Model I. The three stochastic equations are defined in rows 4, 5, and 6 and the three identities in rows 13, 14, and 15. The formulas in these rows do not actually play any functional part in the operation of the worksheet; they are given here only for the edification of the user.

The rows between the equations and the identities (rows 8-11) contain behavioral coefficients for the model, based on the historical performance of the economy being modeled (see the note in the Listing). Cell I15 contains a delta value to control the iteration process. The next few rows (rows 17-20) provide for specification of initial values and growth rates (including zero or even negative growth) for the exogenous variables in the model.

Listing 6.1

```
  |    A     ||B||   C  ||  D  ||  E  ||  F  ||  G  ||  H  ||  I  |
1  MACROECONOMIC MODELING -- KLEIN'S MODEL I          © 1993 by T.J.Cartwright
2
3  - Equations ----
4  Consumption    C(t) = B0 + B1*P(t) + B2*(W+W')(t) + B3*P(t-1) + u1
5  Investment     I(t) = B4 + B5*P(t) + B6*P(t-1) + B7*K(t-1) + u2
6  Labour demand  W(t) = B8 + B9*(Y+T-W')(t) + B10*(Y+T-W')(t-1) + B11*t + u3
7  - Coefficients -
8  Consumption    B0= 16.780    B1=   .020   B2=   .800   B3=   .230
9  Investment     B4= 17.790    B5=   .230   B6=   .550   B7=  -.150
10 Labour demand  B8=  1.600    B9=   .420   B10=  .160   B11=  .130
11 Stochastic     u1=  1.110    u2=  -.270   u3= -1.040
12 - Identities ---
13 Taxes          T(t) = C(t) + I(t) + G(t) - Y(t)         Delta value for
14 Income (net)   Y(t) = W'(t) + W(t) + P(t)               Iteration
15 Capital stock  K(t) = K(t-1) + I(t)                     Control:    .001
16 ─────────────────────────────────────────────────────────────────────────
17 Starting year        1940       Capital stock   204.500         Growth
18 Consumption        65.000       Govt expenditure 15.400          30.0%
19 Investment          3.300       Govt wage bill    8.000          15.0%
20 Priv wage bill     45.000       Profits          21.100           2.0%
21 ─────────────────────────────────────────────────────────────────────────
22        Year         1940     1941     1942     1943     1944     1945     1946
23 US Economy                       Billions of 1934 US Dollars
24 Consumption    C  65.000   70.014   75.887   83.182   92.434  104.222  119.276
25 Investment     I   3.300    3.400    3.221    3.076    2.959    2.866    2.795
26 Govt Expend    G  15.400   20.020   26.026   33.834   43.984   57.179   74.333
27 Pvt Wage Bill  W  45.000   49.350   55.180   62.577   72.180   84.676  100.936
28 Govt Wage Bill W'  8.000    9.200   10.580   12.167   13.992   16.091   18.504
29 Profits        P  21.100   21.522   21.952   22.391   22.839   23.296   23.762
30 Taxes          T   9.600   13.361   17.422   22.956   30.366   40.205   53.201
31 Income (Net)   Y  74.100   80.072   87.713   97.136  109.011  124.063  143.202
32 Capital Stock  K 204.500  207.900  211.121  214.197  217.156  220.022  222.817
33 ─────────────────────────────────────────────────────────────────────────
34 Source:  Adapted from Jan-Henrik Johansson, "Simultaneous Equations with
35    Lotus 1-2-3," BYTE (February 1985), pp. 399-405.  The original model is
36    from L.R. Klein, Economic Fluctuations in the United States, 1921-1941
37    (New York: Wiley, 1950).  Coefficients and initial values are those used
38    by Klein in his original formulation of the model (see ibid., pp. 68
39    and 135).  Estimates for the three stochastic variables are taken from
40    Klein's calculations for 1940 (ibid. p. 69).
41
42
43
```

The rest of the model (rows 24-32) contains formulas for calculating the endogenous variables for each year of the simulation. The formulas in column D, for example, are as follows:

D22	Year	C22+1
D24	Consumption	$C$8+$E$8*D29+$G$8*(D27+D28)+$I$8*C29+$C$11
D25	Investment	$C$9+$E$9*D29+$G$9*C29+$I$9*C32+$E$11
D26	Gov't. expenditure	C26*(1+$I$18)
D27	Private wage bill	$C$10+$E$10*(D31+D30-D28)+$G$10*(C31+C30-C28)+$I$10*(D22-1931)+$G$11
D28	Gov't. wage bill	C28*(1+$I$19)
D29	Profits	C29*(1+$I$20)
D30	Taxes	D24+D25+D26-D31
D31	Income (net)	D27+D28+D29
D32	Capital stock	C32+D25

Note that each of the formulas here is a more or less direct translation of the corresponding algebraic equation discussed in the previous section and displayed in rows 4-6 and 13-15. Note, too, that Klein used a base year of 1931 for his estimates of the private wage bill (cells D27:I27).

Operation

Using the model is as simple as entering the appropriate data—coefficients in rows 8-11, initial values in cells C17:C20 and G17:G20, and assumed growth rates in cells I18:I20. (The source of the data shown in Listing 6.1 is given in the note at the bottom.) The performance of the economy can readily be summarized by means of line and bar graphs, such as those shown in Figures 6.1 and 6.2 below (which are based on the data in Listing 6.1).

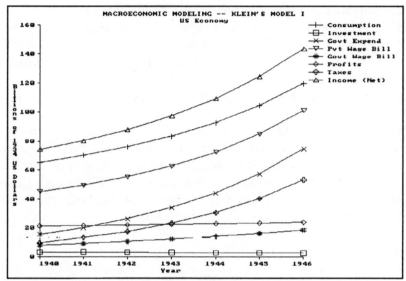

Fig. 6.1. Simulated US Economy, 1940-46, Using Klein's Model I

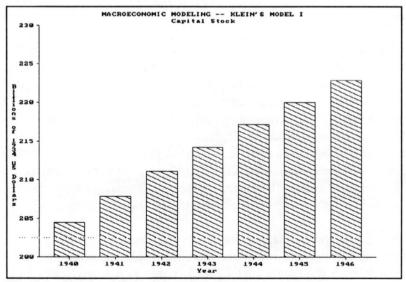

Fig. 6.2. Simulated Growth in US Capital Stock, 1940-46

After that, we can adjust the growth rates and/or the initial values as desired, to reflect the policy options we want to explore. For example, we can model the effects of higher or lower initial values for consumption, investment, labor demand, taxation, net income, or capital stock. The results can be viewed directly or by means of graphs. Similarly, we can model the effects of increases or decreases in rates of growth of public expenditure, the civil service, and corporate profits.

We can also change the behavioral coefficients. This may be done to test the sensitivity of the model to such changes, as a means of assessing its robustness. Or the behavioral coefficients may be altered to simulate structural change in the economy or to represent the economy in a fundamentally different era. Coefficients appropriate for these latter purposes may not be readily available and may have to be estimated rather than calculated from historical data.

Interpretation and Use

Klein's Model I is probably too simple to be used as the basis for public policy. But what the model can do is illustrate how bigger and more sophisticated macroeconomic models work—and such models certainly have been and still are used to influence and inform public policy making. Butt what has been the impact of such models and how that impact is achieved remain matters of considerable debate.

Studies in the United States (e.g., Klein and Burmeister 1976), in the United Kingdom (Britton 1989), and in developing countries (Khayum 1991) all illustrate the extent to which macroeconomic models are currently used. But there is also controversy over how models like these actually achieve their influence. Nearly 20 years ago, Douglass Lee (1973) wrote about the "seven sins" of large-scale models in planning. More recently, Kenneth Kraemer (1987) looked at a number of cases in the United States where large-scale models have played a part in policy making in the federal government, and concluded that their impact has been complex and mixed. Similar conclusions are reached by George Chadwick (1987, chapters 6 and 7) in the context of urban and regional planning in developing countries.

In the end, however, macroeconomic models (like most other models) reflect beliefs about what things are important and about causal relationships between and among those things. Building a model, whether it is a relatively simple model (like Klein's Model I) or a more complex one (like some of its successors), merely makes those beliefs explicit.

Models of all kinds—as well as money, technology, charisma, and many other resources—can be used and misused in making public-policy decisions. It is not the resources themselves so much as the context of their use that determines how well or badly they are used. In the end, there is only one intellectual sin and that is to draw a conclusion without being able to explain how you reached it. That kind of sin can be committed as easily without a computer model as with one.

Thus, the value of macroeconomic models like those of Klein and his successors lies not in whether they give correct answers but rather in what they do to encourage users to explore their problems before rushing in to "solve" them. In other words, the appropriate criteria are: does the model encourage users to think about its assumptions and omissions? to question its results? to examine alternative scenarios? to adjust its parameters? to think about how to design new models? and so on. By this standard, Model I has indeed done well.

Evaluation and Extension

Spreadsheets are capable of accommodating quite large models of this kind. Johansson (1985, p. 404) reports that he has been able to implement spreadsheet models with as many as 450 endogenous variables and 100 exogenous variables, and a time span of 20 years. Such a model, he adds, has a total of almost 9,000 unknowns! Moreover, spreadsheets turn out to be particularly efficient for this kind of modeling. According to Johansson, even with the largest of his models,

> One complete iteration [on his Compaq microcomputer, vintage circa 1985] still takes only about 22 seconds, and [after an initial run] only 4 to 10 iterations are required for convergence [to the solution]. This is fully adequate for practical work with realistic models and makes our simple spreadsheet approach surprisingly competitive when compared even to large mainframe software packages specifically designed to solve macroeconomic models. . . . Iteration is an easy and slow method to solve equations. But because a spreadsheet by default makes a smart guess, we see a drastic improvement in efficiency, allowing us to attack much tougher problems than before [loc. cit. p. 404].

In fact, spreadsheets are so well suited to iteration that Johansson describes the relationship between them as one of "synergism". It is almost as though spreadsheets were born to iterate.

Bibliography and References

Barten, A.P. 1988. "The History of Dutch Macroeconomic Modeling (1936-1986)." In *Challenges for Macroeconomic Modeling*, edited by W. Driehuis, M.M.G. Fase, and H. den Hartog, chapter 3. Amsterdam: North Holland.

Braybrooke, David, and Charles E. Lindblom. 1963. *A Strategy of Decision*. New York: Free Press.

Britton, Andrew, ed. 1989. *Policymaking with Macroeconomic Models*. Aldershot, U.K.: Gower.

Brown, T.M. 1951. "Canadian Experience in Forecasting for Econometric Models." Paper presented at a meeting of the Econometric Society, Boston, in December. Cited in Klein et al. 1961, p. 2.

Chadwick, George. 1987. *Models of Urban and Regional Systems in Developing Countries: Some Theories and Their Application in Physical Planning*. Oxford: Pergamon.

Challen, D.W., and A.J. Haggar. 1983. *Macroeconometric Systems, Constructions, Validations, and Applications*. New York: St. Martin's.

Damus, Sylvester. 1986. *Micro-Computer Simulation with a General Equilibrium Model of Canada*. Economic Council of Canada Discussion Paper No. 311. Ottawa: Economic Council of Canada.

Epstein, Roy J. 1987. *A History of Econometrics*. Amsterdam: North Holland.

Evans, Michael K., and Lawrence R. Klein. 1968. *The Wharton Econometric Forecasting Model*. Programmed by George Schenk. Studies in Quantitative Economics, No. 2. Philadelphia: Economic Research Unit, University of Pennsylvania.

Fair, Ray C. 1984. *Specification, Estimation, and Analysis of Macroeconometric Models*. Boston: Harvard University Press.

Gujarati, Damodar. 1988. *Basic Econometrics*. 2nd ed. New York: McGraw-Hill.

Hudson, John, and Marina Dymiotou-Jensen. 1989. *Modeling a Developing Country: A Case Study of Cyprus*. Avebury, U.K.: Gower.

Ichimura, Shimichi, and Mitamo Ezaki. 1985. *Econometric Models of Asian Link.* Tokyo: Springer-Verlag.

Johansson, Jan-Henrik. 1985. "Simultaneous Equations with Lotus 1-2-3." *BYTE*, 10.2 (February), pp. 399-405.

Khayum, Mohammed F. 1991. *Macroeconomic Modeling and Policy Analysis for Less Developed Countries.* Boulder, Colo.: Westview.

Klaassen, L.H., L.M. Koyck, and H.J. Witteveen, eds. 1959. *Jan Tinbergen: Selected Papers.* Amsterdam: North Holland.

Klein, Lawrence R. 1950. *Economic Fluctuations in the United States, 1921-1941.* New York: Wiley.

———. 1965. *The Keynesian Revolution.* 2nd ed. New York: Macmillan.

———. 1974. *A Textbook of Econometrics.* 2nd ed. Englewood Cliffs: Prentice-Hall.

———, R.J. Ball, A. Hazlewood, and P. Vandome. 1961. *An Econometric Model of the UK.* Oxford: Blackwell.

———, and Edwin Burmeister, eds. 1976. *Econometric Model Performance: Comparative Simulation Studies of the U.S. Economy.* Philadelphia: University of Pennsylvania.

———, and A.S. Goldberger. 1955. *An Econometric Model of the United States, 1929-1952.* New York: North-Holland.

Kraemer, Kenneth L. 1987. *Datawars: The Politics of Modeling in Federal Policymaking.* New York: Columbia University Press.

Lee, Douglass B. 1973. "Requiem for Large-Scale Models." *Journal of the American Institute of Planners*, 39.3 (May), pp. 163-78.

Maddala, G.S. 1977. *Econometrics.* New York: McGraw-Hill.

McGuire, Patrick E. 1983. "A Gauss-Jordan Elimination Method Program." *BYTE*, 8.8 (August), p. 394.

Narasimham, N.V.A. 1957. *A Short-Term Planning Model for India.* Amsterdam: North Holland.

Oshikoya, Temitope Waheed. 1990. *The Nigerian Economy: A Macroeconometric and Input-Output Model.* New York: Praeger.

Schink, George R. 1976. "The Brookings Quarterly Model as an Aid to Longer Term Economic Policy Analysis." In Klein and Burmeister 1976, chapter 14.

Smith, V. Kerry. 1973. *Monte Carlo Methods: Their Role for Econometrics.* Lexington, Mass.: D.C. Heath.

Tinbergen, Jan. 1937. *An Econometric Approach to Business Cycle Problems.* Paris: Hermann.

Uebe, G., G. Huber, and J. Fischer. 1986. *Macro-Econometric Models: an International Bibliography.* Aldershot, U.K.: Gower.

Chapter 7

URBAN FORM: THE LOWRY MODEL OF POPULATION DISTRIBUTION

> *The Lowry model has probably generated more interest*
> *. . . than any other single urban model. It has been*
> *quite widely used and extensively written about.*
>
> *Colin Lee,* Models in Planning

Planners need to be able to forecast not just the growth of population but also its spatial distribution. It is all very well to predict that a city or a region is going to need four more schools or two new health facilities in the next ten years. But it is equally important to know where in the city or region, in precise spatial terms, this demand is going to occur, so that schools and health facilities can be located nearby. One of the most familiar models for this purpose is the so-called Lowry model, named after its inventor, Ira S. Lowry.

At the time he developed the model, Lowry was working on an economic study of the Pittsburgh region. His aim, he said, was to create

an analytical model capable of assigning urban activities to sub-areas of a bounded region in accordance with those principles of locational interdependence that could be reduced to quantitative form. The model is not designed to project regional aggregates, such as total employment or population, but rather to allocate such aggregates to locations within the region. Properly adapted, [the model] should be useful for the projection of future patterns of land development and for the testing of public policies in . . . transportation planning, land-use controls, taxation, and urban renewal [1964, p. 2].

Once his model had been developed, Lowry tested it on 420 zones (each approximately 1 square mile in area) using data that had been gathered for the Pittsburgh Area Transportation Study (PATS) during 1958-60. There have been numerous other applications of the model since then. Moreover, the model has been the object of several major extensions (e.g., Garin 1966; Wilson 1970; and Macgill 1977) and exhaustive reviews (e.g., Mackett and Mountcastle 1977; and Webber 1984).

In retrospect, the Lowry model can be seen to represent at least two major innovations in urban and regional modeling (Lee 1973, pp. 90-91): the model linked quantitative forecasting to spatial distribution; and it combined three hitherto disparate aspects of development (population, employment, and transportation) in a single model. As Michael Webber explains,

> The publication of [the Lowry model] constituted a break-through in attempts to model urban systems. Many of the urban modeling efforts in the early 1960s had floundered or were floundering in a mass of detailed analyses which lacked a clear, comprehensive view of the way a city operates and which were ambitious far beyond the then available computational capacities. By contrast, Lowry offered an extremely simple and elegant insight into the way an urban economy works. . . . The model was computable—its data needs could be met and it could be solved—yet it managed to combine several of the then current macro-geographic ideas about urban areas. . . . Urban geographers, economists, and planners have continued to be fascinated by the Lowry model because it can be operationalised, has a simple yet comprehensive structure, and offers opportunities for further development [1984, p. 27; echoing Goldner 1971].

Purpose of the Model

The purpose of the Lowry model is to allocate the population of a region among a number of smaller areas or zones into which the region is divided. There may be anywhere from a half a dozen to many hundreds of such zones. Lowry divides "activities occupying space" into three categories, which he describes in the following terms (Lowry 1964, pp. 2-3):

1.   A *basic sector*, including industrial, business and administrative establishments whose clients are predominantly nonlocal. These "export" industries are relatively unconstrained in local site-selection by problems of access to local

markets, and their employment levels are primarily dependent on events outside the local economy. Consequently, they have been treated as exogenous to the model, as activities whose locations and employment levels must be assumed as given.

2. A *retail sector*, including those businesses, administrative, and other establishments which deal predominantly and directly with the local residential population. Because these establishments have local clients, site-selection is assumed to be powerfully constrained by problems of access to local residents, and employment levels are assumed to be closely tied to local growth of population. The location and levels of employment of establishments in this sector (which include most schools and local government agencies as well as retail trades and services) are treated as endogenous variables whose values are determined within the model.

3. A *household sector*, consisting of the resident population. It is assumed that the level of employment in the retail sector depends directly on the number of resident households, and that the number of resident households in turn depends upon the number of basic and retail jobs available at any given time. Furthermore, it is assumed that residential site-selection is subject to powerful influence by the location of the resident's place of work. Thus, the number and locations of households are also determined within the model.

The second component of the Lowry model is the assumption that what he called the "powerful" influence between residential location and the location of business and industry could be expressed in terms of transportation costs.[1] In brief, the higher the transportation costs, the more powerful the links between place of residence and place of work are likely to be.

Thus, the Lowry model works like this. Take a particular region and divide it into a number of zones. Then, for a given distribution of *basic sector* jobs across

---

1. This "gravity model" approach appears not to have been Lowry's first choice. It seems he would have preferred a formulation based on the potential of each zone to attract jobs and housing. But the gravity model was forced on him by the availability of data from the PATS study (Lowry 1964, chapter III, and Lee 1973, p. 91). See also Carrothers 1956 for a review of the two approaches.

those zones and a given set of travel costs among the zones, the model computes the total population and employment to be found in each zone.

Conceptual Basis

The version of the Lowry model presented here is adapted from a spreadsheet version developed by Richard Brail (1987). While most of the functions of Brail's version of the model have been preserved here, its form has been substantially altered to take more advantage of the style and operation of a spreadsheet.

As noted above, the Lowry model is based on three fundamental assumptions: (1) that population is a direct function of employment; (2) that total employment is a ultimately a function of the number of jobs in the *basic sector* (which is determined exogenously or outside the model); and (3) that people choose to live in areas where they have reasonable access to their place of work. Although oversimplified, these assumptions are by no means unrealistic.

At the heart of the Lowry model lies a link between total population and the number of workers. This link is defined as a multiplier or set of multipliers. Thus, once the number of jobs is computed for any zone, the total number of people in that zone can be calculated simply by applying the multiplier to the number of jobs.

The second key assumption of the Lowry model is that there are two principal types of employment: "basic" jobs (in what Lowry calls the "basic sector") and "service" jobs (in what Lowry calls the "retail sector"). Thus, basic jobs are jobs whose location is determined exogenously (that is, by factors outside the model); such jobs tend to be in basic industries such as manufacturing and processing. Service jobs, on the other hand, are jobs whose location is determined endogenously (that is, by the spatial distribution of the population within the region) because they are intended to meet the direct needs of people in the region; such jobs tend to be in the retail or service sector.

The number of basic jobs and their distribution among the zones in a region are part of the input data that must be provided to run the Lowry model. The number and distribution of service jobs, on the other hand, is calculated by the model. The Lowry model uses another multiplier or set of multipliers (service workers per capita) to compute the number of service workers required to support a given population. It should be noted, however, that service workers and their families will themselves require the support of still more service workers. So the Lowry model embodies a certain degree of circularity here.[2]

The third main component of the Lowry model is a matrix showing the "generalized cost" (i.e., the cost in terms of both time lost and money spent) of traveling between each pair of zones in the region we are modeling (Black 1981, p. 57). These travel costs may be expressed in monetary terms, but the more common practice is to use travel time or distance as an index of the cost. Thus, for example, if a region had six zones, we would need a six-by-six table showing the average travel time (usually in minutes) from each zone to each other zone. Such a table might look like this:

	Destination Zone					
Origin Zone	1	2	3	4	5	6
1	8	12	15	20	22	21
2	12	6	7	12	20	18
3	15	7	5	8	10	12
4	20	12	8	7	10	20
5	22	20	10	10	8	12
6	21	18	12	20	12	6

According to this table, from an origin in zone 1, it takes an average of 8 minutes to travel to a destination in the same zone, 12 minutes to a destination in zone 2, 15 minutes to a destination in zone 3, 20 minutes to a destination in zone 4, 22 minutes to a destination in zone 5, and 21 minutes to a destination in zone 6. From an origin in zone 2, it takes 12, 6, 7, 12, 20, and 18 minutes, respectively, to travel to the six zones. And so on.

It should be noted that the above matrix is symmetrical about the diagonal, reflecting the fact that travel from any given origin to any given destination takes exactly the same time as travel between the same points in the opposite direction. While this is often the case, it need not be—as in cases where there are one-way streets. Thus, it is possible to create a travel-time matrix that is *not* symmetrical about the diagonal, where travel time between two points does depend on which is

---

2.   See also chapter 6, above, for another example of circularity. As long as the service worker multiplier is less than unity, the iteration will converge on a solution.

the origin and which is the destination. On the other hand, the Lowry model assumes that, for any particular trip, the return part always takes the same length of time as the outbound part.

In order to be able to use the travel-time data, the Lowry model converts them into what is called a "location-probability matrix." (Details of the transformations involved are discussed in a technical note at the end of this chapter.) The probability matrix derived from interzonal travel times is used to determine in which zones people are likely to live, given the location of their places of work. From this distribution, the total population in each zone is calculated using the population-per-worker multiplier discussed above. The Lowry model has done its job.

To summarize, the Lowry model works more or less like this:

1. It assumes an initial distribution of basic workers among the zones in the region.

2. It calculates which zones these basic workers are likely to live in, using the location-probability matrix.

3. It calculates a preliminary population distribution for basic workers only, using a population-per-worker multiplier.

4. It calculates the number of service workers required in each zone to meet the needs of the population in that zone, using a service-worker-per-capita multiplier.

5. It calculates which zones these service workers are likely to live in, using the location-probability matrix again.

6. It calculates a revised population distribution, using the population-per-worker multiplier again.

7. It calculates the total number of workers and the total population, by adding together the basic and service components of each one.

8. It goes back to step 4 and recomputes the number of service workers required, based on the revised total population; then repeats the subsequent steps (steps

5 to 8) until the revised total population in step 7 is no different from what is was after the previous calculation.

When the model converges on a solution like this, it has done its job.

Data Requirements

There are three sets of data required to run the Lowry model:

- The model requires an initial distribution of basic workers among the zones into which the region is divided. This distribution is determined exogenously. So we need to provide such data before we can run the model.

- The model also requires some multipliers. First, we need a multiplier or set of multipliers that reflects the average number of dependants per worker, so the model can calculate total population from the number of workers. The model may use a single multiplier for basic and service jobs in all zones, or a set of multipliers for different types of jobs in different zones. Second, we need a multiplier or set of multipliers for calculating the number of service workers required to meet the needs of a given population. Again, the model may use one multiplier for all zones or different multipliers for different zones. Note that the value of the first multiplier will always be greater than unity, while the value of the second multiplier will always be less than unity.

- In addition, the model requires a matrix of travel times among each pair of zones in the region being modeled. Data on travel times can be gathered in various ways, including empirical analysis, sample measurement, or questionnaire survey (Black 1981, chapter 3). Nor does it matter how the travel occurs —whether it is private automobile, public transport, or even bicycle—as long as it is the means commonly used for that particular route. In fact, it does not even matter how accurate the estimates are, as long as they are a reflection of *perceived* travel times, since (to the extent that behavior is rational) people usually act on the basis of their perceptions of reality (which may not always be objectively accurate).

- Finally, the model also requires a suitable friction factor for use with the travel-time data to calculate a willingness-to-travel matrix and (ultimately) a location-probability matrix. This friction factor is normally a positive value greater than zero.

Overview

The model presented here consists of 11 columns and 46 rows (see Listing 7.1). In this form, the model can simulate up to six zones; however, it can be expanded to accomodate as many zones as your computer memory and software will allow.

The model is divided into two parts. The top twenty rows provide the user interface for data entry and display of the results. The remaining rows contain labels for the graphics and calculation of the two matrices derived from the travel-time data: that is, the willingness-to-travel matrix and the location-probability matrix. The top (visible) part of the model is further divided into three parts. First, there are three adjustable parameters: the friction factor (cell B4) and two multipliers, one to compute overall population from the number of workers (cell B7) and one to compute the number of service workers required to support a given population (cell B10). To the right of these parameters is the generalized travel-time matrix (cells E5:J10). All these data are inputs to the model and must be entered by the user for each region that is to be modeled. Below this is a profile of the zones used in the model. The first line (row 13) contains input data that must be provided by the user; the remaining rows are calculated by the model, using the data entered above. The results are shown in the last two rows (rows 19 and 20).

As soon as the travel-time data are entered in cells E5:J10, the model computes two corresponding matrices: i.e., a willingness-to-travel matrix (in cells E32:J37) and location-probability matrix (in cells E39:J44). Both calculations follow the formulas described above. Thus, the travel time in cell E5 is converted to an index of the willingness to travel in cell E32 by means of the following formula:

```
1/E5^$B$4
```

Then the willingness to travel is, in turn, converted to a location probability in cell E39 by means of the following formula:

Listing 7.1

```
     |   A      ||B ||   C    |  |E ||F  || G  || H  || I  || J  || K  |
1    | THE LOWRY MODEL OF URBAN DEVELOPMENT              © 1993 by T.J. Cartwright
2
3    Generalized Travel Times to > > Zone 1    2    3    4    5    6
4    Friction Factor      .8        ------------------------------------------
5                         From 1    8    12   15   20   22   21
6    Population/worker         2    12   6    7    12   20   18
7    multiplier        2.5      3    15   7    5    8    10   12
8                               4    20   12   8    7    10   20
9    Service worker/pop         5    22   20   10   10   8    12
10   multiplier        .2       6    21   18   12   20   12   6
11   ----------------------------------------------------------------------
12                            Zone1 Zone2 Zone3 Zone4 Zone5 Zone6  Total
13   Number of basic workers    100   300   400    50   200  1000   2050
14   Residential distribution   251   331   410   294   329   436   2050
15   Population distribution    627   827  1025   734   822  1090   5125
16   Number of service workers  263   343   418   318   327   381   2050
17   Residential distribution   275   356   426   342   326   325   2050
18   Population distribution    688   889  1065   855   815   813   5125
19   TOTAL WORKERS              363   643   818   368   527  1381   4100
20   TOTAL POPULATION          1315  1716  2090  1589  1637  1903  10250
21   ----------------------------------------------------------------------
22
23
24   Graph labels:
25   Main Heading      THE LOWRY MODEL OF URBAN DEVELOPMENT
26   Subheading        For a City of Six Zones
27   Variable 1        Basic Jobs
28   Variable 2        Population
29
30                            Zone 1    2    3    4    5    6  Total
31                            -------------------------------------------
32   Willingness-to-Travel Matrix  1  .19  .14  .11  .09  .08  .09   .70
33   (converts travel times into   2  .14  .24  .21  .14  .09  .10   .91
34   an index of the willingness   3  .11  .21  .28  .19  .16  .14  1.09
35   to travel using an inverse    4  .09  .14  .19  .21  .16  .09   .88
36   power function)               5  .08  .09  .16  .16  .19  .14   .82
37                                 6  .09  .10  .14  .09  .14  .24   .79
38
39   Location Probability Matrix   1  .27  .19  .16  .13  .12  .12  1.00
40   (shows probabilities of       2  .15  .26  .23  .15  .10  .11  1.00
41   travel from one zone to       3  .11  .19  .25  .17  .15  .13  1.00
42   another, given the above      4  .10  .16  .22  .24  .18  .10  1.00
43   willingness to travel)        5  .10  .11  .19  .19  .23  .17  1.00
44                                 6  .11  .13  .17  .12  .17  .30  1.00
45   ----------------------------------------------------------------------
46   Source:  Adapted from a version in Richard Brail, Microcomputers in Urban
47       Planning and Management (New Brunswick, N.J.: Center for Urban Policy
48       Research), chapter 5.
```

```
E32/$K$32
```

Similar formulas are used to convert all the other travel times in rows 5-10 first to an index of the willingness to travel, and then to a location probability.

Once this is done, the spreadsheet model follows exactly the eight steps described above (pp. 170-71). That is,

1.   In row 13, the user enters the number of basic jobs in each zone.

2.   In row 14, the model calculates in which zones workers in these basic jobs are likely to live, given the location-probability matrix in rows 39-44. Thus, the number of basic workers living in zone 1, for example, consists of the number of basic jobs in each zone (row 13) times the probability that people in those jobs will live in zone 1 (column E in the probability matrix). Thus, the formula for zone 1 (in cell E14) is as follows:

```
$E$13*E39+$F$13*E40+$G$13*E41+$H$13*E42+
       $I$13*E43+ $J$13*E44
```

Similar formulas for the other zones are found in cells F14:J14.[3]

3.   In row 15, the model calculates a basic population distribution by zones, using the multiplier in cell B7. Thus, the formula in cell E15 is as follows:

```
E14*$B$7
```

4.   In row 16, the model calculates the number of service workers needed in each zone to meet the needs of the current population distribution, using the multiplier in cell B10. Thus, the formula in cell E16 is as follows:

```
E20*$B$10
```

---

3.   Technically speaking, this procedure is called matrix multiplication, because the result in cells E14:J14 is the product of a vector (cells E13:J13) and a matrix (cells E39:J44). Some of the latest versions of spreadsheet programs (e.g. SuperCalc5) have commands specifically for matrix multiplication, but the procedure shown here will still work.

Note that the current population is taken from row 20 rather than row 15, in order to reflect the fact that both basic and service workers and their families require support from service workers. This is what gives the Lowry model its circularity.

5.   In row 17, the model calculates in which zones these service workers are likely to live, again using the probability matrix in rows 39-44. Thus, as in step 2, above, the number of service workers living in zone 1, for example, consists of the number of service jobs in each zone (row 16) times the probability that people in those jobs will live in zone 1 (column E in the probability matrix). The formula for zone 1 (in cell E17) is as follows:

```
$E$16*E39+$F$16*E40+$G$16*E41+$H$16*E42+
        $I$16*E43+$J$16*E44
```

Similar formulas for the other zones are found in cells F17:J17.[4]

6.   In row 18, the model calculates a revised population distribution, again using the multiplier in cell B7. Thus, the formula in cell E18 is as follows:

```
E17*$B$7
```

7.   In rows 19 and 20, the model computes the total number of workers and the total population by adding the basic component to the service component in each zone. The formulas are in cells E19 and E20, respectively:

```
E13+E16
E15+E18
```

8.   Because of the circularity inherent in steps 4 and 7, the spreadsheet will continue to recalculate until the results in rows 19 and 20 converge. As noted, convergence is assured as long as the parameter in cell F10 is less than unity.

---

4.   Again, this is a case of matrix multiplication—a vector (row 17) is the product of another vector (row 16) and a matrix (cells E39:J44).

Operation

The model is easy to use. It is necessary only to enter the data on travel times, the initial distribution of basic workers, and the three parameters—and the model produces the results. You can go back and alter any of the parameters or other input data at any time, and the model will happily recalculate the results each time you do.

Moreover, the graphing capabilities of the spreadsheet can readily be exploited to produce a graph of basic jobs and total population for each zone, such as that shown in Figure 7.1. In this particular case (the one depicted in Listing 7.1), it is easy to see that, because of the dynamics of the Lowry model, zone 4 contains disproportionately more of the total population, and zone 6 disproportionately less, than might be expected given the number of basic workers in each zone.

Interpretation and Use

According to Michael Webber,[5]

Lowry intended that his model be used as a device for evaluating the impact on urban form of such public decisions as urban renewal, tax policy, land-use control, and transport investment; and for predicting the changes in metropolitan form which are caused by the changing location of industry, the improving efficiency of transport, and the growth of population [1984, p. 6].

To illustrate how the model might be used in this way, let us suppose that the six zones in the model just presented are arranged along both sides of a river, thus:

Zone 3	Zone 1	Zone 5
River		
Zone 4	Zone 2	Zone 6

Suppose too that it takes 5 minutes to travel within any zone and 10 minutes to travel

---

5.   Cf. Lowry's own explanation cited on p. 165, above.

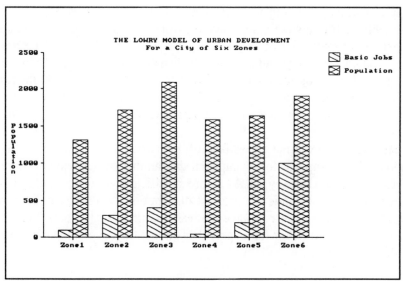

Fig. 7.1. Distribution of Population and Basic Jobs from Listing 7.1

from any zone to an adjacent one. Suppose finally that there is a bridge across the river between two pairs of zones but not the other: that is, between zones 1 and 2 and between zones 3 and 4, but not between zones 5 and 6. Thus, travel between zones 1 and 2 and between zones 3 and 4 also takes 10 minutes; but travel between zones 5 and 6 takes 30 minutes, because of the need to travel via zones 1 and 2.

Given this scenario (which is easily entered into the model), there are at least three different things we can do:

● we can analyze the effect of changing some of the model's parameters;

● we can simulate the effect of industrial development (and industrial decline) on population distribution; and

● we can look at the effect of changes in the transportation system on either population distribution or industrial development.

Let us start by providing an initial distribution of basic workers (say, 1,000 per zone), suitable multipliers (say, a population/worker multiplier of 3.5 and a service-worker-per-capita multiplier of .8), and a friction factor of unity. The model will quickly calculate the total population and its distribution by zones. Then we can try increasing and decreasing the population/worker ratio; then we can do the same with the service-worker-per-capita multiplier. What are the effects? How sensitive does the model seem to be to changes in these parameters?

Next, we can examine the effect of industrial development on population distribution. What happens, for example, if the number of basic jobs in some of the zones goes up or down by, say, 10%? More specifically, in which zones do changes in employment appear to have the greatest distributional effect on other zones, and in which zones do they appear to have the least distributional effect? Questions like these are obviously critical for (amongst other things) the development of an effective urban industrial policy.

A third use of the model is to examine the effect of changes in the transportation network on population distribution and/or on industrial development. What would be the effect of (for example) adding a bridge between zones 5 and 6? Alternatively, what would be the effect of restricting the bridge between zones 3 and 4 to "one-way south," thereby adding 20 minutes to the trip from zone 4 to zone 3?

Finally, once several interesting policy options have been identified, the model can be used to test for sensitivity to changes in the friction factor. The friction factor is, at best, a crude estimate. So it is always wise to examine the effect of adjusting this factor slightly, to see if it makes any significant difference to your conclusions.

Evaluation and Extension

Since it was first proposed nearly 30 years ago, the Lowry model has attracted a lot of comment and criticism. At the end of his exhaustive review of the model, Webber complains that the model explains nothing and predicts little and that it "ignores causes and . . . is not grounded in a suitable theory of society" (pp. 164, 180). Yet, he admits, the model it remains very popular. Others are more generous. Michael Batty, for example, has written:

At present, the [Lowry] model provides a reasonably good simulation device, and it can be extremely useful if used with care. It is manageable in terms of both technique and cost, and if it is integrated with objective techniques of design and evaluation of the system, the planner can have high confidence in the model's prediction [1970; cited in Lee 1973, pp. 101-2].

Moreover, in its spreadsheet version at least, the model is very flexible. The number of zones can be increased up to the limits of the available hardware and software. Although only two multipliers and one friction factor are used in the version shown here, the model could easily be adapted to use a set of multipliers (for different types of jobs and/or different zones) and a set of friction factors (e.g., for different zones).

Finally, the Lowry model is appealing for the simple reason that it links typically disparate features of a region in a meaningful way. As Tomas de la Barra put it recently (1989, p. 56), "The main contribution of Lowry is that several sub-models are linked to each other within an iterative system, thus allowing for a more complex structure." Intuitively, of course, we know that form and function are related. Yet, in practice, few urban models have emphasized that relationship as effectively as the Lowry model.

Technical Note

The Lowry model converts travel times to location probabilities by means of two transformations. First, travel times are converted into an index of the willingness to travel; then the willingness to travel is converted into a location probability.

Travel times are converted into a willingness to travel by making the entirely reasonable assumption that people are less willing to travel long distances than short ones. In other words, the willingness to travel is assumed to vary inversely with travel time. The precise relationship is mediated by some constant—often called a "friction factor"—that links travel time in minutes to the willingness to travel. One popular transformation (see Black 1981, chapter 2)—which is the one used here (following Brail 1987, chapter 5)—relies on an inverse power function:

$$W = \frac{1}{T^f}$$

where W is the willingness to travel, T is the travel time or distance, and f is the friction factor. In this case, the friction factor, f, must have a value of at least zero (which makes all zones equally attractive, regardless of travel time). The higher the friction factor, on the other hand, the greater the tendency for the zone with the lowest travel time to become overwhelmingly attractive relative to all the others.

However, a problem arises when this kind of transformation is made. The problem is that the base of the index is not the same for all zones. If you sum the willingness to travel from each zone to all of the zones, you do not get the same total for each zone. In other words, it appears that people in some zones are *fundamentally* (in some sense) more likely to travel than people in other zones. Yet such is not necessarily the case.

So a second transformation is required, in order to convert all the indices to a common base—that is, to standardize the indices. This is achieved by simply dividing each index by the total of all indices for that zone of origin. The results of this transformation can be interpreted as a matrix of probabilities, showing the relative likelihood that people in any zone are prepared to travel to each of the zones. This is called a location probability matrix.

Bibliography and References

Barra, Tomas de la. 1989. *Integrated Land Use and Transport Modelling: Decision Chains and Hierarchies*. Cambridge University Press.

Batty, Michael. 1970. "An Activity Allocation Model for Notts/Derby Subregion." *Regional Studies*, 4.3. Cited in Lee 1973, pp. 101-2.

———. 1976. *Urban Modelling*. Cambridge: Cambridge University Press.

Black, John. 1981. *Urban Transport Planning*. Baltimore: Johns Hopkins University Press.

Brail, Richard. 1987. *Microcomputers in Urban Planning and Management*. New Brunswick, N.J.: Rutgers University Press.

Carrothers, G.A.P. 1956. "An Historical Review of the Gravity and Potential Concepts of Human Interaction." *Journal of the American Institute of Planners*, 22, pp. 94-102.

Cartwright, T.J. 1989. *The Management of Human Settlements in Developing Countries*. London: Routledge.

————. 1989. "Urban Management as a Process, Not a Result". *Review of Urban and Regional Studies*, 2 (July), pp. 107-13.

Chadwick, George. 1987. *Models of Urban and Regional Systems in Developing Countries: Some Theories and Their Application in Physical Planning.* Especially chapter 6. Oxford: Pergamon.

Foot, David. 1981. *Operational Urban Models: An Introduction.* New York: Methuen.

Garin, R.A. 1966. "A Matrix Formulation of the Lowry Model for Intra-Metropolitan Activity Location." *Journal of the American Institute of Planners*, 32, pp. 361-64.

Goldner, W. 1971. "The Lowry Model Heritage." *Journal of the American Institute of Planners*, 37, pp. 100-10.

Lee, Colin. 1973. *Models in Planning: An Introduction to the Use of Quantitative Models in Planning.* Urban and Regional Planning Series. Oxford: Pergamon.

Lowry, Ira S. 1964. *A Model of Metropolis.* Memorandum No. RM-4035-RC. Santa Monica: Rand Corporation.

Macgill, S.M. 1977. "The Lowry Model as an Input-Output Model and Its Extension to Incorporate Full Intersectoral Relations." *Regional Studies*, 11, pp. 337-54.

Mackett, R.L., and G.D. Mountcastle 1977. "Developments of the Lowry Model." In *Models of Cities and Regions*, edited by A.G. Wilson, P.H. Rees, and C.M. Leigh, pp. 209-82. Chichester, U.K.: Wiley.

Webber, Michael J. 1984. *Explanation Prediction and Planning: The Lowry Model.* London: Pion.

Wilson, A.G. 1970. *Generalising the Lowry Model.* Working Paper No. 56. London: Centre for Environmental Studies.

Chapter 8

AFFORDABLE HOUSING: THE BERTAUD MODEL

The importance of housing can hardly be exaggerated.
In human terms, access to satisfactory housing is
vital to health, happiness and civilized living.

John Black and David Stafford, Housing Policy and Finance

Housing is one of the most important social and economic activities that human beings engage in. Indeed, it can be said that housing is a measure of the health, wealth, and happiness of a society. As has been said of the United Kingdom,

> The economic facts about housing bear this out. In the UK in 1985, owner-ship of dwellings gave rise to 5.8 per cent of Gross Domestic Product, and use of dwellings accounted for 9.2 per cent of total consumer expenditure. . . . for 20.2 per cent of gross fixed capital formation . . . . and for 31.6 per cent of net investment. . . . At the end of 1985, the value of the stock of dwellings, at replacement cost, was 30.7 per cent of the total real domestic capital of the UK [Black and Stafford 1988, p. ix].

One of the most intractable problems facing countries around the world is that of ensuring a minimum standard of shelter for everyone. Numerous strategies have been proposed, ranging from almost complete reliance on market forces, through so-called "enabling strategies," to full-scale public housing. Similarly, there have been variations on each theme, from land grants, to site-and-service schemes, to core-housing projects, to upgrading schemes, to rent subsidies, and many others. As time goes by, different types of projects seem to move in and out of public favor. The one clear lesson seems to be that shelter, like most other "metaproblems" societies now seem to face, is unlikely to prove amenable to any "quick fix" (Chevalier 1969; Cartwright 1987). Instead of being able to solve problems, planners may have to settle instead for incremental improvements based on detailed analysis of tradeoffs.

When it comes to detailed analysis, it is natural to think of computers. Thus, this chapter presents and discusses a simul ation model of the cost and affordability of low-cost housing under varying conditions of physical design and consumer financing. The model uses only the most basic arithmetic capabilities of the spreadsheet and makes good use of its graphing commands (especially the pie chart). In spite of its relative simplicity, however, the model provides a useful perspective on the impact of various design and financing options.

Purpose of the Model

When it comes time for implementation, most shelter projects become the object of detailed financial analysis. The purpose of these calculations is usually to show: (a) overall costs and benefits that are expected to accrue from the project, in order to show that the project is a good one; and (b) cash flow over time, in order to show how much money will be required to finance the deficit between expenses and income, as the project is built and comes on stream. If the owner or developer of the project does not require such statements, the financier almost certainly will!

But models like these are designed primarily to serve financial objectives. Planners need something a bit different—a model that can be created at a much earlier stage in the proceedings, when data are still only vague and approximate, and something that can incorporate a broader range of criteria than the merely financial. Planners want to be able to compare the effect of planning at different scales of development, at different densities, and for different standards of housing; and they want to examine the effect of various proposals for subsidies and cross-subsidies. In addition, planners want to be able to assess the *affordability* of their scheme for potential target groups, under varying assumptions of loan terms, interest rates, and progression rates.[1]

One of the most popular cost/affordability models for housing is the Bertaud model developed for the World Bank by Alain Bertaud in the early 1980s (World Bank 1981). The Bertaud model was designed to meet the needs of planners, engi-

---

1.   The criterion of what is "affordable" derives from the nineteenth-century slogan of "a week's wages for a month's rent", although the exact percentage of income that is considered affordable depends a lot on what is included in the cost of housing (Feins and Lane 1981, p. 8 and Appendix B).

neers, economists, and other professionals involved in developing affordable housing schemes for low-income sectors of the population in developing countries. The original Bertaud model was written for use on a programmable calculator, like the Hewlett Packard HP91; subsequently, the model was adapted to the spreadsheet environment by both Bertaud and others working independently. Some versions have been divided into several separate modules, including (in some cases) one that can print a simple sketch plan of the resulting housing scheme. There have also been various modifications and adaptations of the original model, one of which (de Kruijff 1983) serves as the basis for the version presented here.

The Bertaud model requires three kinds of input data:

- project design parameters,
- development and construction costs, and
- financing terms and conditions for the eventual owner.

On that basis, the model calculates how much each housing unit will cost to build and how much that will mean in terms of monthly carrying costs for the consumer. The model is intended primarily for projects involving new housing, such as site-and-service or core-housing schemes, but can also (with a little imagination) be applied to upgrading schemes as well.

The Bertaud model is not a precise accounting tool; for example, there is no allowance for borrowing costs or the rate of inflation. But the model is a useful device for reviewing options and tradeoffs entailed in the design and financing of the project. What happens, for example, if densities are increased? if housing standards are improved? if subsidies are reduced? if cost-recovery is imposed? if interest rates are changed? These are the questions that this model can help planners to answer.

Conceptual Basis

The version of the Bertaud model presented here is adapted from one prepared by G.J.W. de Kruijff in 1983 as part of a United Nations project for the Directorate of Building Research in Bandung (Indonesia). De Kruijff's version followed the form and the function of the original quite closely, whereas the version presented here has been modified considerably to take advantage of the capacity of the spreadsheet to do what-if modeling. To understand how the model works, suppose that a piece of land has been identified as the potential site for a new housing scheme. In due course, accountants and developers will produce detailed cost analyses and cash-flow summaries. But, in the meantime, planners want to examine some of the potential tradeoffs in design, financing, and so forth. To this end, the Bertaud model can be understood in terms of five basic issues: site design, development costs, unit costs, planning profile, and affordability.

• **Site Design:** The first question is how much land (if any) is to be allocated to nonhousing uses (i.e., commercial uses, industrial uses, community facilities, and open space)? Second, how many plots, and of what size, are to be made available for housing, and what area needs to be set aside for vehicular and pedestrian circulation (i.e., streets and roads)? Naturally, the sum of all these allocations cannot exceed the total size of the site; if, on the other hand, there is land left over after all allocations are made, it is just classified as unplanned.

• **Development Costs:** The next issue is that of cost. For the site itself, there are costs associated with buying the land, preparing the site, and building the basic infrastructure or services (water, sewers, electricity, and roads). For the housing units there are: the construction costs; the cost of fixtures and equipment, connection to services, and common facilities (if any); and costs for on-site services.

• **Unit Costs:** In order to estimate unit housing costs, we must find a way of allocating the site development costs. To do this, we first calculate an average site development cost per square meter for the entire project site. Since the entire site can never be developed—some space is required for roadways, and some will probably be left undeveloped as open space—we need to adjust the raw site development cost to reflect the maximum possible (but less than total) development of the site. Moreover, if we impute a market value to land on the site which has been serviced by the proposed project but which will not be used for housing purposes, we can compute a net site development cost per square meter for housing: that is, a cost that is adjusted for the revenue that might be derived from sale of the nonhousing serviced land. Finally, if there is more than one type of housing unit in the proposed project, we can (if we like) provide for cross-subsidies between them. That is, we can adjust the net site development costs for each type of plot in such a way that some plots become more expensive than others—presumably in pursuit of some policy regarding the affordability of different types of housing.

- **Planning Profile:** If we provide an estimate of mean household size, we can have the model present a profile of the character of the proposed project from a planning point of view: e.g., its maximum population, its likely density, some statistics on land use per capita, the percentage of land used for particular purposes, and so on.

- **Affordability:** Finally, if we enter data on the terms and conditions of financing available to potential owners, we can calculate the monthly payments required for each type of housing unit, based on their respective unit costs.[2] Then, on the basis of cost and terms of financing, we can assess the affordability of different units in the project.

These are the five aspects of housing reflected in the Bertaud model.

Data Requirements

Compared to most of the other models in this book, the Bertaud model requires quite a lot of data. These are listed in Table 8.1.

Note that, in the last section (section I), the user has the option of providing either global numbers or more detailed calculations for each of the four types of infrastructure considered here (i.e., water supply, sewerage, electricity, and roadways). In fact, estimated or approximate data may be adequate in many cases, at least for an initial run of the model.

Overview

The Bertaud model is contained entirely within a matrix of 10 columns and 77 rows. The model is divided into two parts: the main part of the model, which is in the top 49 rows, and the rest which is used for entering estimates of the costs of on-site services (as specified in Table 8.1 above). As noted, the estimate for each kind of infrastructure can be either a single, global amount or a detailed calculation.

---

2. In the form presented here, the model provides only for varying the minimum down payment, the term of the loan in years, the rate of interest (or administration charge), and the rate at which interest payments may be graduated (progressive payments).

Table 8.1: Data Requirements for the Bertaud Model

A. Site

Overall size: _____ ha
Allocation to non-housing uses
        Commercial: _____ ha
        Industrial: _____ ha
        Community facilities: _____ ha
        Open-space: _____ ha

B. Housing Area

Frontage: _____ m   Depth: _____ m
Site plan   Quantity   Frontage   Depth
        Plot #1:   _____   _____ m   _____ m
        Plot #2:   _____   _____ m   _____ m
        Plot #3:   _____   _____ m   _____ m
Allocation for roadways
        Footpaths: _____ m^2
        Access streets: _____ m^2
        Collector roads: _____ m^2

C. Site Acquisition and Development Costs

Land acquisition costs: $ _____
Site preparation costs: $ _____
Off-site services   water supply: $ _____
                                    sewerage: $ _____
                                    electric power: $ _____
                                    roadways: $ _____
                                    other costs: $ _____

D. Housing and On-Site Costs

Floor area   Plot #1:   _____ m^2
                          Plot #2:   _____ m^2
                          Plot #3:   _____ m^2
Construction costs: $ _____ per m^2
Fixtures and features: $ _____ per unit
Connection costs: $ _____ per unit
Common-services costs: $ _____ per unit
On-site services   water supply: $ _____ all units
                                    sewerage: $ _____ all units
                                    electricity: $ _____ all units
                                    roadways: $ _____ all units

E. Site Potential

Market value of commercial land: $ _____ per ha
Market value of industrial land: $ _____ per ha
Market value/community facilities: $ _____ per ha

Table 8.1: Data Requirements for the Bertaud Model *(continued)*

F. Cross-Subsidy Factor: _____ %

G. Mean Household Size: _____ persons

H. Unit Costs and Affordability

Minimum down payment: _____ %
Term of financing: _____ years
Interest rate, etc. _____ %
Progression rate (if applicable): _____ %

I. Detailed On-Site Infrastructure Costs

Note: enter either global estimates in the right column or detailed calculations in the left column (but not both).

Water Supply:	Primary	_____ m	
	Cost	$ _____ /m	
	Secondary	_____ m	
	Cost	$ _____ /m	
	Other costs	$ _____	
	Total cost	$ _____	$ _____
Sewerage:	Primary	_____ m	
	Cost	$ _____ /m	
	Manholes		
	Unit cost	$ _____	
	Secondary	_____ m	
	Cost	$ _____ /m	
	Manholes		
	Unit cost	$ _____	
	Other costs	$ _____	
	Total cost	$ _____	$ _____
Electricity	Primary	_____ m	
	Cost	$ _____ /m	
	Secondary	_____ m	
	Cost	$ _____ /m	
	Other costs	$ _____	
	Total cost	$ _____	$ _____
Roadways:	Footpaths	_____ m	
	Cost	$ _____ /m	
	streets	_____ m	
	Cost	$ _____ /m	
	Collectors	_____ m	
	Cost	$ _____ /m	
	Total cost	$ _____	$ _____

For convenience, the top part of the model is divided into nine sections, as shown in Listing 8.1. Of these nine sections, the four on the left deal with the site or the project as a whole, while the five on the right deal with the individual housing units. At the end of the worksheet is a tenth section which performs some detailed calculations for one of the upper sections. The model is set up here for use with metric units of measurement and a dollar currency.

Section I contains basic site data. In cell B3, the user enters the total amount of available land in cell B3 and, in cells B7:B10, the amount to be used for non-housing land uses. The percentages in column C are calculated by the computer. When data on housing are entered in section II, the computer converts the total area in cell I3 from square meters to hectares and displays it in cell B5; then it computes the total unallocated space and displays it in cell B11 as net unplanned space.

Section II contains basic housing data. The user enters the length and width of the area designated for housing, and the computer calculates the area. The user enters the number and size (frontage and depth) of each of up to three plots, and the computer calculates the total area allocated to each type of plot.[3] The user also enters an estimated total area for each of up to three types of roadways, and the computer displays a total in cell I10. Finally, the computer adds up the area allocated to housing and roadways and displays the unallocated area in cell I11.

In sections III and IV, the user enters basic cost data pertaining to the site and to housing and infrastructure costs, respectively. Note that (for the sake of convenience) data in section III are shown in thousands of dollars. In section IV, data are entered for the floor area of the units to be constructed on each type of plot, as well as for the costs of construction, fixtures, service connections, and common services. Costs for on-site services for water, sewers, electricity, and roads are copied from section X, while costs of other services are entered directly in cell I19. The computer calculates the percentage distribution of all the on-site service costs.

---

3.   The model does not require that there be more than one type of plot, but allows for up to three different types. If more than one type of plot is defined, it is important to ensure that plots of type 1 be smaller or inferior in some other way to those of type 2 and plots of type 2 be smaller or inferior in some other way to those of type 3. This is because any "cross-subsidy factor" that is entered in section VI is applied (in section VIII) so as to favor plots of type 1 more than those of type 2 and those of type 2 more than those of type 3.

Listing 8.1

```
  | A    !! B   !! C  !!D!! E   !!  F  !!  G  !! H  !!  I  !!  J  !
1 BERTAUD AFFORDABLE HOUSING MODEL              © 1993 by T.J. Cartwright
2
3 I. SITE    105.00 ha        II. HOUSING AREA       146      76  11,096 m2
4                                         Qtty.  Front  Depth    Area
5 Housing     1.11      1%    Plot #1      48    5.00  12.00    2,880      26%
6                             Plot #2      52    6.00  12.00    3,744      34%
7 Commerce    5.25      5%    Plot #3      18    8.00  15.00    2,160      19%
8 Industry     .00      0%    Total       118                  8,784      79%
9 Community   5.25      5%    Roadways  Paths Streets   Roads   Total
10 Open spac 10.50     10%       m2       864    552      608   2,024      18%
11 Unplanned 82.89     79%    Unallocated net space (m2)          288       3%
12
13 III. SITE COSTS ($000)      IV. HOUSING AND ON-SITE COSTS
14 Land acquisition  $4,200   Plot No   FlArea  Const Fixture Connect  Common
15 Site preparation    $315      1       15.00   $40   $306    $225       $3
16 Water supply         $21      2       18.00   $40   $306    $225       $3
17 Sewers            $1,000      3       30.00   $40   $306    $225       $3
18 Electric power        $0    Water  Sewers ElecPwr Roadway   Other   Total
19 Roadways             $29   $3,317 $19,300     $0  $13,074     $0  $35,690
20 Other                $13      9%     54%     0%      37%      0%     100%
21
22 V. SITE POTENTIAL           VI. AVERAGE SITE DEVELOPMENT COSTS ($/m2)
23 Commercial $/m2   $45.00           Land    Prep OffSite  OnSite   Total
24 Industrial $/m2    $.00    Cost/m2:  $4.00   $.30  $1.01   $3.22   $8.53
25 Community $/m2    $15.00   Adjusted cost based on total max devt   $12.11
26 Max housing (ha)  63.47    Net site devt cost, housing only        $9.15
27 Max other devt    10.50    Cross-subsidy factor                       0%
28 Total max devt    73.97    Adjusted net devt cost, housing only     $9.15
29
30 VII. PROJECT PROFILE        VIII. UNIT COSTS AND AFFORDABILITY      Dollars
31 Mean household     6.00
32 Maximum plot #1   3,468                 Plot #1 Plot #2 Plot #3 Wt Mean
33 Maximum plot #2   3,757
34 Maximum plot #3   1,301    Const & fixtures $1,134 $1,254 $1,734 $1,278
35 Total             8,526    Site development   $549   $658 $1,097   $681
36 Residents/ha        487    Total cost       $1,683 $1,912 $2,831 $1,959
37 Plots/ha             81    Downpaym    5%      $84    $96   $142    $98
38 Plots/res ha        106    Term/yrs    15
39 Commerce m2/pers   1.03    Payment     8%   $15.56 $17.69 $26.19 $18.12
40 Industry m2/pers    .00    ProgRate    0%
41 Communit m2/pers   1.03
42 Openspac m2/pers   2.46    IX. OVERALL PROJECT COSTS               Unit
43 Net residential     60%                                   $000     Mean
44 Commercial           5%    Land acquisition             $4,200    $493
45 Industrial           0%    Off-site services            $1,062    $125
46 Community            5%    On-site services             $3,692    $433
47 Open space          12%    Housing/facilities          $10,899  $1,278
48 Roadways            18%    Total costs                 $19,853  $2,329
49
```

	A	B	C	D	E	F	G	H	I	J
50	X. ON-SITE INFRASTRUCTURE COSTS (copied up into cells E19, F19, G19 & H19)									
51	Note: Detailed costing is replaced by estimated, if latter is nonzero.									
52										
53	Water Supply		$0		Electricity			$0		< Nonzero
54										values
55	Primary pipe (m)		76		Primary network (m)			0		here
56	Cost ($/m)		$10.00		Cost ($/m)			$.00		replace
57	Second pipe (m)		367		Second network (m)			0		computed
58	Cost ($/m)		$6.50		Cost ($/m)			$.00		totals
59	Other costs		$171		Other costs			$0		│
60	Total cost		$3,317		Total cost			$0		< here.
61										
62	Sewerage		$0		Roadways			$0		< Nonzero
63										values
64	Primary pipe (m)		73		Footpaths (m)			864		here
65	Cost ($/m)		$50.00		Cost ($/m)			$6.50		replace
66	No manholes		3		Access streets			552		computed
67	Unit cost		$700.00		Cost ($/m)			$5.80		totals
68	Second pipe (m)		380		Collector roads			608		│
69	Cost ($/m)		$30.00		Cost ($/m)			$7.00		│
70	No manholes		10		Total cost			$13,074		< here and
71	Unit cost		$250.00							│
72	Other costs		$0							│
73	Total cost		$19,300							< here.
74										
75	Source: Adapted from G.J.W. de Kruijff, "Urban Cost Model," United Nations									
76	Project Working Paper (Bandung, Indonesia: Directorate of Building Re-									
77	search, September 1983).									

In section V, the user enters data on current market values for serviced land (i.e., land already supplied with urban services, such as water, sewage, road access, etc.) of various kinds in cells C23, C24, and C25. Then the computer calculates the overall *development potential* of the site. The development potential is based on a simple assumption of "more of the same" in the rest of the site.[4] Thus, the computer applies the percentage in cell J8 to the total of all land that could be used for housing (i.e., cells B5 and B11)—less only an allowance for roads, streets, and paths for other land uses that is based on the percentage computed for housing in cell J10.[5] The result is then displayed in cell C26. The "other development" given in cell C27 means the sum of land allocated to commercial, industrial, or community uses (cells B7:B9), excluding land set aside for use as open space (cell B10). Total maximum development (cell C28) is simply the sum of the land available for housing (cell C26) and the land available for other kinds of development (cell C27). This represents the overall development potential of the site.

In section VI, the computer calculates average site development costs by dividing the total costs shown in section III (in cells C14:C20) by the total size of the site (cell B3), and then allocating them to the categories shown for cells F24, G24, and H24.[6] The average cost of on-site services (cell I24) is computed by dividing total costs (cell J19) by total housing area (cell I3). These four average costs (cells F24:I24) are shown as a total in cell J24.

This calculation of average site development costs per square meter is based on the assumption that the entire site could be developed. But, as shown in section V, the maximum development potential of the site (cell C25) is less than the total size of the site (cell B3). Accordingly, we really ought to make a couple of adjustments:

---

4. In other words, the computer assumes that the ratio computed in cell J8—which is the percentage of land allocated to housing (in section I) that is actually used for housing (in section II)—can be applied to any land on the site that might be allocated to housing in the future.

5. In fact, land uses other than housing will probably require a lesser allowance for roadways than land used for housing. So the blanket use of a ratio appropriate to the first is likely to prove conservative, if anything.

6. The results are adjusted in each case by 0.1 to reflect the conversion from thousands of dollars in cells C14:C20 and hectares in cell B3 to dollars per square meter here.

●   First, it would be misleading to allocate development costs to land which we know cannot or will not be developed. So we should inflate the average site development costs shown in cell J24 by the ratio of the total size of the site (cell B3) to the maximum development potential of the site (cell C28). This yields the result shown in cell J25.

●   Second, it is really housing costs that we are interested in. So we should try to deflate the site development costs attributable to housing by whatever benefits are derived by nonhousing uses. One way of doing that is to adjust the costs shown in cell J25 by the revenue that could be derived from sale (at the prices shown in cells C23:C25) any of the (now serviced) land allocated to nonhousing uses (shown in cells B7:B9). This yields the result shown in cell J26.

Finally, site development costs can also be adjusted by means of a "cross-subsidy factor" in cell J27. This factor determines the extent (if any) to which the costs of providing lower-quality units are allocated to higher-quality units. Specifically, the model assumes that plots of type 1 are subsidized by plots of type 2 to the extent of the cross-subsidy factor and by plots of type 3 to the extent of double the cross-subsidy factor; and that plots of type 2 are subsidized by plots of type 3 to the extent of the cross-subsidy factor. The magnitude of the cross-subsidy factor is set by the user; it can, of course, be zero (as it is in Listing 8.1)

In fact, the cross-subsidy factor appears the model at two points. Here in section VI, it is built into calculation of adjusted site development costs; then, in section VIII, it is used to calculate unit costs for the three different types of plots. Thus, in cell J28, the adjusted net site development costs (ANSDC) in dollars per square meter are calculated by means of the following formula:

$$ANSDC = \frac{NSDC * (P_1 + P_2 + P_3)}{P_1 + (1 + XDF) * P_2 + (1 + 2 * XDF) * P_3}$$

where $NSDC$ = net site development cost ($/m^2),
                   P_1 = area of plot #1 (m^2),
                   P_2 = area of plot #2 (m^2),
                   P_3 = area of plot #3 (m^2), and
                   XSF = cross-subsidy factor (%).

Section VII provides a summary "profile" of how the site as a whole could be expected to develop on the basis of the costs and choices made so far. For this, however, the user must provide (in cell C31) an estimate of the average number of persons living in a household. Note too that all of the calculations are based on the assumption that any net "unplanned space" shown in cell B11 will eventually be used for housing, in accordance with the calculations in section V (cells C26:C28).

Section VIII calculates unit costs and affordability. Costs of construction and fixtures for each type of plot are derived from data in section IV. Site development costs are calculated from plot sizes entered in section II, multiplied by the adjusted net site development cost in cell J28, and adjusted by the cross-subsidy factor (if any) in cell J27. The two costs are added together in row 36 (cells G36:J36). Financial terms and conditions are entered in column F (cells F37:F40). Actual downpayments and monthly payments are calculated and displayed in rows 37 and 39.

Monthly payments (MP) in cells G39:J39 are calculated like this:

$$MP = \frac{P * (1 + i)^n * (i - j)}{12 * [(1 + i)^n - (1 + j)^n]}$$

where

	P	= the principal of the loan (less any down payment)
	i	= the rate of interest or administration (%),
	n	= the term of the loan (years), and
	j	= the rate of progression (%).

For each of these items, the computer also calculates a weighted mean, based on the number of units of each type (as specified in cells F5, F6, and F7) and displays the means for each plot in column J.

Note that the calculation of monthly payments with the formula above is for repayment of interest only, not the principal. In other words, monthly payments in this model are not what are called "blended payments." However, the model does allow the user to specify progressive, or graduated (rather than equal), monthly payments over the term of the loan. In other words, some of the interest in the early years is deferred to later years, so that monthly payments are lower than they should be at the beginning of the term of the loan and higher than they should be at the end. Assuming that inflation will reduce the value of money and/or that the debtor's real income will rise over time, progressive payments may help make housing more affordable to more people. On the other hand, the use of progressive interest payments effectively increases the loan principal, since it amounts to borrowing additional funds to cover the portion of the interest payments that is deferred in the early years of the term. This means that the total amount of interest paid over the full term of the loan will be greater than it would without progressive payments.[7]

Section IX summarizes overall project costs and calculates the mean costs per unit of housing. In order to assign overall costs on a realistic basis, mean costs are based on the development potential of the entire site (as computed in section V of the model), rather than the actual number of units proposed at the present time (as entered in section II). The difference between average unit costs in cell J36 and average unit costs in cell J48 is explained by the fact that the former is adjusted (through the formula in cell J26) for potential revenue to be derived from the sale of serviced land for nonhousing purposes (cells B7, B8, and B9), while the latter is not adjusted in that way.

Finally, section X at the bottom of the worksheet provides an area for entering and computing the costs of construction for on-site services. The section is divided into four quadrants, one each for water supply, sewerage, electric power, and roadways. Here the user has a choice. If a nonzero value is entered into the cell at the top of a quadrant (i.e., in cell C53, H53, C62, or H62), this nonzero value is automatically copied to the bottom cell in the same quadrant (i.e., cell C60, H60, C73, or H70, as the case may be)—regardless of any more detailed data that may be contained in any intervening cells. If a more detailed calculation is desired for any of the services, a zero should be entered in the cell at the top of the appropriate quadrant—in which case, the corresponding total at the bottom of the quadrant will be computed from the data immediately above it. In either case, the values in the bottom cell of each quadrant (whether copied or computed) are in turn copied up to cells E19, F19, G19, and H19 respectively in section IV of the model, where they are used for calculation of housing and on-site development costs.

---

7. Thus, it is important not to let the progression rate get too high. The higher the rate, the lower the initial payments but the higher the later payments. It is possible to reach a state where the principal grows at such a rate (through the addition of deferred interest) that there is no realistic prospect of its ever being paid off. In other words, the debtor faces "runaway" debt. Thus, in the United States, for example, the maximum allowable rate of progression under federal law is 7.5% on a loan with a 5-year term and 3% on a loan of 10-year term (Rosen 1984, p. 145).

Operation

The ordinary way of using a model like this is to start with a preliminary design and some rough data on costs and finance. Once these data are all entered, they can be adjusted to see how they interact with each other in defining the eventual profile of a project.

Ultimately, users of this model are likely to have three basic concerns, as far as housing projects are concerned. These are: the quality of life to be provided by the project as a whole, and the standard of housing to be built; the overall financial feasibility of the project; and the affordability of the housing units for their intended target groups. For example, starting from the data shown in Listing 8.1, we might ask some of the following kinds of questions:

- What would be the effect of increasing or decreasing the overall size of the site? or of changing the extent of the proposed use of the site for nonhousing purposes?

- What would be the effect of scaling the size of the proposed housing project up or down? What if the plots are made larger or smaller? What if the balance among plots of different types is altered?

- How critical are the estimates of major capital costs? Could moderate cost overruns lead to unacceptable increases in unit costs of housing?

- What would be the effect of increasing or decreasing floor area or improving fixtures in each housing unit? How critical are the estimates of construction, connection, and servicing costs?

- How vulnerable is the project to changes in the market price of serviced land for nonhousing purposes?

- What is the effect of adjusting the cross-subsidy factor?

- What is the effect of changing the financing terms and conditions to be made available to consumers?

The windowing capability of the spreadsheet can easily be deployed to facilitate investigation of many of these "what-if" scenarios. For example, in order to examine the impact of design changes on affordability, a lower window can be fixed on section VIII and an upper window moved to wherever there are variables to be adjusted. For example, we might move the upper window to section I, in order to see the effect on affordability of increasing or decreasing the amount of land allocated to housing. We might move the upper window to section IV, in order to see what would be the effect of higher construction costs or fewer fixtures and features. Or we might move the upper window to section VI, in order to see the effect of changing the cross-subsidy factor. And so on.

Finally, simple graphs can be constructed to illustrate the results obtained. For example, a user might construct the following graphs, each keyed to a particular section of the model:

- Initial land use on the site (pie chart) from section I,

- Land use within the housing area (bar chart) from section II,

- Site costs (pie chart) from section III,

- On-site servicing costs (pie chart) from section IV,

- Potential site development (pie chart) from section V,

- Average site development costs (pie chart) from section VI,

- Potential land use on the site (pie chart) from section VII,

- Unit costs and affordability, by type of plot (barchart) from section VIII, and

- Overall project costs (pie chart) from section IX.

The first two and the last two of these graphs are presented in Figures 8.1 to 8.4 on the next two pages; they were produced from the empirical data shown in Listing 8.1.

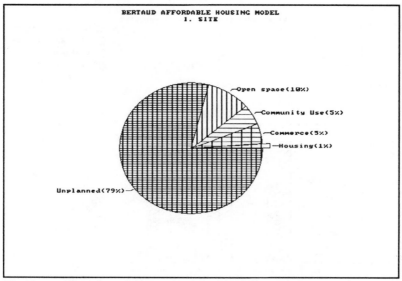

Fig. 8.1. Allocation of Site to Various Land Uses

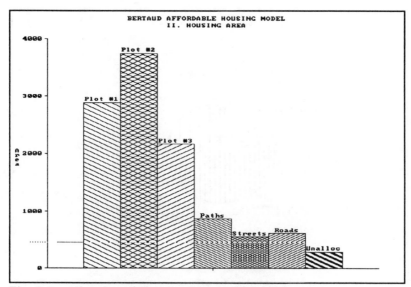

Fig. 8.2. Allocation of Housing Area to Various Uses

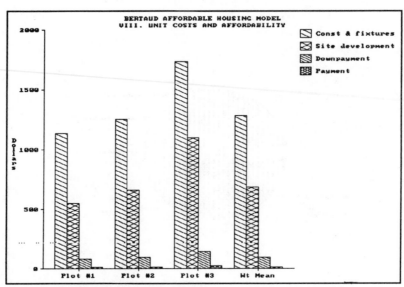

Fig. 8.3.   Unit Costs and Affordability

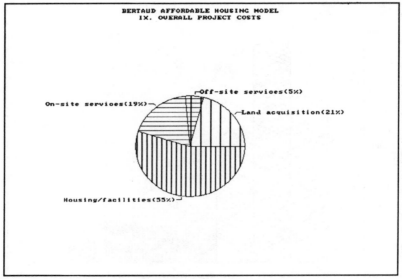

Fig. 8.4.   Distribution of Total Project Costs

Interpretation and Use

Planning rarely yields definitive answers; so there is no reason to think that a model like the Bertaud model is going to provide them. For one thing, there is no way that the results produced by the model can ever be better or more accurate than the data from which they are derived. Second, and even more fundamentally, good planning requires judgment as well as computation. There is no arithmetically correct housing project. In the end, all we can ever say is that a project optimizes certain variables. What these variables *really* represent, or whether it is desirable to optimize them in the first place, are questions of judgment as well as calculation. We cannot expect to be able to replace a planner's experience and insight with the calculations of a computer.

But we can use the computer to supplement the planner's skills. What the Bertaud model and others like it can do is to illuminate the choices and tradeoffs that are involved when the planner makes his or her judgments. For example, the model can help to identify areas where relationships among variables are particularly sensitive and other areas where they are not. If housing costs are relatively insensitive to construction costs, for example, then we may be able to afford to increase the gross floor area of our designs. Similarly, if monthly payments are highly sensitive to the cost of providing on-site services, then perhaps we should consider delaying the construction of some of these services, until some of the serviced land for nonhousing purposes is sold off—or alternatively whether more of their cost could be allocated to the site rather than to the individual housing units. The model cannot advise which of these options is desirable—but it can help make planners more aware of the implications.

Evaluation and Extension

The Bertaud model has been used in numerous international projects and elsewhere (World Bank 1981). Its use is clearly more appropriate at a design stage, rather than once implementation has begun. But there is a lot of scope in the model for improvement and extension of existing projects. One area where this has occurred is in providing more capability for varying the parameters of the physical design of the housing site (see Ast 1981 and Sanders 1984). Bertaud himself has created versions of his model that provide for printing a sketch plan of the plots and roadways drawn to scale.

Another way in which the model could be extended is by providing for more variety in financing. This is an aspect of housing where there is scope for innovation, in both industrial and developing countries. As Kenneth Rosen has pointed out (1984, p. 142), "The fixed-payment, fixed-interest rate mortgage (FPM) has been the mainstay of the housing finance system for nearly thirty years. In the present environment of volatile, high interest and inflation rates . . . the FPM serves well [the interests of] neither the borrower nor the lender." Others have reached the same conclusion in other countries (Boleat and Coles 1987, especially part II). Rosen goes on to describe at least five major alternatives to the fixed-payment mortgage (FPM) that most of us are familiar with:

- The *graduated-payment mortgage (or GPM)*, which is based on a progression rate governing the rate at which monthly payments increase over time. In other words, instead of monthly payments being all the same, they are slightly lower at the beginning of the term and slightly higher at the end.

- The *variable-rate mortgage (or VRM)*, in which the interest rate is pegged to some market-based index of borrowing costs. This means that the borrower never pays more (or less) than current market rates, which leave him or her entirely at the mercy of market forces. If interest rates rise to a point where monthly payments are no longer affordable, the borrower risks losing the property and any equity in it.

- The *equity-adjusted mortgage (or EAM)*, in which the borrower is permitted to increase the principal of the loan during its term in some proportion (e.g., 50%) to the rate of inflation, and to apply this sum to reducing monthly interest payments. This amounts to borrowing against the future value of the property, rather than just its present value.

- The *shared-appreciation mortgage (or SAM)*, in which the borrower is charged a reduced rate of interest in return for assigning to the lender an agreed-upon share of any increase, or appreciation, in the value of the property over the term of the loan.

- The *dual-interest mortgage (or DIM)*, in which the borrower is charged one rate of interest (usually a reduced or graduated rate of some kind), while interest on the loan actually accrues at a second (often variable) rate of interest, with the difference being added to or subtracted from the principal.

The first variation (the graduated-payment mortgage based on progressively increasing payments) is already provided for in the model shown here. The other four variations are not in the present version, but it would be easy to build them in.

In summary, the Bertaud model is not an especially sophisticated model. Perhaps its greatest value lies in the convenient way it provides for examining the interaction between design standards (such as site density and basic services), financing choices, and (ultimately) affordability to the consumer. These are complex issues that are too often left to specialists—architects, accountants, and social workers —who are too often to be found working more or less in isolation from (not to say, conflict with) each other. The great merit of this model is that it shows clearly how these issues are in the end related to each other, and that gains in one area may involve tradeoffs in other areas. Housing is not just a matter of good design and quality control; nor is it just an exercise in internal rates of return and cost-recovery ratios; nor is it even just a matter of public involvement and community development. In the end, housing is all these and more. The Bertaud model provides a framework for bringing some of these disparate issues together in a way that all of us can understand.

Bibliography and References

Ast, Guido. 1981. *Space Standards for Urban Low-Cost Housing in Kenya: A Proposed Method for the Definition of Living Spaces, Dwelling Sizes, Plot Sizes and Plot-Use Heights and Densities.* Housing Research and Development Unit, University of Nairobi. Nairobi: HRDU.

Black, John, and David C. Stafford. 1988. *Housing Policy and Finance.* London: Routledge.

Boleat, Mark, and Adrian Coles. 1987. *The Mortgage Market: Theory and Practice of Housing Finance.* London: Allen and Unwin.

Caminos, Horacio, and Reinhard Goethert. 1980. *Urbanization Primer: Project Assessment, Site Analysis and Design Criteria for Sites and Services.* Boston: MIT Press.

Cartwright, T.J. 1991. "Planning and Chaos Theory." *Journal of the American Planning Association,* 57.1 (Winter), pp. 44-56.

————. 1989. *The Management of Human Settlements in Developing Countries*. London: Routledge.

————. 1987. "The Lost Art of Planning." *Long Range Planning*, 20.2, pp. 92-99.

————. 1973. "Problems, Solutions, and Strategies: A Contribution to the Theory and Practice of Planning." *Journal of the American Institute of Planners*, 39.3 (May), pp. 179-86.

Chevalier, Michel. 1969. *Social Science and Water Management: A Planning Strategy*. Policy Planning Branch, Department of Energy Mines and Resources. Ottawa: Queen's Printer.

Feins, Judith D., and Terry Saunders Lane. 1981. *How Much for Housing? New Perspectives on Affordability and Risk*. Cambridge, Mass.: Abt Books.

Joergensen, N.O. 1980. *Housing Finance for Low-Income Groups, with Special Reference to Developing Countries*. Housing Research and Development Unit, University of Nairobi. Nairobi: HRDU.

Kruijff, G.J.W. de. 1983. "Urban Cost Model." United Nations Project Working Paper. September. Bandung, Indonesia: Directorate of Building Research. Includes source code and documentation.

Rosen, Kenneth T. 1984. *Affordable Housing: New Policies for the Housing and Mortgage Markets*. A Twentieth Century Fund Report. Cambridge, Mass.: Ballinger.

Sanders, Welford. 1984. *Affordable Single-Family Housing: A Review of Development Standards*. Chicago: American Planning Association.

World Bank. 1981. *The Bertaud Model: A Model for the Analysis of Low-Income Shelter in the Developing World*. Urban Development Department Technical Paper no. 2. Washington, D.C.: World Bank.

Chapter 9

TRAFFIC ON THE ROADS: MODELING TRIP GENERATION AND TRIP DISTRIBUTION

. . . any color you like, as long as it's black.

Attributed to Henry Ford,
on being asked the color of his cars.

It may be that the single biggest influence on the environment in this century has been the invention of the automobile. Quite apart from its direct impact on personal mobility, energy consumption, air quality, and even the landscape, the automobile has had an indirect impact on almost every aspect of our lives, from cultural development to economic growth, and from education to warfare. Henry Ford cannot have conceived of the revolution he was starting!

Moreover, the automobile is ubiquitous. Human settlements in almost every country of the world are characterized by road networks and patterns of transport. Transportation is at once a cause and an effect of urban and regional development. On the one hand, development creates the capital necessary for creating a transportation system. On the other hand, development is reinforced by the ability to move people and goods quickly and efficiently from place to place. Beyond doubt, transportation is a vital factor in the modern world.

Purpose of the Model

Planners often want to know about the relationship between the characteristics of a city or region (e.g., its income and its land use) and the pattern of road transportation that takes place within it. In particular, they often want to know what the impact on traffic flows is likely to result from a given pattern of residential, commercial, or industrial development.

The most widely used model for linking human settlements to patterns of road transportation is based on a sequence of four steps:[1]

1. the *trip generation* phase, in which an estimate is made of the number of trips (usually per day) produced by, or having their origin in, each zone within a city or region;

2. the *trip distribution* phase, in which these trips are allocated among the various zones, according to destination;

3. the *modal split* phase, in which the trips are divided among the various modes of transport that may be available (the automobile is typically the dominant mode); and

4. the *traffic assignment* phase, in which trips between each pair of zones are assigned to the alternative road and highway links between these zones.

In this chapter, we will concentrate on modeling only the first two steps in this process.

The purpose of the model presented here is to provide planners with a dynamic view of likely traffic flows generated in a city or region characterized by a given residential distribution (based on household income) and a given set of nonresidential land uses (based on area).

In reality, of course, there must be many factors that influence travel patterns; models like this one are a crude simplification of real life (Jones et al. 1983). But the model does allow planners to see what kind of effects are likely to result from specific changes in residential patterns and/or land use, and how serious these changes are likely to be.

---

1. Credit for inventing this model is usually given to Douglas Carroll, Jr., who directed traffic studies in Detroit, Chicago, and Pittsburgh in the 1950s and 1960s. See Lowry 1964, p. 1; Detroit MATS 1953; and Carroll and Bevis 1957. For criticisms and an alternative approach ("continuous modeling"), see Vaughan 1987; for a review of relevant microcomputer software, see Marshment 1985 and Suchorzewski et al 1986.

Conceptual Basis

The trip distribution model presented here is derived from one developed by Richard Brail (1987, chapter 5).[2] This model depends on at least four important assumptions:

- **Zones**: The model assumes that a city or region can be divided into a fixed number of zones, which can, in turn, be defined in terms of a unique travel-time or distance from each of the other zones. Clearly, the more of these zones there are, the more accurate the model will be. On the other hand, more zones mean more calculations and slower performance (the relationship between the number of zones and the number of calculations is geometric, not linear!)

- **Trip Purposes**: The model assumes that trips can be divided into discrete categories according to purpose. Typically, planners use three categories:

  > Homebased-work (HBW) trip: "a trip, for the purpose of work, with one end at the residence of the trip-maker" (Sosslau et al. 1978b, p. 227);

  > Homebased-nonwork (HBNW) trip: "a trip, for the purpose of shopping, or for a social-recreational purpose, or for any other purpose other than work, with one end at the residence of the trip-maker" (ibid.); and

  > Nonhomebased (NHB) trip: "a trip that takes place between two points, neither of which is the home end of the trip-maker" (ibid.).

- **Trip Generation**: The model assumes that the number of trips produced, or generated, by each zone is a function of the residential characteristics of that zone. In other words, all trips are assumed to be generated by households,

---

2.   In addition to the trip generation and trip distribution phases shown here, Brail provides what he calls a "primitive" traffic assignment phase; but that is not included here. See also Blunden and Black 1984, chapter 3.

even if the physical origin and/or destination of the trip is elsewhere. For example, a trip to and from work is counted as two trips, but both are assumed to have been generated by the residence. Similarly, a trip from, say, a factory to a shopping center is assumed to have been generated by the household whose member or members made the trip.

• **Trip Distribution**: The model assumes that the allocation or distribution of these trips between pairs of zones is a function of the activities or land uses (residential and nonresidential) occurring in each zone, adjusted by the travel time or distance between each pair of zones. For this latter purpose, a form of gravity model is typically used.

In short, the *number* of trips people make is assumed to be a function of their lifestyle (as measured by income). *Where they go* on these trips (i.e., what zones they travel to) is assumed to be a function of what is going on in each zone (as measured by land use) and how willing people are to travel there.

Three more assumptions are required in order to help us get the input data required to run the model. These additional assumptions concern:

• **Willingness to Travel**: The model assumes that the greater the travel time or distance between two zones, the more reluctant (or the less willing) people will be to travel between those zones. In order to define the relationship between physical time or distance and the willingness to travel for that time or over that distance, planners typically use something called a "friction factor." There are various ways of describing this link; one is to use an inverse power function (Black 1981, chapter 2):[3]

$$W = \frac{1}{T^f}$$

where W is the willingness to travel, T is the travel time or distance, and f is the friction factor.

---

3.   The same technique is used and explained in chapter 7, above, on the Lowry model; for further details, see the technical note to that chapter on pages 179-80.

- **Trip Production**: The model assumes that trip production in one city or region can be estimated on the basis of evidence from "similar" cities or regions elsewhere. For example, if we have a breakdown of households by income level for the city we are studying, and if we know from studies elsewhere how many trips are typically produced by households in these same income categories, then we can estimate trip production for our study area on that basis.

- **Trip Attraction**: The model assumes that trip attraction can likewise be estimated on the basis of evidence from similar cities or regions elsewhere. For example, if we have a breakdown of land use by area for the city we are studying and we know from studies elsewhere how many trips are typically attracted by these same land uses, then we can estimate trip attraction for the study area on that basis.

Data Requirements

From the foregoing, it will be apparent that we need three kinds of data in order to be able to model the transportation pattern for a given city or region (see Sosslau et al. 1978a and 1978b for details). These are (a) data on travel times or distances, (b) data on residential characteristics, and (c) data on land use.

Time-Distance Matrix

We need a generalized time-distance matrix describing the time or distance involved in travel between all the different pairs of zones in our study area. Such a matrix might look like the one illustrated in Table 9.1.[4]

Associated with this matrix, we need a *friction factor*, so that we can convert the times or distances between pairs of zones in the time-distance matrix into an index of the willingness to travel between those zones. In some cases, it might be more appropriate to have a set of several different friction factors (e.g., for different types of trips of for trips with different purposes).

---

4. See also discussion of the travel-time matrix in chapter 7, above (pages 169-70).

Table 9.1:   Sample Time-Distance Matrix for a City or Region of Six Zones

Origin Zone	Destination Zone					
	1	2	3	4	5	6
1	5	14	7	11	13	15
2	14	5	21	15	21	18
3	7	21	5	18	14	22
4	11	15	18	5	12	8
5	13	21	14	12	5	5
6	15	18	22	8	5	3

Source: Based on R.K. Brail, <u>Microcomputers in Urban Planning and Management</u> (New Brunswick, N.J.: Center for Urban Policy Research, 1987), Exhibit 5-10, Table 16.

Residential Characteristics

We also need data on the residential characteristics of the city or region in our study area. Typically, this might be a table like Table 9.2, showing the distribution of households by income level for each zone. Associated with this table, we need data on trip production in similar cities or regions. For example, in the above case, we would need data on the average number of trips per household by income level and how these trips were distributed according to the purpose of the trip. Such data might look like those shown in Table 9.3, which are based on experience in the United States. If the available data pertain to *person-trips* (as they do here), then we also a means of converting them to *vehicle-trips*. One way of doing this (see Sosslau et al. 1978b, chapter 5) is to use a set of automobile occupancy rates (e.g., one rate for each type of trip according to purpose),

Finally, there is one more set of residential data required, although it is for estimating trip attraction rather than trip production. These data have to do with houses as destinations for, rather than origins of, automobile trips. We need to know how many such "social" trips (homebased-nonwork and nonhomebased) are attracted by each household, so that we can add these trips to estimates of trip attraction based on nonresidential land uses (see below).

Table 9.2:   Sample Distribution of Households by Zone and Income Level

Income Level	Number of Households by Zone						
	1	2	3	4	5	6	Total
Under $25,000	50	20	10	10	10	20	120
$25,000-39,999	20	0	0	40	0	10	70
$40,000-54,999	10	100	120	150	200	100	680
$55,000-69,999	50	0	0	0	10	0	60
$70,000 and over	60	300	120	230	100	50	860
Total	190	420	250	430	320	180	1790

Source: Adapted from R.K. Brail, <u>Microcomputers in Urban Planning and Management</u> (New Brunswick, N.J.: Center for Urban Policy Research, 1987), Exhibit 5-10, Table 3.

Table 9.3:   Sample Distribution of Trips per Household by Income Level and Percentage Distribution according to Trip Purpose

Income Level	Person Trips per Household	Trip Purpose			Total
		Home- Work	Home- Nonwork	Non- Home	
Under $25,000	6	.22	.58	.20	1
$25,000-39,999	25	.20	.57	.23	1
$40,000-54,999	30	.19	.56	.25	1
$55,000-69,999	20	.18	.56	.26	1
$70,000 and over	21	.18	.55	.27	1
Total	102				

Source: Adapted from 1970 data in Sosslau et al., <u>Quick-Response Urban Travel Estimation Techniques and Transferable Parameters: User's Guide</u>, National Cooperative Highway Research Program, Report no. 17 (Washington, D.C.: Transportation Research Board, 1978), Table 2; as cited in Brail, op. cit., Exhibit 5-10, Table 1.

Nonresidential Characteristics

Finally, we need data on the nonresidential characteristics of the city or region in our study area. Typically, this will be in the form of a table showing the quantity of land in each zone that is allocated to different types of uses. Typically, these data are measured in either spatial terms (e.g., hectares or square kilometers) or institutional terms (e.g., census tracts or city blocks). Thus, there might be a table showing the distribution of land use in each zone across a series of specific uses, such as shopping centers, other retail outlets, light industry, manufacturing, office space, restaurants, fast-food outlets, and so on. Such a table might look like the one shown in Table 9.4.

Associated with this, we need data on trip attraction in similar cities or regions. For example, in the above case, we would need data on the extent to which land uses like those shown above attract trips per unit area. Such data might look like those shown in Table 9.5. In the United States, data such as these can be obtained from a handbook called *Trip Generation*, which is published by the Institute of Transportation Engineers (ITE 1983).

Overview

Let us begin by examining what the model does; then we can discuss how that is accomplished. The model can be understood in terms of seven basic steps:

1. The model begins by computing the number of person-trips *produced* by each zone, which is based on a comparison of its residential characteristics (typically, the distribution of households by income level) with those of similar cities or regions for which we have data on trip generation.

2. Then, the model converts the number of generated person-trips per zone to vehicle-trips per zone and distributes them according to the purpose of the trip, again on the basis of data for similar cities or regions.

3. Next, the model computes the number of vehicle-trips *attracted* to each zone on the basis of the activities going on in each zone (typically, the distribution of nonresidential land use by area) compared to those in similar cities or regions for which we have data on trip attraction.

Table 9.4:   Sample Distribution of Activities by Type of Activity in Unit Area per Zone

Type of Activity	1	Unit Areas of Activity per Zone 2	3	4	5	6	Total
Shopping Center	4	2	1	1	1	1	10
Other Retail	4	3	3	2	1	1	14
Light Industry	0	0	0	1	1	5	7
Manufacturing	0	0	0	0	2	4	6
Industrial Park	0	0	0	0	0	9	9
Small Offices	1	1	1	1	1	1	6
Large Offices	1	1	1	0	0	1	4
Quality Restaurant	2	0	0	0	0	0	2
Other Restaurant	1	1	1	1	1	1	6
Fast Food	1	1	1	1	1	1	6
Local Park	0	0	4	1	1	1	7
Total	14	9	12	8	9	25	77

Source: Based on illustrative data provided in R.K. Brail, <u>Microcomputers in Urban Plan-</u><u>ning and Management</u> (New Brunswick, N.J.: Center for Urban Policy Research, 1987), Exhibit 5-10, Table 7. The unit area can be any convenient unit of measurement, such as square kilometer, census tract, or city block.

Table 9.5:   Sample Distribution of Trips per Unit Area by Type of Activity and Percentage Distribution according to Trip Purpose

Type of Activity	Trips per Area	Trip Purpose HBW	HBNW	NHB	Total
Shopping Center	889	.05	.60	.35	1
Other Retail	1000	.05	.60	.35	1
Light Industry	52	.90	.05	.05	1
Manufacturing	39	.90	.05	.05	1
Industrial Park	63	.90	.05	.05	1
Small Offices	137	.20	.45	.35	1
Large Offices	166	.20	.45	.35	1
Quality Restaurant	200	.05	.70	.25	1
Other Restaurant	932	.05	.54	.41	1
Fast Food	1825	.05	.54	.41	1
Local Park	6	.00	.90	.10	1
Total	5309				

Source: Based on illustrative data provided in R.K. Brail, <u>Microcomputers in Urban Plan-</u><u>ning and Management</u> (New Brunswick, N.J.: Center for Urban Policy Research, 1987), Exhibit 5-10, Table 6. The unit area can be any convenient unit of spatial measure-ment, such as a square kilometer or a city block. Trips are divided into homebased-work (HBW) trips, homebased-nonwork (HBNW) trips, and nonhomebased (NHB) trips.

4.   Then, the model distributes the total number of vehicle-trips per zone according to the purpose of the trip, again based on data for similar cities or regions. (Normally, it is not necessary to convert data on *attracted* trips from person-trips to vehicle-trips, because such data are normally collected in terms of vehicle-trips rather than person-trips.)

5.   Because the previous calculations of trip attraction are based on what are essentially nonresidential land uses, it is necessary to add to these calculations the number of trips attracted to each zone by residential land uses (i.e., by what might be called *social trips*).

6.   Since total trip *attraction* for each purpose must be the same as total trip *production* (since origins and destinations must be exactly equal for the city or region as a whole), the former must be adjusted until they equal the latter. In other words, trip attractions must be scaled up or down, until they match trip productions for each type of trip according to purpose.

7.   In the final (and most complicated) step, trip production is matched to trip attraction in an interzonal trip distribution matrix. This calculation is performed separately for each type of trip according to purpose, and the results are then added together to get a total trip distribution. In each case (that is, for each trip purpose), the procedure is as follows:

   •   using the appropriate friction factor, the time-distance matrix is converted to a corresponding willingness-to-travel matrix;

   •   using the willingness-to-travel matrix, trip production is distributed by zone of destination according to the relative attractiveness of each zone; and

   •   the results are adjusted (as before) to ensure that total trip distribution corresponds to total trip production.

Operation

   The model presented here (Listing 9.1) consists of 14 columns and 172 rows. The model is divided into a series of nine parts, but these can be grouped into three main sections.

Listing 9.1

```
     | A|| B|| C ||D ||E || F ||G|| H ||  I ||  J ||  K || L || M ||  N |
 1   TRAFFIC PLANNING MODEL                              © 1993 by T.J. Cartwright
 2
 3                          To:     1      2      3      4      5      6   Total
 4   TRIP DISTRIBUTION       ---------------------------------------------------
 5                          From: 1    715    281    395    242    187    255   2074
 6   By Zone of Origin  and       2   1271   1913    674    687    476    765   5786
 7   By Zone of Destination       3   1190    449   1020    366    369    424   3817
 8                                4   1242    782    626   1170    574   1065   5459
 9   No of Vehicle-Trips          5   1115    626    732    663    991   1274   5401
10                                6    506    345    272    432    494    856   2906
11                             Tot   6039   4396   3719   3560   3091   4639  25444
12
13   1. TIME-DISTANCE MATRIX   To:     1      2      3      4      5      6
14      AND FRICTION FACTORS   ---------------------------------------------------
15                          From: 1      5     14      7     11     13     15
16      Friction Factors:         2     14      5     21     15     21     18
17                HBW   .4         3      7     21      5     18     14     22
18                HBNW  .7         4     11     15     18      5     12      8
19                NHB   .7         5     13     21     14     12      5      5
20                                6     15     18     22      8      5      3
21
22   2. RESIDENTIAL DATA
23          Person                                     Zones
24   Income  Trips  Trip Purpose     1      2      3      4      5      6   Total
25   Level   per HH HBW HBNW NHB Tot
26   ---------------------------------------Number of Households---------
27   <15     6.4   .22 .58 .20  1     50     20     10     10     10     20    120
28   15-25    25   .20 .57 .23  1     20      0      0     40      0     10     70
29   25-35    30   .19 .56 .25  1     10    100    120    150    200    100    680
30   35-45  19.6   .18 .56 .26  1     50      0      0      0     10      0     60
31   >45    20.5   .18 .55 .27  1     60    300    120    230    100     50    860
32   Total  101.5                    190    420    250    430    320    180   1790
33                                --------Number of Person-Trips-------
34   Auto occupancy                  320    128     64     64     64    128    768
35   (pass/veh)       1.37 1.81 1.43 500      0      0   1000      0    250   1750
36                                   300   3000   3600   4500   6000   3000  20400
37   Trip attractions                980      0      0      0    196      0   1176
38   per household         1    .5  1230   6150   2460   4715   2050   1025  17630
39                                  3330   9278   6124  10279   8310   4403  41724
40
41   3. NON-RESIDENTIAL DATA
42                                                     Zones
43   Type of  Trip/  Trip Purpose    1      2      3      4      5      6   Total
44   Activity Area  HBW HBNW NHB Tot
45   ---------------------------------Unit Area of Activity---------
46   ShopCent  889  .05 .60 .35  1      4      2      1      1      1      1     10
47   Oth Reta 1000  .05 .60 .35  1      4      3      3      2      1      1     14
48   Light In   52  .90 .05 .05  1      0      0      0      1      1      5      7
49   Manufact   39  .90 .05 .05  1      0      0      0      0      2      4      6
50   Ind Park   63  .90 .05 .05  1      0      0      0      0      0      9      9
51   Sma Offi  137  .20 .45 .35  1      1      1      1      1      1      1      6
52   Lar Offi  166  .20 .45 .35  1      1      1      1      0      0      1      4
53   Qual Res  200  .05 .70 .25  1      2      0      0      0      0      0      2
54   Oth Rest  932  .05 .54 .41  1      1      1      1      1      1      1      6
55   Fast Foo 1825  .05 .54 .41  1      1      1      1      1      1      1      6
56   Loc Park    6  .00 .90 .10  1      0      0      4      1      1      1      7
57   Total    5309                     14      9     12      8      9     25     77
```

	A|| B|| C || D || E || F ||G||	H ||	I ||	J ||	K ||	L ||	M ||	N |
58				-------Vehicle-Trip Attractions-------					
59	ShopCenter	3557	1779	889	889	889	889	8893
60	Oth Retail	4000	3000	3000	2000	1000	1000	14000
61	Light Industry	0	0	0	52	52	262	367
62	Manufacturing	0	0	0	0	78	156	233
63	Ind Park	0	0	0	0	0	565	565
64	Sma Office	137	137	137	137	137	137	822
65	Lar Office	166	166	166	0	0	166	664
66	Qual Restaurant	400	0	0	0	0	0	400
67	Oth Restaurant	932	932	932	932	932	932	5592
68	Fast Food	1825	1825	1825	1825	1825	1825	10950
69	Loc Park	0	0	24	6	6	6	42
70	Total	11017	7839	6973	5842	4920	5938	42528

71									
72	4. TRIP PRODUCTION				Zone			
73		1	2	3	4	5	6	Total
74	---							
75	HBW	456	1245	833	1400	1137	608	5679
76	HBNW	1030	2838	1882	3160	2560	1359	12830
77	NHB	588	1704	1103	1847	1481	776	7498
78	Total	2074	5786	3817	6407	5179	2743	26007

79									
80								
81	5. TRIP ATTRACTION				Zone			
82		1	2	3	4	5	6	Total
83	Residential	-----------------------------------						
84	HBW	0	0	0	0	0	0	0
85	HBNW	190	420	250	430	320	180	1790
86	NHB	95	210	125	215	160	90	895
87	Total	285	630	375	645	480	270	2685
88	Non-Residential							
89	HBW	596	437	393	357	377	1177	3338
90	HBNW	6439	4492	3980	3292	2696	2813	23713
91	NHB	3981	2909	2600	2193	1847	1947	15477
92	Total	11017	7839	6973	5842	4920	5938	42528
93	All sources							
94	HBW	596	437	393	357	377	1177	3338
95	HBNW	6629	4912	4230	3722	3016	2993	25503
96	NHB	4076	3119	2725	2408	2007	2037	16372
97	Total	11302	8469	7348	6487	5400	6208	45213
98	All sources, adjusted to TP/TA							
99	HBW 1.70	1015	744	668	607	641	2003	5679
100	HBNW .50	3335	2471	2128	1872	1517	1506	12830
101	NHB .46	1867	1428	1248	1103	919	933	7498
102	Total .58	6501	4871	4227	3731	3106	3571	26007

103									
104	6.1 HOMEBASED-WORK (HBW) TRIPS							
105	Willingness-to-Travel To:	1	2	3	4	5	6	Total
106	---							
107	From: 1	.5253	.3480	.4592	.3832	.3584	.3385	2.4126
108	2	.3480	.5253	.2959	.3385	.2959	.3147	2.1182
109	3	.4592	.2959	.5253	.3147	.3480	.2904	2.2334
110	4	.3832	.3385	.3147	.5253	.3701	.4353	2.3671
111	5	.3584	.2959	.3480	.3701	.5253	.5253	2.4230
112	6	.3385	.3147	.2904	.4353	.5253	.6444	2.5486
113								
114	Trip Distribution To:	1	2	3	4	5	6	Total
115	---							
116	From: 1	109	54	65	47	47	135	456
117	2	223	254	130	129	120	389	1245
118	3	190	93	150	77	91	232	833
119	4	240	160	135	195	146	524	1400
120	5	171	107	114	105	158	483	1137
121	6	79	55	47	60	77	289	608
122	Tot	1012	722	641	612	640	2051	5679
123								
124	Adjusted Total Number of Trips:	1017	767	698	602	642	1956	5679
125								

```
      | A|| B||  C  || D ||E || F ||G|| H || I || J || K || L || M || N |
126   6.2 HOMEBASED-NONWORK (HBNW) TRIPS
127   Willingness-to-Travel        To:     1      2      3      4      5      6    Total
128                                     ---------------------------------------------------
129                        From: 1    .3241  .1577  .2561  .1866  .1661  .1502 1.2408
130                              2    .1577  .3241  .1187  .1502  .1187  .1322 1.0016
131                              3    .2561  .1187  .3241  .1322  .1577  .1149 1.1037
132                              4    .1866  .1502  .1322  .3241  .1756  .2333 1.2021
133                              5    .1661  .1187  .1577  .1756  .3241  .3241 1.2663
134                              6    .1502  .1322  .1149  .2333  .3241  .4635 1.4182
135
136   Trip Distribution           To:     1      2      3      4      5      6    Total
137                                     ---------------------------------------------------
138                        From: 1     390    144    208    125     88     75    1030
139                              2     664   1036    337    351    220    231    2838
140                              3     638    224    545    183    173    119    1882
141                              4     555    339    266    534    230    288    2213
142                              5     577    313    370    338    495    468    2560
143                              6     258    172    133    222    245    330    1359
144                              Tot  3081   2229   1859   1752   1450   1512   11882
145
146   Adjusted Total Number of Trips:   3610   2740   2437   2001   1588   1500   13853
147   ──────────────────────────────────────────────────────────────────────────────────
148   6.3 NONHOMEBASED (NHB) TRIPS
149   Willingness-to-Travel        To:     1      2      3      4      5      6    Total
150                                     ---------------------------------------------------
151                        From: 1    .3241  .1577  .2561  .1866  .1661  .1502 1.2408
152                              2    .1577  .3241  .1187  .1502  .1187  .1322 1.0016
153                              3    .2561  .1187  .3241  .1322  .1577  .1149 1.1037
154                              4    .1866  .1502  .1322  .3241  .1756  .2333 1.2021
155                              5    .1661  .1187  .1577  .1756  .3241  .3241 1.2663
156                              6    .1502  .1322  .1149  .2333  .3241  .4635 1.4182
157
158   Trip Distribution           To:     1      2      3      4      5      6    Total
159                                     ---------------------------------------------------
160                        From: 1     216     83    122     71     52     45     588
161                              2     384    623    206    208    136    146    1704
162                              3     361    132    326    106    105     73    1103
163                              4     447    283    226    441    198    252    1847
164                              5     367    207    248    220    337    323    1703
165                              6     170    118     93    150    173    237     939
166                              Tot  1945   1445   1220   1195   1001   1076    7883
167
168   Adjusted Total Number of Trips:   1791   1412   1277   1017    844    809    7131
169   ──────────────────────────────────────────────────────────────────────────────────
170   Source:  adapted from Richard Brail, Microcomputers in Urban Planning and
171            Management (New Brunswick: Center for Urban Policy Research, 1987),
172            chapter 5.
```

The first 11 rows contain a trip distribution table, which displays the results of running the model. Below this, there is a set of three parts (labeled 1, 2, and 3) in which the various input data are entered and a few preliminary calculations are made. Below this are five more parts (labeled 4, 5, 6.1, 6.2, and 6.3) that are used for converting the input data in parts 1, 2, and 3 into the results displayed at the top of the model.

Part 1 (just below the table of results at the top) is a time-distance table that is used for entering data on travel times or distances between each pair of zones. Beside the table are three friction factors, one for each type of trip according to purpose.

Next, part 2 contains data on the residential characteristics of the study area. On the left are data on trip generation by income level derived from "standard" regions; below this are data on automobile occupancy rates and residential attractiveness. On the right is a distribution of households in the study area by zone and by income level; beneath it is a table that converts the number of households into an estimate of the number of person-trips generated by those households, using the multipliers in cells C27:C31. Thus, the formula in cell H34 (for example) is

    H27*$C27

Totals on the right (column N) and at the bottom of both tables (rows 32 and 39) are computed using the SUM() function.

The third part of the model contains data on the nonresidential characteristics of the study area. On the left are data (taken from "standard" regions) on trip attraction by type of activity. On the right is a distribution of land use (measured in whatever spatial or institutional units may have been chosen for this purpose) by zone and by type of activity in the study area. Below this is a table that converts the land-use data into an estimate of the number of vehicle-trips attracted by that land use, using the multipliers in cells C46:C56. Thus, the formula in cell H59 (for example) is

    H46*$C46

Totals on the right (column N) and at the bottom of both tables (rows 57 and 79) are computed using the SUM() function.

The remaining data tables are interim calculations required to get from the three parts of the model just discussed (i.e., the time-distance matrix and the data on the residential and nonresidential characteristics of the study area) to the results right at the top of the worksheet. Thus, in the fourth part, the model deals with trip production. For each zone, the model calculates the total number of person-trips of each type (i.e., HBW, HBNW, and NHB) for all income levels. For zone 1, for example, the following formulas are used in cells H75, H76, and H77 respectively, and then copied across the rest of their respective rows:

```
(H34*$D27+H35*$D28+H36*$D29+H37*$D30+H38*$D31)/$D35

(H34*$E27+H35*$E28+H36*$E29+H37*$E30+H38*$E31)/$E35

(H34*$F27+H35*$F28+H36*$F29+H37*$F30+H38*$F31)/$F35
```

Again, the totals on the right (column N) and at the bottom (row 78) are calculated using the SUM() function.

In the fifth part, the model deals with trip attractions and reconciles them with trip production. First, trip attractions by residential areas are computed. There are, by definition, no homebased-work (HBW) trips that are attracted by residential areas; so cells H84:M84 are all zero. Other types of trips that are attracted by residential areas are computed using the trip attraction data entered in cells E38 and F38. Thus, the formulas in cells H85 and H86 (for example) are respectively as follows:

```
$E38*H32

$F38*H32
```

Next, nonresidential trip attractions are computed by taking vehicle-trip attractions and allocating them by type of trip. Thus, the formulas in cells H89, H90, and H91 are respectively as follows:

```
H59*$D46+H60*$D47+H61*$D48+H62*$D49+H63*$D50+H64*$D51+
    H65*$D52+H66*$D53+H67*$D54+H68*$D55+H69*$D56

H59*$E46+H60*$E47+H61*$E48+H62*$E49+H63*$E50+H64*$E51+
    H65*$E52+H66*$E53+H67*$E54+H68*$E55+H69*$E56

H59*$F46+H60*$F47+H61*$F48+H62*$F49+H63*$F50+H64*$F51+
    H65*$F52+H66*$F53+H67*$F54+H68*$F55+H69*$F56
```

Then, residential and nonresidential trip attractions are added together in rows 93-97. Finally, total trip attractions from all sources are adjusted in rows 98-102 to reconcile them with trip production.[5] For this purpose, a conversion factor (trip production over trip attraction) is calculated in cells F99:F101 and applied to the raw data in cells H94:M96. Thus, the formulas in cells F99 and H99 (for example) are respectively

```
N75/N94
```

```
H94*$F99
```

with the latter being copied into cells H99:M101.

The last three parts of the model (labeled 6.1, 6.2, and 6.3) compute the willingness to travel and the consequent interzonal trip distribution for each different type of trip. As noted above, the willingness to travel is calculated by means of an inverse power function. Thus, the formula in cell H107 (for example) is as follows:

```
IF(H15=0,0,1/H15^$E$17)
```

or, if your spreadsheet can cope with Boolean operations,[6]

```
(H15>0)/H15^$E$17
```

This formula is copied into all the other cells in the table. The corresponding formulas in parts 6.2 and 6.3 of the model (cells H129 and H151) are similar, except that the reference to cell E17 has to be replaced by a reference to cell E18 and cell E19, respectively. Again, the formulas can be copied into the rest of each table.

Beneath each willingness-to-travel table is a trip-distribution table. In this table, the model determines what proportion of the total number of trips produced of each type (that is, what proportion of each cell in the table in part 4 of the model)

---

5.   In other words, trip production data are treated as empirical, while trip attractions are regarded essentially as an index. Consequently, the latter are adjusted so they correspond to the former.

6.   Boolean operations are discussed further in appendix A, below.

can be assigned to each origin-destination pair. This is calculated by multiplying the total number of trips produced in each zone by the *proportion* to be assigned to each destination zone, based on the willingness to travel to that zone. Thus, cell H116 (for example) is defined as the product of the total number of HBW trips produced in zone 1 (cell H75) times the product of the willingness-to-travel to zone 1 (cell 107) and the total number of trips to zone 1 (cell H124), all over the sum of the products of the willingness-to-travel and the total number of trips from zone 1 to all of the zones. That is, cell H116 contains the following formula:

$$\$H75*H107*H124/(\$H107*H124+\$I107*\$I124+\$J107*\$J124+$$
$$\$K107*\$K124+\$L107*\$L124+\$M107*\$M124)$$

This formula is copied into all the other cells in the table. The corresponding formulas in parts 6.2 and 6.3 of the model (cells H138 and H160) are similar (see appendix 2, below, for precise details of the adjustments that need to be made in each case.

An important point to note is that parts 6.1, 6.2, and 6.3 of the model are in fact circular, which means that iteration occurs. This is because there is both a total and an adjusted total for the number of trips made to each zone at the bottom of the trip-distribution table. The total is simply the sum of the trips from each of the zones, whereas the adjusted total represents an adjustment of this total to the trip-attraction capacity of each zone (as calculated in part 5 of the model). Thus, the formulas in cells H122 and H124 are respectively as follows:

SUM(H116:H121)

IF(H99=0,0,H99^2/H122)

or, if your spreadsheet can cope with Boolean operations, the second formula can be written:

(H99>0)*H99^2/H122

From this last formula, it can be seen that cell H124 depends on cell H122; however, cell H122 depends on cell H116 (as well as other cells) and cell H116 depends on cell H124 (again, among others). So the calculations are circular and therefore iterate. This ensures that any adjustment made to reflect the interzonal distribution of trip destinations is in turn adjusted to the correct interzonal distribution of trip origins.

The final step in the model occurs when the table right at the top of the model (cells H5:M10) is calculated from the three trip-distribution tables at the bottom. Thus, cell H5 (for example) is simply the sum of the trips within zone 1 for all three types of trip purposes (HBW, HBNW, and NHB) and contains the formula

```
H116+H138+H160
```

The results of using the model with the input data shown in Listing 9.1 are shown in Figures 9.1 and 9.2 below. Figure 9.1 shows the distribution of vehicle-trips by zone of origin, with different destinations stacked on top of each other to show the total. Figure 9.2 shows the distribution of vehicle-trips by zone of destination, with different origins stacked on top of each other to show the total.

Interpretation and Use

First, of course, the model allows us to get some insight into the current state of affairs. Using the data shown in Listing 9.1. (as depicted in Figures 9.1 and 9.2), for example, it is immediately apparent that most of the traffic *originates in* zones 2, 4, and 5, whereas most of it *goes to* destinations in zones 1 and 2. For a typical city in an industrial country, such a result would imply that "work" tends to be located in zones 1 and 2 and "homes" tend to be located in zones 3, 4, and 5.

Second, we can examine the sensitivity of the model to its parameters. Chief among these are the "standard" data on the distribution of trip production by income level and trip purpose, and on the distribution of trip attraction by type of urban activity and trip purpose. As noted, the data shown here are derived from experience in the United States. Naturally, you should give some thought to how far these ratios are applicable to the case you may be dealing with (see Suchorzewski et al. 1986), whether your own study area is also in the United States or in some other country.

You can also use the model to test its sensitivity to the parameters you have used. You can change the parameters by small amounts and examine what effect that has. In some cases, the results may be highly sensitive to small changes in the parameters; in other cases, even quite large changes in the parameters may not make much difference to the results. The same thing can be done with friction factors, automobile occupancy rates, and assumed trip attraction rates per household. Naturally, it is a good idea to know how sensitive your results are to certain parameters, before worrying about how accurate those parameters may be.

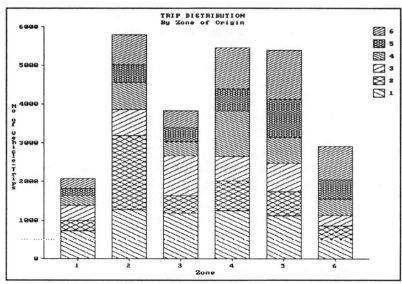

Figure 9.1: Trip Distribution by Zone of Origin

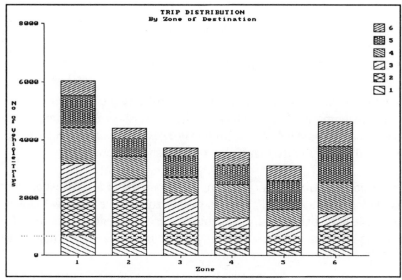

Fig. 9.2. Trip Distribution by Zone of Destination

But the model lets you do more than just understand the present. You can also model the future. There are three obvious points of attack: namely,

- What if travel times were to change substantially? Suppose existing roads were improved and new roads were built, thereby reducing travel times: how would that alter the trip distribution? Suppose, on the other hand, that travel times increased due to introduction of flow-control measures, such as one-way streets or exclusive lanes: what effect would that have?

- What if the residential characteristics of some zones were to change? Suppose a zone were to improve economically: how would that alter the current trip distribution? Alternatively, suppose a zone were to decline economically: how would that affect the overall trip distribution?

- What if land-use patterns were to change? Suppose the number of acres or hectares devoted to a given land use in a particular zone were to double: what effect on trip distribution would that have? Suppose it were to decline by 25%? What effect might the following kinds of developments have: a new shopping center in one zone? a local park in another zone? or a switch from light industry to manufacturing in a third zone? And so on.

All these and other scenarios can be simulated using the model shown here.

Evaluation and Extension

The trip distribution model shown here is probably the most widely used of all transportation planning models. Naturally, the spreadsheet environment in which it is presented here is somewhat limiting. For example, most real-life planning situations would require more than the six zones shown here.

It is possible to build bigger versions of this model in a spreadsheet by making use of linked worksheets. Models of up to twenty zones have reportedly been created in this way. The spreadsheet is also capable of a more extended versions of the trip-distribution model, that is, one containing all four of the stages discussed at the beginning of the chapter.[7] In Calgary (Canada), for example, a four-stage spread-sheet model has been developed in which

> The trip generation model is a regression-type model while the trip distribution model is of the gravity category. . . . The modal split model used is an origin-destination normative type. Only AM [i.e., morning], peak-hour, homebased-work (HBW) trips are modeled. Total AM, peak-hour trips are derived by factoring the modeled trips by origin and destination land-use type. Manual and all-or-nothing assignments are used to obtain screen-line/corridor and desire-line volumes [Fung 1985, p. 637].

Indeed, the spreadsheet is not just a good modeling environment; it can even be thought of as a metaphor for the city as a spatial matrix in its own right!

Nevertheless, traffic planning models written in conventional programming languages (e.g., Fortran, Pascal, etc.) can typically handle hundreds of zones rather than six or twenty. This means a city can be divided into much smaller zones. Such models will certainly require a great deal more data, but they are also likely to increase confidence in the model's results.

The chief value of the model in the form shown here is probably educational. Thanks to the transparency of the spreadsheet as a programming environment, the workings of the model are apparent even to users with very little understanding of computers and computer programming. Spreadsheets, in other words, can help to make traffic modeling seem much less of a "black box" than it frequently does.

---

7.   Some other spreadsheet applications to transportation are described in the Proceedings of the National Conference on Microcomputers in Urban Transportation held in San Diego in June 1985 (Abkowitz 1985): e.g., see the paper by Peter Eakland (1985) on transit scheduling. The Microcomputer Center on Transportation (MCTRANS) at the University of Florida maintains a collection of microcomputer-based transportation models.

Bibliography and References

Abkowitz, Mark D., ed. 1985. *Microcomputer Applications within the Urban Transportation Environment*. Proceedings of the National Conference on Microcomputers in Urban Transportation, San Diego. New York: American Society of Civil Engineers.

Batty, Michael. 1976. *Urban Modeling*. Cambridge: Cambridge University Press.

Black, John. 1981. *Urban Transport Planning*. Baltimore: Johns Hopkins University Press.

Blunden, W.R., and J.A. Black. 1984. *The Land-Use Transport System*. 2nd ed. Urban and Regional Planning Series, vol. 2. Sydney: Pergamon.

Brail, Richard K. 1987. *Microcomputers in Urban Planning and Management*. New Brunswick, N.J.: Center for Urban Policy Research.

Carroll, J.D., and H.W. Bevis. 1957. "Predicting Local Travel in Urban Areas." *Papers and Proceedings of the Regional Science Association*, 3, pp. 183-97.

Cartwright, T.J., and I.I. Gabbour. 1975. "Graph Theory and Managing Urban Change." *Socio-Economic Planning Science*, 9 (Summer), pp. 118-32.

Detroit Metropolitan Area Traffic Study (MATS). 1953. *Report on the Detroit Metropolitan Traffic Study*. J.D. Carroll, Jr., Study Director. Detroit: n.p.

Eakland, Peter. 1985. "Transit Scheduling with Spreadsheets." In Abkowitz 1985, pp. 465-72.

Fung, Yu Han. 1985. "Use of Lotus 1-2-3 for Transportation Planning." In Abkowitz 1985, pp. 636-44.

Institute of Transportation Engineers (ITE). 1983. *Trip Generation*. 3rd. ed. Washington, D.C.: Institute of Transportation Engineers.

Jones, P.M., M.C. Dix, M.I. Clarke, and I.G. Heggie. 1983. *Understanding Travel Behaviour*. Aldershot, U.K.: Gower.

Lowry, Ira S. 1964. *A Model of Metropolis*. Memorandum No. RM-4035-RC. Santa Monica: Rand Corporation.

Marshment, Richard S. 1985. "Transportation System Forecasting Using Microcomputers." In Abkowitz 1985, pp. 107-15.

Sosslau, Arthur, A.B. Hassam, M.M. Carter, and G.V. Wickstrom. 1978a. *Travel Estimation Procedures for Quick Response to Urban Policy Issues*. Comsis Corporation for the National Cooperative Highway Research Program (NCHRP). Report No. 186. Washington, D.C.: Transportation Research Board.

———. 1978b. *Quick-Response Urban Travel Estimation Techniques and Transferable Parameters: User's Guide*. Comsis Corporation for the National Cooperative Highway Research Program (NCHRP). Report No. 187. Washington, D.C.: Transportation Research Board.

Suchorzewski, Wojtek, T.J. Cartwright, and M.R. Brown. 1987. "Transportation Software." *Habitat Microcomputer Users Directory*, 4 (July). Special Issue. Nairobi: United Nations Centre for Human Settlements (Habitat).

United States Department of Transportation. 1984. *Microcomputers in Transportation: Quick Response System (QRS) Documentation*. Comsis Corporation for the Federal Highway Administration. Washington, D.C.: Department of Transportation.

———. 1985. *Microcomputers in Transportation: Software and Source Book, February 1985*. Prepared by the Methods Division, Urban Mass Transportation Administration. Washington, D.C.: Department of Transportation.

Vaughan, Rodney. 1987. *Urban Spatial Traffic Patterns*. London: Pion.

Chapter 10

THROWING THINGS AWAY:
A MODEL FOR
WASTE MANAGEMENT

*A society in which consumption
has to be artificially stimulated
in order to keep production going
is a society founded on trash and waste,
and such a society is a house built upon sand.*

Dorothy L. Sayers, Creed or Chaos

A presentiment like this may have seemed far-fetched when it was made in 1949, but it hardly seems so today.[1] According to the U.S. Environmental Protection Agency (USEPA 1988; cited in Erwin and Healey 1990, p. 14), municipal solid waste in the United States in 1988 amounted to a staggering 160 million tons. That works out to about 1.5 kilogram per person per day! One third of that garbage is estimated to be packaging (Franklin Associates 1988; cited in Erwin and Healey 1990, p. 25).

Nor is there any obvious cure. In fact, Vance Packard argued in *The Waste Makers* (1960) that waste is inherent in development, at least as we practise it today. With development, he argued, the capacity to produce eventually exceeds the ability to consume. In order to sustain the demand for goods and services, therefore, we have to effectively reduce their capacity to meet our needs. That is, we have to build in greater and greater proportions of waste. Packard identified nine strategies that he felt were already in use for this purpose, including unnecessary model changes,

---

1.   Vance Packard uses the same quotation in his prescient study *The Waste Makers* (1960).

"planned obsolescence" (a term Packard invented), throw-away items, consumption for the sake of consumption, and others. His views aroused a lot of controversy when they were published—but they hardly seem extreme now.

Even for nontoxic and nonhazardous wastes, simply throwing things away (i.e., disposal by burying or dumping) is becoming more and more problematical, due to the sheer volume of these wastes. Incineration—and the prospect of killing two birds with one stone, so to speak, by turning waste into energy—has reduced the problem, but only by creating another one: namely, how to dispose of the ash. A few years ago, a city in the United States became involved in a ludicrous but tragic case of the "politics of garbage" when the city tried to dispose of ash from its municipal incinerators.

With [all] its land-based disposal options under attack, the city finally arranged . . . to ship the ash abroad. . . . [and] a seventeen-year old freighter named the *Khian Sea* . . . laden with more than 14,000 tons of . . . ash . . . wandered through the Caribbean, first stopping in the Bahamas, where the cargo was rejected. Next stop was Panama, where a hastily arranged deal collapsed after . . . Greenpeace released EPA memos detailing the heavy metals and toxic dioxins present in the ash.

The *Khian Sea* then left Panama. . . . [for] Haiti . . . where another deal was arranged . . . to dump the ash on a beach near the port of Gonaives. . . . But when that country's political opposition—alerted once again by Greenpeace— found out . . . the action was brought to a halt, although possibly as much as several thousand tons [of ash] remained behind [in Haiti].

The *Khian Sea*, subsequently renamed twice, hopefully to lessen attention to its mission, once again took to the oceans. . . . through 1987 and 1988 . . . travelling to Africa, then Sri Lanka, and finally Singapore, where its cargo mysteriously disappeared. The freighter's last owner . . . sent a message, ultimately forwarded to [city] officials [in the United States], claiming the ash had been discharged but. . . . "Owners will not advise location. . . ." Court documents later suggested that . . . [the ship] had actually dumped the ash in the Indian Ocean [Blumberg and Gottlieb 1989, pp. 4-5].

Meanwhile, the United States alone has another half million tons of municipal solid waste to dispose of every day.

Purpose of the Model

The purpose of this chapter is to discuss a much more modest case than this. Here we are dealing with a small waste-disposal site, such as a neighborhood dump or a toxic-waste site. We have some idea of how many customers and how much waste we want the site to be able to handle in the long term (say, over a year), but we do not know how big a site to build or how often we should plan to empty it. The problem is that the flow of waste to the site is going to be uneven: there will be times of feast and times of famine, so to speak—times of heavy demand and times of low demand. In the long term, the highs and lows will probably balance each other, and the number of customers will reflect a fairly stable mean. But in the short run, we have no idea how many customers will arrive tomorrow or next week. Thus, a facility designed for little more than the average number of customers may well break down under the pressure of a few days or a few weeks of heavy demand. At the same time, we do not want to "overbuild" the facility either, just to make sure it can cope with the kind of demand that occurs only once or twice in several hundred days or weeks. So what do we do?

In some cases, it is reasonable to just ignore random influences or at most to make some general allowance for them.[2] But, in other cases, we want to things that are fundamentally unpredictable: that is, randomness is not just one of several explanatory influences; it is the main one. At first glance, it might appear that there is, by definition, no way of modeling such events; however, that is not the case. It is possible to simulate random events by using random (or pseudorandom) numbers. This technique is usually attributed to John von Neumann (1951) and Stanislaw Ulam (Metropolis and Ulam 1949), who called their technique the "Monte Carlo" method because it seemed to rely on little more than gambling. Since then, the Monte Carlo method has found many applications as a random sampling technique for multidimensional problems in both the natural and the social sciences.[3]

---

2.   For example, the Klein model discussed in chapter 6, above, includes a "stochastic factor" to allow for random disturbances.

3.   See, for example, Hammersley and Handscomb 1964; Smith 1973; Kleijnen 1974, 1975; and Rubinstein 1981.

One process in which Monte Carlo simulation has been successfully used is modeling the arrival of "customers" at some point, to pick up or deliver a product or service of some kind. In terms of environmental management, for example, we might think of people arriving at a licensing authority to pick up new licenses or at a hospital to donate blood; of vehicles arriving at an intersection or campers arriving at a camp site; or of lightning strikes in a forest or wolf attacks on a herd of elk. In all these cases, the key condition for using Monte Carlo simulation is that there be many, uncoordinated customers whose effects are intermittent and independent of each other. It is also assumed that there are no cyclical or seasonal patterns and that the probability of any particular customer being served in any given period of time is relatively low. Such, in fact, is the character of our waste-disposal site.

Conceptual Basis

The Monte Carlo method is any method for the solution of a model that uses random or pseudorandom numbers (Hammersley and Handscomb 1964; cited in Kleijnen 1974, p. 6). We will apply this technique to the waste-disposal site by assuming that the arrival of customers can be simulated by a set of random numbers. However, while we have no idea how many customers will turn up at any specific time (e.g., tomorrow or next week), it is reasonable for us to assume that very large or very small numbers of customers are less likely than near-average numbers (whatever the average may turn out to be). In other words, the likelihood or probability of different numbers of customers is not the same. Some numbers of customers are clearly going to be more likely or more probable than other numbers. To summarize, the number of customers arriving in any single period of time can be simulated by a random number; but over many periods of time the distribution of the numbers of arrivals is unlikely to be either constant or random. Instead, it is likely to form a probability distribution of some kind.

Thus, we can visualize a curve showing the likelihood or probability (on the y-axis) of different numbers of customers (on the x-axis). If numbers of customers get progressively less likely as we move away from the mean, or average, the probability distribution would be highest at the mean and would slope downwards on either side. In other words, the further we move from the mean, the less likely we are to get that particular number of customers. This does not mean that we will *usually* get the mean number; usually, in fact, we will not. But it does mean that near-average numbers of customers are more likely than numbers further away from the average.

If we make one further assumption—that the probability distribution has the shape of what statisticians call a Poisson curve—we can infer an exact probability for the arrival of each number of customers. In assuming a Poisson distribution, we are assuming that:

- customers arrive individually and not in groups;

- the arrival of each customer is independent of the arrival of all other customers (e.g., customers are not deterred by the presence or absence of other customers); and

- the arrival of any customer is independent of time (e.g., customers are neither more nor less likely to arrive in the morning or the afternoon, on Mondays rather than Fridays, etc.)

Thus, according to one source (Law and Kelton 1982, p. 206), "the Poisson process is the most commonly used model for the arrival process of customers to a queuing system."

Once we have both a mean and an exact distribution for the number of arrivals at our site over time, then we can calculate the exact probability of any specific number of arrivals. In the case of a Poisson distribution, the probability of a given number of customers arriving at a particular time, P_n, is given by the formula:

$$P_n = \frac{A^n}{\exp(A) * n!}$$

where   A   = the average or mean number of arrivals,
             $\exp(A)$ = the exponent of A (or e^A, where e is approximately 2.7183),
and   n!   = factorial n.[4]

Thus, the probability of getting, say, 5 customers when the mean is 10 is given by:

$$P_5 = 10^5 / (2.7183^{10} * 5 * 4 * 3 * 2 * 1)$$

---

4.   Factorial n means the product of all the integers between 1 and n inclusive. There is a discussion of exponents and the base e in chapter 1 (page 37), above.

which works out to just under 2% (.0189). In short, once we know the mean of a Poisson distribution, we can calculate the probability of any particular number in it. Naturally, we can use a spreadsheet to do these calculations for us (see Table 10.1).[5] If we plot the results on a bar graph (see Figure 10.1), we can see the characteristic shape of the Poisson distribution: sloping down on either side of the mean but slightly skewed towards the right or non-zero side.

The next step is to use this distribution to simulate the pattern of demand at the waste-disposal facility. If the pattern of demand were entirely random, we could just select a random number between 1 and, say, 25 and assume that it represented the number of customers at a particular time. But, as noted above, the long-run pattern of demand is not random; it reflects a Poisson distribution. So we have to be able to select a random number from a scale that reflects a Poisson distribution. In other words, we have to divide the range over which the random numbers occur (usually from zero to one) into units that are *not* equal in size but vary in size in accordance with a Poisson distribution. Thus, the probability that a random number will fall into a given part or unit on our scale will vary.

This may sound more complicated than it really is. Suppose that random numbers are generated within the range from zero to one and that we want to simulate a Poisson distribution with a mean of 10, just as we did in the illustration above. This means we are assuming that, in the long run, an average of 10 people will arrive at the waste-disposal site each day (or whatever unit of time we may be using). In order to be able to simulate this kind of pattern, we would have to ensure that just under 2 percent of the random numbers we select yield values of 5, just under 4 percent yield values of 6, just over 6 percent yield values of 7, and so on—all precisely in accordance with the probabilities shown in Table 10.1.[6]

---

5.   Some of the latest spreadsheet versions (e.g., Excel 4.0) have a built-in function to generate random Poisson numbers.

6.   If this still sounds a bit fishy, another way of understanding what is going on is to imagine collecting rain water in a series of buckets. If the buckets are all the same size and assuming the rain falls in a random manner, then presumably the same amount of water will collect in each bucket. But if we adjusted the size of the buckets (strictly speaking, the area of their openings) so that that they were proportional to the probabilities of a Poisson distribution, then the amount of water collected in the various buckets would reflect a Poisson distribution, even though the rain was still falling in an entirely random manner.

Table 10.1: Poisson Distribution about a Mean of 10

Customers n	Probability P(n)	Cumulative P(n)	Customers n	Probability P(n)	Cumulative P(n)
1	.0000	.0000	14	.0729	.8645
2	.0005	.0005	15	.0521	.9165
3	.0023	.0028	16	.0347	.9513
4	.0076	.0103	17	.0217	.9730
5	.0189	.0293	18	.0128	.9857
6	.0378	.0671	19	.0071	.9928
7	.0631	.1301	20	.0037	.9965
8	.0901	.2202	21	.0019	.9984
9	.1126	.3328	22	.0009	.9993
10	.1251	.4579	23	.0004	.9997
11	.1251	.5830	24	.0002	.9999
12	.1137	.6968	25	.0001	1.0000
13	.0948	.7916			

Note: probabilities for n < 2 and n > 25 are shown as zero because they are less than
.00005 (or 1/200th of 1%), which is the smallest number that can be specified with
the four figures of accuracy used here.

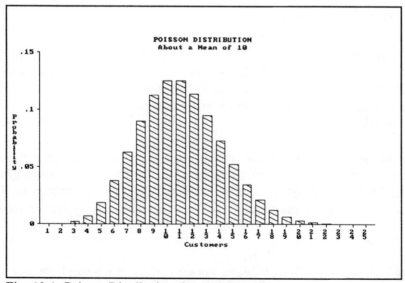

Fig. 10.1. Poisson Distribution about a Mean of 10

One way of doing this is to calculate the cumulative probability associated with each number of customers, and use these cumulative probabilities to "mark off," or calibrate, a scale for interpreting the random numbers.[7] Thus, referring to Table 10.1, we can say that, if a random number is less than or equal to .0293 (which is the cumulative probability for 5 customers) but more than .0103 (which is the cumulative probability for 4 customers), then it means 5 customers will arrive. We can see from Table 10.1 that there is a probability of .0189 that a random number will have a value within this range. Similarly, if a random number falls between .0671 and .0293, it means 6 customers; if it falls between .2202 and .1301, it means 7 customers; if it falls between .3328 and .2202, it means 8 customers; and so on. In other words, a cumulative probability distribution provides a convenient scale for converting random numbers in the same range (zero to one) into a value appropriate for a Poisson distribution.

Let us now return to the case of the waste-disposal facility. If we know that we will have an average of 10 customers per day or per week over the long term, we can simulate demand in the short term (i.e., each day or each week) by choosing a random number in the range zero to one and then "looking it up" in a table of cumulative probabilities like the one in Table 10.1.

Data Requirements

In order to run the model described here, only four items of input data are required, two pertaining to the facility and two to the demand. All units in the model as presented here are kilograms, although they could as well be tonnes, cubic meters, liters or any other appropriate unit of measurement (including the corresponding Imperial units).

As far as the facility is concerned, we need to know its capacity and the manner in which it is emptied from time to time. In the model presented here, we assume that the capacity of the facility is limited only by the capacity of the site. Similarly, we assume (for the sake of argument) that the site is emptied in accordance

---

7.   The cumulative probability for given number of customers is simply the sum of all the probabilities of numbers up to and including that given number. In other words, the cumulative probability of, say, 5 customers means the probability that 5 or fewer customers will arrive.

with the following procedure: the level of waste in the facility is checked at the beginning of each period (i.e., each day, week, or whatever unit of time is in use) and that, if the amount of waste is greater than or equal to the capacity of the truck used to empty the facility, that amount is removed from the facility at the end of the period. Naturally, there are other assumptions that could have been made, but this is the procedure that has been built in to the model presented here.

As far as the demand is concerned, we need to know the average number of customers in the long term and the average amount of waste they want to dump at the facility each time they come. For example, if you expect the facility to service 5,000 customers per year, it is easy to see that this means an average of 96 customers per week or 14 per day (assuming the facility is open every day). Similarly, if you expect the facility to handle 100,000 kilograms of waste per year, that means an average of 20 kg per customer.

Some of these data are likely to be only rough estimates; so one of the uses of the model will be to see how sensitive the operations of the facility are to changes in some of these parameters. What if demand is higher than expected? What if the average customer brings more waste than expected? What if the faclity is a bit bigger or a bit smaller than proposed? What if it is emptied a bit more frequently or a bit less frequently than proposed? And so on.

Overview

The model consists of 13 columns and 83 rows (see Listing 10.1). It is designed to simulate 31 time periods because that is the number of days in a typical month. But it could easily be modified to provide for more or fewer periods, as might be required to simulate hours in a day or days in a year.

The model is divided into three parts. At the top on the left is the user interface for the model, where the input parameters are entered and the results of running the model are displayed in summary form. To the right of the user interface are the detailed results of running the model through as many time periods as are provided for. The third part of the model is a lookup table of cumulative probabilities for a Poisson distribution. In this version of the model (with 31 time periods), the table begins at row 38 and extends to the end of the model at row 80.

The model works in five steps:

1.　The amount of waste in the facility at the beginning of each day is computed in column G.　On day 1, the level of waste is arbitrarily set at half the total capacity of the facility.　On subsequent days, the level is determined by taking the level at the beginning of the previous period, adding new waste brought in during the previous period (column J), and subtracting waste removed since then (if any, as indicated in column L).　Thus, the formula in cell G5 (for example) is

```
G4+J4-L4*$D$5
```

2.　Then the number of customers arriving each day is simulated in column H by means of random numbers in column M and the lookup table at the bottom of the model.　Thus, the formula in cell H4 (for example) is

```
VLU(M4,$J$41:$L$80,2)
```

This formula means: get the value in cell M4, look it up in the left-hand column of the block J41:L80, and return the value in the corresponding cell 2 columns to the right.　Cell M4, in turn, contains the RANDOM function that returns a random number.　The block J41:L80 is part of a five-column lookup table found at the bottom of the model, which matches specific ranges of random numbers to specific numbers of customers.　The five columns and their contents are as follows:

Column F	consecutive integers, beginning with 0;
Column G	factorials of the integers in column F; thus, cell G42 (for example) contains the formula: F42*G41;
Column I	Poisson probabilities of the integers in column F; thus, cell I41 (for example) contains the following formula, derived from the equation on page 233 above: ($D$6^F41)/(EXP($D$6)*G41);
Column J	cumulative Poisson probabilities of the integers in column F; thus, cell J41 (for example) contains the formula I41+J40; and

Listing 10.1

	A	B	C	D		F	G	H	I	J	K	L	M
1	WASTE MANAGEMENT MODEL USING MONTE CARLO SIMULATION								© 1993 by T.J. Cartwright				
2													
3	Model Parameters					Day	Level	Cust	Waste	Dumped	Excess	Full	Random
4	Site capacity (kg)			2000		1	1000	20	500	500	0	1	.47039
5	Removal capacity (kg)			1000		2	500	24	600	600	0	0	.79907
6	Mean no of customers			20		3	1100	25	625	625	0	1	.88326
7	Mean waste dumped			25		4	725	19	475	475	0	0	.40767
8	————————————					5	1200	28	700	700	0	1	.95839
9	Summary of Results					6	900	24	600	600	0	0	.79862
10	Mean site level (kg)			1043		7	1500	24	600	500	100	1	.82702
11	Mean no of customers			21		8	1000	18	450	450	0	1	.37805
12	Min no of customers			11		9	450	17	425	425	0	0	.25564
13	Max no of customers			28		10	875	28	700	700	0	0	.95557
14	No of removals			16		11	1575	19	475	425	50	1	.44577
15	No of periods			31		12	1000	23	575	575	0	1	.72146
16						13	575	16	400	400	0	0	.18189
17	Excess waste (kg)			150		14	975	23	575	575	0	0	.73641
18	Customers affected			6		15	1550	13	325	325	0	1	.05157
19	As % of total			.9%		16	875	19	475	475	0	0	.44620
20	————————————					17	1350	23	575	575	0	1	.77461
21						18	925	24	600	600	0	0	.82595
22						19	1525	17	425	425	0	1	.24322
23						20	950	19	475	475	0	0	.40243
24						21	1425	18	450	450	0	1	.30415
25						22	875	11	275	275	0	0	.01919
26						23	1150	19	475	475	0	1	.39883
27						24	625	25	625	625	0	0	.85417
28						25	1250	23	575	575	0	1	.77104
29						26	825	18	450	450	0	0	.37989
30						27	1275	20	500	500	0	1	.48681
31						28	775	25	625	625	0	0	.85078
32						29	1400	18	450	450	0	1	.34712
33						30	850	19	475	475	0	0	.38951
34						31	1325	22	550	550	0	1	.68861
35													
36													
37													
38	Lookup table for Poisson distribution:												
39										Cum		No of	
40						n	n!		P(n)	P(n)		Customers	
41						0	1		2.1e-9	2.1e-9		1	
42						1	1		4.1e-8	4.3e-8		2	
43						2	2		4.1e-7	4.6e-7		3	
44						3	6		2.7e-6	3.2e-6		4	
45						4	24		.00001	.00002		5	
46						5	120		.00005	.00007		6	
47						6	720		.00018	.00026		7	
48						7	5040		.00052	.00078		8	
49						8	40320		.00131	.00209		9	

	A	B	C	D	F	G	H	I	J	K	L	M
50					9	362880		.00291	.00500		10	
51					10	3.63e6		.00582	.01081		11	
52					11	3.99e7		.01058	.02139		12	
53					12	4.79e8		.01763	.03901		13	
54					13	6.23e9		.02712	.06613		14	
55					14	8.7e10		.03874	.10486		15	
56					15	1.3e12		.05165	.15651		16	
57					16	2.1e13		.06456	.22107		17	
58					17	3.6e14		.07595	.29703		18	
59					18	6.4e15		.08439	.38142		19	
60					19	1.2e17		.08884	.47026		20	
61					20	2.4e18		.08884	.55909		21	
62					21	5.1e19		.08461	.64370		22	
63					22	1.1e21		.07691	.72061		23	
64					23	2.6e22		.06688	.78749		24	
65					24	6.2e23		.05573	.84323		25	
66					25	1.6e25		.04459	.88782		26	
67					26	4.0e26		.03430	.92211		27	
68					27	1.1e28		.02541	.94752		28	
69					28	3.0e29		.01815	.96567		29	
70					29	8.8e30		.01252	.97818		30	
71					30	2.7e32		.00834	.98653		31	
72					31	8.2e33		.00538	.99191		32	
73					32	2.6e35		.00336	.99527		33	
74					33	8.7e36		.00204	.99731		34	
75					34	3.0e38		.00120	.99851		35	
76					35	1.0e40		.00069	.99920		36	
77					36	3.7e41		.00038	.99958		37	
78					37	1.4e43		.00021	.99978		38	
79					38	5.2e44		.00011	.99989		39	
80					39	2.0e46		.00006	.99995		40	
81												

82 Source: Adapted from a model in James L. Conger, "Using Financial Tools for
83   Nonfinancial Simulations," BYTE (January 1988), pp. 291-96.

Column L numbers of customers corresponding to each Poisson probability (starting with 1).[8]

3. The third step is to estimate the daily demand by multiplying the number of customers (column H) by the average amount of waste dumped (cell D7). Thus, the formula in cell I4 (for example) is:

    H4*$D$7

4. The amount of waste arriving each day is compared to the remaining capacity of the facility—that is, the capacity shown in cell D4 minus the level at the beginning of the period, as shown in column G. If there is sufficient capacity, all the waste is accepted into the facility; if not, the excess has to be turned away. Thus, the formula in cell J4 (for example) might be

    IF(G4+I4<=$D$4,I4,$D$4-G4)

    However, the same effect can be achieved more elegantly like this:

    MIN($D$4-G4,I4)

    The excess (if any) is recorded in column K; thus, the formula in cell K4 (for example) is

    I4-J4

5. The final step is to compare the level of waste in the facility at the beginning of the period (column G) to the capacity of the mechanism used for emptying or removing waste from the facility (cell D5). If the former is larger than the latter, then emptying is presumed to be warranted and a "flag" is set in column L; if not, a zero is displayed in column L. To achieve this effect, the formula in cell L4 (for example) might be:

    IF(G4>=$D$5,1,0)

---

8. Note that the number of customers corresponds to one more than the integer used to determine the probabilities. This is because of the way lookup tables work in SuperCalc and most other spreadsheets: that is, when an exact match is not found in the lookup table, the function returns a value corresponding to the next *lower* value in the table.

or, more elegantly,

```
G4>=$D$5
```

which is evaluated as a Boolean expression returning a one if true and a zero if false. This is exactly what we want in this column, so it can be used to compute the waste level at the beginning of the following period: see step 1 above.

The results of a series of such calculations—31 of them in the case of the model shown here, one for each possible day of a month—are displayed in an area labeled "Summary of Results" in the upper left corner of the worksheet. Altogether, nine different indicators are shown, each being calculated by means of the following formulas:

Row 10	Mean site level (kg)	AV(G4:G34)
Row 11	Mean no. of customers	AV(H4:H34)
Row 12	Minimum no. of customers	MIN(H4:H34)
Row 13	Maximum no. of customers	MAX(H4:H34)
Row 14	No. of removals required	SUM(L4:L34)
Row 15	No. of periods	MAX(F4:F34)
Row 17	Excess waste (kg)	SUM(K4:K34)
Row 18	No. of customers affected	D17/D7
Row 19	As a % of total customers	D18/SUM(H4:H34)

With these indicators, it should be clear how well the waste-disposal facility is meeting the demand. Naturally, any final assessment of the adequacy of the facility will have to weigh the cost of providing additional capacity against the cost of occasional shortages.

Operation

The model is easy to use. Once the four items of input data are provided, the model automatically displays a set of results for 31 days. Of course, these results are the product of a set of 31 random numbers; so you may want to run the model more than once to see how much the results tend to vary. It is easy to make a graph of the results of any monthly simulation: Figure 10.2 illustrates the results shown in Listing 10.1.

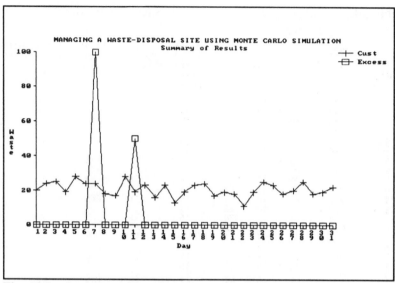

Fig. 10.2. Summary of Results over a 31-Day Run

To simulate an entire year or more, the user need only run the model (i.e., press the recalculation key) the appropriate number of times. A sample of the results that might be obtained over a 24-month period (given the input data shown in Listing 10.1) is presented in Table 10.2 and illustrated in Figure 10.3.[9]

The key point to note from the graph in Figure 10.3 is that problems related to lack of capacity at the facility (as shown in the lower line) need not correspond with peak loadings (as shown in the upper line). This is because the facility is designed to be able to cope with *occasional* peak loadings. Problems arise when there are *sustained* periods of abnormally high loadings. In such cases, the "spare" capacity built in to the site simply gets "used up".

---

9.   Note that these results were compiled manually. The model (as presently constituted) is not designed to record the results from a series of runs.

Table 10.2: Managing a Waste Disposal Site over 24 Months

Months	Mean Monthly Waste (kg)	-- Customers ---			No. of Times Emp- tied	Total No. of Daily Periods	Total Excess Waste (kg)	Total Customers Refused Service	
		Mean	Min.	Max.				No.	% Total
1	1002	19	13	28	15	31	125	5	.8
2	987	19	13	31	15	31	125	5	.8
3	936	19	10	27	15	31	300	12	2.0
4	937	19	13	27	15	31	0	0	.0
5	943	20	11	31	15	31	325	13	2.1
6	1030	20	13	29	15	31	225	9	1.4
7	932	20	12	27	15	31	125	5	.8
8	910	20	14	30	15	31	275	11	1.8
9	981	21	12	27	15	31	350	14	2.2
10	965	21	11	27	16	31	225	9	1.4
11	907	21	12	31	16	31	275	11	1.7
12	1090	19	10	28	15	31	0	0	.0
13	978	21	12	28	16	31	175	7	1.1
14	905	20	12	32	15	31	75	3	.5
15	923	21	9	31	16	31	75	3	.5
16	932	21	12	30	16	31	200	8	1.2
17	938	20	9	27	15	31	0	0	.0
18	1012	20	8	29	16	31	50	2	.3
19	951	20	9	29	15	31	125	5	.8
20	986	21	10	31	16	31	400	16	2.5
21	942	20	11	27	16	31	75	3	.5
22	990	20	11	32	15	31	150	6	1.0
23	949	20	15	31	16	31	225	9	1.4
24	974	20	13	29	15	31	0	0	.0

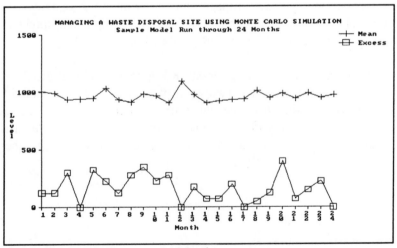

Fig. 10.3. Sample Run over 24 Months

Interpretation and Use

Whether the facility illustrated here is adequate depends on how seriously you regard the fact that some 1% or 2% of customers are refused service. That is, there are a few customers who arrive at the waste-disposal site and have to be told to "come back later" because the site is full. If these customers can go elsewhere or can come again another day without undue inconvenience, then the cost of occasional lack of capacity is probably not going to be high. If, on the other hand, disappointed customers tend to dispose of their waste in illegal or dangerous ways and/or have significant political clout, then you may feel that even a little lack of capacity is a big problem.[10]

In any case, you may want to try adjusting some of the parameters of the model in the upper left corner of the worksheet. For example, you could alter the capacity of the truck or other mechanism used to remove waste from the site from time to time. Clearly, if you *reduce* its capacity, the truck will have to empty the site more frequently, whereas if you *increase* its capacity, the truck will empty the site less frequently. This leads to what may seem at first to be a somewhat surprising result —what Jay Forrester (1968) has called a "counter-intuitive" result. This is that increasing the capacity of the truck makes the situation worse, not better, whereas reducing the capacity of the truck does help, at least up to a point.

If we run the model repeatedly, it soon appears that reducing the capacity of the truck and increasing the frequency of emptying the site does help. That is, it leads to fewer customers being refused service. But, of course, this works only up to a point. Make the capacity of the truck too small and any practical increases in frequency may not be enough to make up for the reduced capacity. On the other hand, increasing the capacity of the removal mechanism and reducing the frequency of emptying the site makes things worse. This is because the site itself now has to get closer to its own limits before emptying takes place. In effect, we have created the classic kind of situation in which nothing gets done until a crisis is imminent.

A second way in which we can use the model is to examine the effect of increasing and decreasing the mean number of customers, leaving the other para-

---

10. Brown and Lebeck 1976 provide a case study of how rural communities dispose of waste in the absence of adequate public services and facilities.

meters as they were. In this case, the effect is not so surprising but its extent may be. Repeated runs of the model show that fairly small changes (e.g., 10%) in the mean number of customers make a major difference to the adequacy of the site. In other words, given the situation defined in Listing 10.1 and illustrated in Figure 10.2, the capacity of the site appears to be very sensitive to the number of customers it is intended to serve.

A third possibility is to increase or decrease the average amount of waste dumped by each customer. Not surprisingly, more waste per customer means the site reaches capacity sooner and that tends to cause more frequent refusals of service. Here, too, the site seems to be quite sensitive to this parameter.

On balance, therefore, you may not be too happy with this particular design for a waste-disposal facility. In the short term, you might be prepared to accept the occasional lack of capacity. But in the long term, you might be worried about the ability of the facility to cope with potential growth in the amount of waste production or even changes in the pattern of waste disposal. After all, there are few signs of an imminent decline in the amount of waste we produce.

Evaluation and Extension

The model shown here could be elaborated in a number of ways. One obvious possibility is to build into the model the ability to run itself a predetermined number of times, record the results of each run, and then present a summary table of results. In fact, this is what was done manually to produce Table 10.2 and Figure 10.3. It would not be difficult to write a macro to issue the recalculate command (say) 12 or 24 times, enter the results in a remote part of the spreadsheet, and then generate an appropriate graph. For the enterprising reader, the macros used in chapters 3 and 11 can serve as a guide for extending the model in this way.

Another way of extending the model would be to change the rules governing the emptying of the site. For example, we could modify the model to provide for a fixed schedule of emptying (e.g., every so many days, regardless of whether this meant a full or partial load for the truck). We could even program the model for various hybrid arrangements, such as one in which the truck had a fixed schedule but could adjust the schedule in cases of emergency (i.e., where the capacity of the site reached a certain level). In short, there are numerous ways of adapting the model presented here so that it suits the specific situation in which it is to be applied.

Thus, Monte Carlo simulation can be a very useful modeling technique. With this technique, random numbers are turned into a kind of "engine" that drives a model through endless runs. Such runs do not necessarily provide answers, of course, because there is nothing optimum about random numbers and there is always the temptation to run the model "one more time." But that is the nature of simulation modeling and, perhaps, of life itself.

Bibliography and References

Blumberg, Louis, and Robert Gottlieb. 1989. *War on Waste: Can America Win Its Battle with Garbage?* Washington, D.C.: Island Press.

Brown, F. Lee, and A.O. Lebeck. 1976. *Cars, Cans, and Dumps: Solutions for Rural Residents.* Baltimore: Johns Hopkins University Press for Resources for the Future.

Conger, James L. 1988. "Using Financial Tools for Nonfinancial Simulations." *BYTE*, 13 (January), pp. 291-96.

Erwin, Lewis, and L. Hall Healey, Jr. 1990. *Packaging and Solid Waste: Management Strategies.* New York: American Management Association.

Forrester, Jay. 1968. *Principles of Systems.* Cambridge, Mass.: MIT Press.

Franklin Associates Ltd. 1988. *Characterization of Municipal Solid Waste in the United States, 1960-2000 (Update 1988).* For the Environmental Protection Agency. Washington: EPA.

Hammersley, J.M., and D.C. Handscomb. 1964. *Monte Carlo Methods.* New York: Wiley.

Kaufman, A. 1968. *Introduction to Operations Research.* Troy, Mich.: Academic Press.

Kleijnen, Jack P.C. 1974, 1975. *Statistical Techniques in Simulation.* 2 vols. New York: Marcel Dekker.

Knuth, Donald E. 1981. *The Art of Computer Programming: Semi-Numerical Algorithms,* vol. 2. Reading, Mass.: Addison-Wesley.

Law, Averill M., and W. David Kelton. 1982. *Simulation Modeling and Analysis*. New York: McGraw-Hill.

Metropolis, Nicholas, and Stanislaw Ulam. 1949. "The Monte Carlo Method." *Journal of the American Statistical Association*, 44, pp. 335-41. Reprinted in *Stanislaw Ulam: Sets, Numbers, and Universes; Selected Works*, edited by W.A. Beyer, J. Mycielski, and G.-C. Rota, paper no. 38, pp. 319-25. Cambridge, Mass.: MIT Press, 1974.

Neumann, John von. 1951. "Various Techniques Used in Connection with Random Digits." Reprinted in *John von Neumann Collected Works*, edited by A.H. Taub, vol. 5, no. 23. New York: Pergamon, 1961.

Packard, Vance. 1960. *The Waste Makers*. New York: McKay.

Rubinstein, Reuven Y. 1981. *Simulation and the Monte Carlo Method*. New York: Wiley.

Russel, Edward C. 1985. *Building Simulation Models with Simscript II.5*. Consolidated Analysis Centers Inc. (CACI), vol. 4. Los Angeles: CACI.

Schriber, Thomas J. 1974. *Simulation using GPSS*. New York: Wiley.

Smith, V. Kerry. 1973. *Monte Carlo Methods: Their Role for Econometrics*. Lexington, Mass.: D.C. Heath.

United States Environmental Protection Agency (USEPA). 1988. *The Solid Waste Dilemma: an Agenda for Action*. 2 vols. EPA/530-sw-89-019. Washington, D.C.: Environmental Protection Agency.

Chapter 11

MULTI-CRITERIA ANALYSIS: AN ENVIRONMENTAL IMPACT ASSESSMENT MODEL

*A Commission was set up to examine the environmental impacts of
the Wealden iron mills and furnaces in Kent and Sussex;
this body consisted of four chief assessors and 16 other investigators. . . .
Most of their evidence was collected from representatives of
the towns and districts: the number of mills, how much wood
they consumed yearly, how much the cost of wood had increased
due to shortages; which towns would suffer economically
from further development; what trades would suffer
from timber shortage; how many jobs would be lost; and
why the price of iron was higher than it was when there were fewer mills.*

C.A. Fortlage, Environmental Assessment: A Practical Guide

We tend to think of concern for the environment as a relatively recent phenomenon, but this is not so. While the scale and scope of environmental impacts may be greater than ever before, the problem itself is far from new. For example, the English commission described above was appointed more than four hundred years ago, in 1548! However, its findings were much like those you would expect from an environmental impact assessment today:

Each mill used about 1,500 loads of wood yearly and made no effort to renew the woodland . . . ; the trades depending on timber for their raw material were in distress . . . ; the fishermen had insufficient wood to build boats . . . ; and more importantly there was not enough large timber for repairing the harbours and houses.

The Commissioners . . . predicted that job losses and community decay would eventually be enormous if no steps were taken to mitigate the effects of further development. Their recommendations included restrictions on tree felling and a reduction in the number of mills; though whether these were actually carried out is not certain [Fortlage 1990, pp. 1-2].

What was untypical about the inquiry in 1548 is that it took little more than two months to complete its work and make its report (Cartwright 1975)!

Nevertheless, in spite of how long we have been grappling with the problem of reconciling public and private interests, appropriate tools remain elusive. In the case of environmental impact assessment, some of the tools are so general as to seem little more than applied common sense; while others are so complex that their results are hard to interpret (Draggan et al. 1984, pp. 9 and 16). The purpose of this chapter is to present something of a compromise. The model presented here is more than just common sense, but thanks to the fact that the computer does all the detailed calculations, the model is relatively easy to understand and use.

The model shown here—which is variously called "optimum pathway matrix analysis" or "probabilistic linear vector analysis"—is not presented as the ultimate in environmental impact assessment tools. But it is a model that is intuitively appealing and, in one form or another, quite widely used.[1] The purpose of this discussion is to show that, in cases where the method is appropriate, it has some important advantages. These advantages are: (a) the model is easy to adjust—to add new impacts or delete old ones, to make some things a bit more important and others a bit less important—and to see what effect this has on the overall assessment; and (b) the computer can do some random "fudging" of the data itself, to see if this has much effect on the overall result. Thus, as well as providing a more dynamic work-space than pencil and paper, the computer can also give some indication of how reliable its results may be.

---

1.   There are also non-probabilistic, non-linear techniques for environmental impact assessment. See, for example, the Multi-Attribute Tradeoff System (MATS) developed by the U.S. Bureau of Reclamation (Brown et al 1986), which is described as a "certainty-based decision analysis program" (page 5) and which relies on "function forms" to describe the different levels of utility (or disutility) that may be derived from different quantities of a given impact.

Purpose of the Model

Environmental impact assessment techniques are typically divided into categories according to the methods they employ (Clark et al. 1980; Biswas and Geping 1987, chapters 1 and 2). These include:

- **Overlay mapping methods**—originally developed by Ian McHarg (1968) to help planners select highway routes and subsequently developed into a general approach to planning (1969)—and other techniques for spatial analysis.

- **Cost-benefit techniques**, adapted from traditional applications in economics to the needs of environmental impact assessment (e.g., Kneese 1984; OECD 1989).

- **Checklists and matrices**, which provide for the systematic examination of a set of causes for a set of possible effects. Among the best known of these methods are the so-called Leopold matrix developed for the U.S. Geological Survey (Leopold et al. 1971), the Environmental Evaluation System developed at Battelle Laboratories (Dee et al. 1973), and a component-interaction matrix used by the Ministry of the Environment in Canada (Environment Canada 1974).

- **Network methods**, which attempt to follow chains of cause and effect in more detail than is possible with two-dimensional checklists or matrices. Invention of this technique is usually credited to Sorensen 1971; the U.S. Forest Service has developed a more sophisticated, computerized version (Thor et al. 1978).

- **Quantitative methods** of various kinds, including so-called "totality indices" (such as those proposed by Odum et al. 1975 and 1976) and impact indices (as developed by Stover 1972).

- **Simulation modeling**, advocated in studies such as Holling 1978 and Munn 1975 as a more dynamic, adaptive, and (therefore) realistic alternative to the encyclopedic approach of so many other techniques. Emphasis in this approach is typically placed on prediction as well as assessment and on the importance of continual, post-assessment monitoring (Holling 1978, p. 1).

There are at least four basic problems involved in trying to assess the environmental impact of any event or set of activities:

Scope   What factors or components should be taken into account in making the assessment?

Measurement   What is the appropriate way to measure the impact of each component or criterion, and how do we forecast what those impacts will be in the future?

Evaluation   What importance or weight should be attached to each component?

Integration   What method is appropriate for combining all these factors into a single scale, so that the overall impact of different events or activities can be calculated and compared.[2]

For some people, these problems are mainly problems of tactics. Do the best you can, and while the results may be less than perfect, they are better than nothing. As one textbook puts it (Midgley and Piachaud 1984; cited in Massam and Skelton 1986, p. 55), "Although it is recognized [that] planning techniques have many limitations . . . we believe they are helpful aids to policy-making which enhance objectivity and efficiency. To reject their use is to deny the need for greater rationality in decision-making." For others, these problems of uncertainty are fundamental. We are dealing with apples and oranges (so to speak), and no matter what we do, the results will vary according to the methods we choose to reconcile their differences.

The model in this chapter aims to provide a reasonably "objective" way of reconciling these two positions, at least to a degree. On the one hand, the model reflects a set of explicit procedures for dealing with each of the four basic sources of complexity set out above. On the other hand, the procedures of this model are so flexible and make so much allowance for the imponderable and the immeasurable that even the most skeptical should be persuaded. Of course, there is a price for this—

---

2.   These four sources of difficulty are similar to the four sources of "uncertainty" proposed by Voogd 1983, pp. 190-91: namely, criterion uncertainty, assessment uncertainty, priority uncertainty, and method uncertainty. See also Cox 1982 for a slightly different trio of uncertainties.

and this is that the model may not always produce an answer. That is, if the alternatives are not sufficiently different from each other in terms of their overall impact, then the model may fail to produce a clear and consistent winner. But perhaps that will add to, rather than detract from, the appeal of this model.

Conceptual Basis

The model presented here is derived from a model developed in the early 1970s by Joseph C. Zieman and his colleagues at the Institute of Ecology of the University of Georgia (Zieman et al. 1971).[3] The model was developed to examine the environmental impact of eight alternative alignments for an extension to an interstate highway (Route I-75) near Marietta, Georgia (see Figure 11.1). The state department of transportation had made a study of the situation and recommended a particular route. However, protests by environmentalists and others led to a review of the decision, and a new study of eight alternative alignments was commissioned from the University of Georgia. In this second study, a new model was developed, which produced a dramatically different result:

A set of 56 variables was used [as criteria] to evaluate each proposed [highway] alignment. The variables addressed four areas of concern:

1. Economics and highway engineering,
2. Environment and land use,
3. Recreation, and
4. Social and human factors.

The study group selected weights that were assigned to each criterion. Using the full set of data and the [new] model, it was concluded that the alternatives could be classified in two distinct sets. The most attractive ones included routes G, G-1, T, and T-1 [all of which lay to the west]. . . . The four easterly alignments (F, F-1, O, and P) were all clearly inferior. . . . [although the] preliminary study [by the state department of transportation] . . . had offered F as the preferred route [Massam and Skelton 1986, p. 55].

---

3.   The use of the model was further described in articles by Eugene Odum and others (1975, 1976); more recently, the model was reviewed in a comparative setting in a short paper by Massam and Skelton (1986).

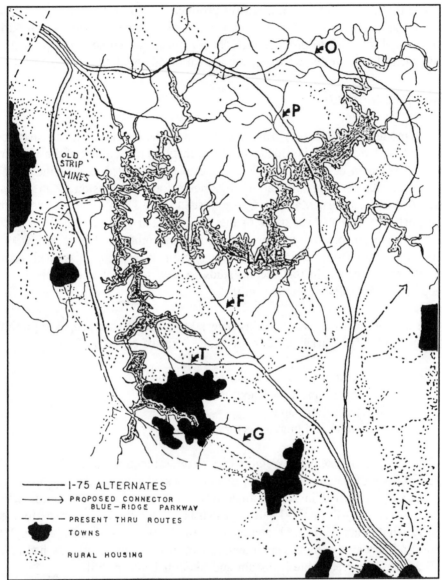

Fig. 11.1. Alternative Routes for Highway I-75 near Marietta, Georgia (Source: J.P. Zieman et al., "Optimum Pathway Matrix Analysis" [1971])

The original model was programmed in PL/1 on an IBM 360/65.[3] The version presented here preserves as much of the original as possible, including use of the data from the I-75 case for illustrative purposes (see below). At the same time, as we shall see, certain features not in the original model have been added to this version, particularly with respect to choosing the technique used for standardization.

The model is based on a simple and flexible method, which can be summarized in terms of four key steps. The first step is to define the problem in three main dimensions:

- There has to be a set of alternative events or options to be assessed. For example, this might be alternative sites for a new waste-disposal site or a new dam; or it might be alternative routes for a highway or electrical transmission line.

- There has to be a list of all the components to be included in the assessment. For example, this might include trees cut down, jobs created, pollution caused, and so on. You do not have to worry at this stage whether the list is complete, since it is easy enough to delete components and/or add new ones later on.

- There has to be a set of "weights" to indicate the relative importance to be attached to the various components in the assessment. For example, loss of agricultural land might be weighted as twice as important as loss of forest land but only half as important as job creation. If desired, several different weights can be specified for each component. For example, there might be present weights and future weights or weights for different interest groups.

The second step is to take each of the options in the first data set and give it a "score" in terms of each of the components in the second data set. For example, option A might be defined as involving the loss of 1,000 ha of agricultural land, the creation of 100 new jobs, and so on. Similar sets of component scores are provided for each of the other options. It does not matter what units or scales are used for measuring these impacts (as long as it is consistent for each component).

---

3. A simulation run with 50 variables on the IBM 360 reportedly took about 3.5 minutes. The spreadsheet version presented here takes about 50 seconds on a 80386 laptop, including the time required to graph the results on the screen!

The third step is to make a couple of adjustments to the component scores and the weights: i.e., they are *standardized* and then *randomized*. The purpose of standardization is clear: it is to remove the differences in units and scales among the various components so that we can, in due course, compare the scores for one component with those for another. There are various techniques of standardization provided in the model, and these are discussed in a technical note at the end of this chapter. The purpose of the second transformation (randomization) is a bit more complicated, but the reason for it is essentially this. Suppose the input data on component scores and weights are unreliable. Suppose we estimate that each datum could be wrong by as much as, say, 50% in either direction. It is not that the input data are systematically wrong; it is just that they may be subject to random errors. In order to simulate the effect of such errors, we can run the model many times, adjusting the component scores and weights each time by a random amount, and average the results. That way, we can calculate not just the average component scores and weights but also a confidence range for each. This confidence range will give us an idea as confident we can be (with, say, 95% probability) of our results, given an estimate of the potential for errors in our input data, no matter how many times we might run the model.

The fourth and final step is to take the component scores for each option, multiply each one by its respective weight, and add up the results.[4] That gives us an overall environmental impact score for each option. If one option has an average weighted component score, standardized and randomized within a 95% confidence range, that lies unambiguously above that of another option, then we can conclude that the first option has a more beneficial (or less harmful) overall impact on the environment than the second. Where the confidence ranges of the two options overlap, no such conclusion is warranted.

Data Requirements

In one sense, the model requires many items of data—but the nature of the data is very flexible. As already discussed, there are four sets of data required:

---

4.   There are alternatives to a weighted summation (see Hobbs et al. 1984, p. 233ff.) and some of them could be built into a model like the one in this chapter.

Options:
a list of the alternatives or options whose impacts are to be assessed and (presumably) compared; in the I-75 case, eight different options were identified.

Components:
a list of the various components to be included in the assessment; in the I-75 case, a total of 56 different components were identified.

Weights:
a set of weights to indicate the relative importance to be attached to each component; in the I-75 case, two weights were provided for each component—"present" or near-term future weights, and longer-term or "future" weights—both of which were derived in various ways and from various sources.

Scores:
a set of scores for each component for each option; in the I-75 case, all the scores were ratio-scale scores, but they represent scales ranging from precise units (such as miles and acres) to educated guesses on an essentially arbitrary scale.[5]

A detailed description of the data used in the I-75 case is provided in the original presentation (Zieman et al. 1971; Odum et al. 1972) and is discussed here no further.

In addition to these four data sets, there are four other choices the user must make in order to run the model. These choices were not required in the original version, since it was programmed to run in one particular way. However, in the version to be presented here, the user must make the following choices:

• the technique used for standardizing the component scores,

• the technique used for standardizing the weights,

• the relative weight attached to "present" and "future" weights, and

• a randomization factor to indicate how much to vary the input data.

---

5.   A ratio scale is a scale in which the units are known and the origin (indicated by the zero value) is known; see Voogd 1983, chapter 5.

The first two parameters involve the selection of a specific technique for standardizing component scores and weights. There are two options for the first and four for the second; as previously noted, these are discussed further in a technical note at the end of the chapter. The third parameter is just a simple way of allowing the user to adjust the relative importance of present versus future interests. For example, giving the present a value of 1 and the future a value of 10 would mean that you regard long-term consequences as ten times more important than short-term ones.

The fourth parameter allows the user to specify how far randomization is to be allowed to affect the weights and scores originally specified. For example, we might begin by specifying a fairly small randomization factor (e.g., .1, or 10%) and seeing which options do best. Then we might increase the factor to, say, .5, or 50% and see how far this puts the results of the first trial in doubt. That is, we can check to see if any of the options that were superior under conditions of low randomization are still clearly superior—even under conditions where the randomization factor is fairly high (i.e., the input data are assumed to be fairly unreliable). To put it another way, the higher the randomization factor that still yields a clear winner, the more confident we can be of choosing that option.

Overview

The model consists of 23 columns (A to W) and 186 rows. It can be divided into five main parts (see Listing 11.1):

- rows 1-67, for data entry, calculation of composite present/future weights, and displaying a summary of the results;

- rows 69-125, for standardization and randomization of component scores;

- rows 127-83 (right-hand side), for multiplication of component scores by composite weights;

- rows 133-60, for holding the results of twenty successive runs of the model; and

- rows 162-72 (left-hand side), that contains a macro to run the model 20 times and graph the results.

In the top part of the model, rows 12-67 provide for entry of the various components of the impact assessment. Component names are entered in column A, their units (if any) in column B, their weights (present and future) in columns D and E, and their scores for each alternative or option in columns G through N. Here, too, the user specifies values for each of the four methodological choices just discussed:

- in cell B5, the technique to be used for standardizing weights (you can specify either technique 1 or 2 from the list of four shown in cells A4:A7);

- in cell B7, the technique to be used for standardizing component scores (you can choose technique 1, 2, 3, or 4 from the list in cells A4:A7);

- in cell E9, the extent to which you want randomization to affect the results (you can specify any value between zero and one); and

- in cells K9 and N9, the relative importance to be attached to short-term and long-term consequences.

Also in the top part of the model, the formulas in cells G5:N7 transfer results obtained elsewhere up to the summary here. Note that the formulas provide for blanking out the error messages that would otherwise be returned by some of the statistical functions during the first few runs of the model (e.g., from trying to calculate the standard deviation of a single observation). Thus, the formula in cell G5 is as follows:

```
IF(ISERROR(G156)," ",G156)
```

which means that, if cell G156 shows an error message, display a blank; otherwise, display the value of cell G156.

Columns P, Q, and R provide for calculation of certain statistical values that are used elsewhere. Similarly, columns U, V, and W provide for calculation of a single, composite (present and future) weight for each component, and then adjust that weight by a random value to the extent specified (in cell E9). Thus, the formula in cell U12 is

$$(D12*\$K\$9+E12*\$N\$9)/(\$K\$9+\$N\$9)$$

Listing 11.1

A	B|D|E	G|H| I |J|K|L|M|N	P|Q|R|S|T|U|V|W

ENVIRONMENTAL IMPACT ASSESSMENT USING PROBABILISTIC LINEAR VECTOR ANALYSIS   © 1993 by T.J. Cartwright

			G	G1	T	T1	F	F1	P	O					
Standardization Method:	Option >>		G	G1	T	T1	F	F1	P	O	Statistical Summary				Present/Future
1 Raw/Max	Wts										of Scores for				Composite Wts
2 Raw/Sum	2	Mean :	.284	.686	.075	.272	-.59	-.31	-1.1	-1.0	All Options in				
3 Raw-Min/Max-Min	Sco	StDev:	.455	.436	.441	.628	.675	.592	.390	.292	Each Component				Raw Max AbSum
4 Raw-Avg/StdDev	1	StErr:	.107	.103	.104	.148	.159	.139	.092	.069					Data: 45.5 48.5
											Max	Min	Sum	Std	
Now iterating . . . 20	Rnd: .5	Relativ Wt -- Pres: 1 Future: 10												Dev	Raw Strd Rand
															Comp ized ized

Components:	Unit	Pres	Fut	G	G1	T	T1	F	F1	P	O	Max	Min	Sum	Std Dev	Raw Comp	Strd ized	Rand ized
Pine lost	ha	-3	-10	1407	1411	1438	1442	1352	1356	811	826	1442	811	1e4	254.	-9.4	-.19	-.21
Mixed lost	ha	-4	-10	1619	1619	1406	1406	1092	1092	1263	1445	1619	1092	1e4	193.	-9.5	-.19	-.29
Hardwood lost	ha	-7	-10	215	226	277	288	344	355	152	164	355	152	2021	71.4	-9.7	-.20	-.21
Agric lost	ha	-3	-10	313	282	305	275	325	294	257	279	325	257	2330	20.9	-9.4	-.19	-.11
Idle lost	ha	5	8	227	237	175	185	147	157	61	51	237	51	1240	64.2	7.73	.159	.173
Water lost	ha	-7	-4	17	27	23	33	22	32	52	19	52	17	225	10.5	-4.3	-.09	-.10
Swamp lost	ha	-4	-4	0	0	0	0	0	0	0	2	2	0	2	.661	-4	-.08	-.06
Mined land lost	ha	8	10	69	69	68	68	0	0	0	0	69	0	274	34.3	9.82	.202	.222
Urban land lost	ha	-6	10	212	199	175	156	68	40	22	30	212	22	902	75.3	8.55	.176	.171
Water sources lost	#	-2	0	1	1	1	1	0	0	0	0	1	0	4	.5	-.18	>>>>	-.01
Unique areas lost	#	-2	-2	0	0	0	0	0	0	1	1	1	0	2	.433	-2	-.04	-.04
Streams crossed	#	-2	-2	29	27	28	36	36	24	23	24	36	23	227	4.82	-2	-.04	-.03
Small abridgements	#	-2	-2	6	4	4	2	2	0	0	0	6	0	18	2.11	-2	-.04	-.03
Major bridges	#	-5	-5	0	0	0	0	1	1	1	0	1	0	3	.484	-5	-.10	-.09
Total length	m	-5	-5	0	0	0	0	1400	1400	3100	0	3100	0	5900	1071	-5	-.10	-.08
Other bridges	#	-3	-3	2	3	2	2	1	2	0	2	3	0	14	.829	-3	-.06	-.04
Total length	m	-3	-3	1600	2100	1700	2300	500	1000	0	1500	2300	0	1e4	736.	-3	-.06	-.06
Composite soil limits		-4	-3	293	293	300	300	320	320	328	303	328	293	2457	12.7	-3.1	-.06	-.07
Max possible sediment		-6	-3	315	313	295	292	350	397	270	290	397	270	2522	38.0	-3.3	-.07	-.10
Min expected sediment		-8	-3	44	44	40	40	48	48	37	40	48	37	341	3.77	-3.5	-.07	-.05
Paved area	km2	-2	-2	675	666	649	640	553	644	519	609	675	519	4955	52.2	-2	-.04	-.04
Major noise	km2	-7	-4	103	103	90	90	63	63	64	73	103	63	649	16.3	-4.3	-.09	-.06
Minor noise	km2	-7	-7	33	33	50	50	84	84	95	66	95	33	495	22.5	-7	-.14	-.14
Total system cost	$m	-20	0	108	101	103	95	106	98	89	82	108	82	782	8.24	-1.8	-.04	-.02
Annual costs	$m	0	-20	8.7	8.3	8.5	7.8	8.6	8	7.2	6.8	8.7	6.8	63.9	.643	-18.	-.37	-.38
Total excavation	m3	-2	0	25.3	25	25.5	25.3	20.6	20.1	20.9	25.5	25.5	20.1	188.	2.33	-.18	>>>>	-.01
Road user costs/yr	$m	-10	-10	90.8	91.9	90	91.1	88.5	89.1	94.2	104.	104.	88.5	739.	4.57	-10	-.21	-.16
Benefit/cost ratio		-10	-10	-17.	-9.6	-9.8	-3.2	0	-.9	-4.1	-8.2	0	-17.	-53.	5.31	-10	-.21	-.19
Trunk roads	km	-2	-2	27.1	27.1	27.1	27.1	24.5	24.5	27.4	31.9	31.9	24.5	217.	2.14	-2	-.04	-.04
Taxbase lost	ha	-8	-8	3955	3995	3737	3736	3039	3084	2557	2808	3995	2557	3e4	521.	-8	-.16	-.24
Public space lost	ha	-8	-8	242	197	200	204	401	338	151	92	401	92	1825	92.9	-8	-.16	-.12
Families displaced	#	-20	0	177	99	177	97	146	66	101	53	177	53	916	44.3	-1.8	-.04	-.03
Day noise, minor	#	-2	-2	3072	3015	3222	3432	1555	1667	1956	2521	3432	1555	2e4	693.	-2	-.04	-.03
Day noise, major	#	-5	-5	789	797	2088	1901	1368	1076	998	975	2088	789	9992	465.	-5	-.10	-.09
Nite noise, minor	#	-4	-4	1431	1072	1185	826	958	599	637	598	1431	598	7306	285.	-4	-.08	-.08
Nite noise, major	#	-10	-10	174	215	470	513	263	306	193	268	513	174	2402	118.	-10	-.21	-.20
Church noise, min	#	-2	-2	10	8	6	7	6	7	5	7	10	5	56	1.41	-2	-.04	-.06
Church noise, maj	#	-5	-5	2	2	2	3	4	4	3	3	4	2	23	.781	-5	-.10	-.10
School noise, min	#	-3	-3	2	1	0	0	0	0	0	3	3	0	6	1.09	-3	-.06	-.04
School noise, maj	#	-10	-10	0	0	1	6	1	1	1	0	6	0	10	1.85	-10	-.21	-.18
Lives saved/yr 1	#	50	0	16	16	42	42	82	82	5	5	82	5	290	29.6	4.55	.094	.057
Lives saved/yr 12	#	0	50	377	377	389	389	385	385	301	305	389	301	2908	35.2	45.5	.936	1.10
Interchanges	#	2	6	16	16	14	14	13	13	10	10	16	10	106	2.17	5.64	.116	.076
Second growth potential		0	25	10	10	10	10	8	8	4	6	10	4	66	2.11	22.7	.468	.332
Second growth suitabilit		0	25	8	9	8	9	-7	-6	-9	-5	9	-9	7	7.70	22.7	.468	.292
Water quality		5	2	-1	-2	-1	-2	-2	-3	-2	-2	-1	-3	-15	.599	2.27	.047	.049
Visual disturbance		2	2	-2	-3	-2	-3	-5	-6	-5	-2	-2	-6	-28	1.5	2	.041	.051
Hunting and game		4	1	0	-1	0	-1	-6	-7	-6	-4	0	-7	-25	2.76	1.27	.026	.021
Loss of nat character		4	4	-1	-2	-1	-2	-4	-5	-3	-2	-1	-5	-20	1.32	4	.082	.100
Safe access		6	6	6	7	6	7	6	7	4	2	7	2	45	1.65	6	.124	.157
Reservoir impact		1	2	2	3	2	3	3	4	2	-3	4	-3	12	2.40	1.91	.039	.057

	A	B	D	E		G	H	I	J	K	L	M	N		P	Q	R	S	T	U	V	W
63	Driving pleasure	2	1			3	4	3	4	8	9	8	7		9	3	46	2.33		1.09	.022	.021
64	Noise on camping	-3	-3			44	44	48	48	89	89	51	23		89	23	436	21.5		-3	-.06	-.07
65	Noise on picnic	-2	-2			40	40	87	87	26	26	17	9		87	9	332	28.0		-2	-.04	-.02
66	Recreatn loss, actual	5	0			-3	-4	-1	-2	-6	-7	-4	-3		-1	-7	-30	1.85		.455	.009	.005
67	Recreatn loss, potential	0	3			-1	-2	-1	-2	-4	-5	-5	-3		-1	-5	-23	1.54		2.73	.056	.059

		G	G1	T	T1	F	F1	P	O		G	G1	T	T1	F	F1	P	O
69		G	G1	T	T1	F	F1	P	O		G	G1	T	T1	F	F1	P	O
70		.976	.979	.997	1	.938	.940	.562	.573		.883	1.07	1.34	.930	1.18	1.25	.482	.599
71		1	1	.868	.868	.674	.674	.780	.893		.687	.878	.598	.884	.825	.597	.507	1.15
72	Left Table:	.606	.637	.780	.811	.969	1	.428	.462		.340	.765	.919	1.04	.792	1.14	.399	.627
73		.963	.868	.938	.846	1	.905	.791	.858		.597	.467	1.22	.805	.941	.621	.668	.494
74	Standardized Component Scores	.958	1	.738	.781	.620	.662	.257	.215		.889	.973	.470	.460	.382	.712	.325	.154
75		.327	.519	.442	.635	.423	.615	1	.365		.192	.308	.492	.673	.267	.600	.877	.250
76	Computed using one of the	0	0	0	0	0	0	0	1		0	0	0	0	0	0	0	.794
77	following calculations:	1	1	.986	.986	0	0	0	0		1.30	1.40	.504	.942	0	0	0	0
78		1	.939	.825	.736	.321	.189	.104	.142		.737	1.06	.529	.863	.200	.164	.079	.150
79	(1) Raw/Max	1	1	1	1	0	0	0	0		.819	1.34	.970	1.30	0	0	0	0
80	(2) Raw/Sum	0	0	0	0	0	0	1	1		0	0	0	0	0	0	1.18	.962
81	(3) (Raw-Min)/(Max-Min)	.806	.75	.778	1	1	.667	.639	.667		.990	.444	.714	.670	.715	.907	.689	.472
82	(4) (Raw-Avg)/Std Dev	1	.667	.667	.333	.333	0	0	0		.574	.476	.771	.191	.264	0	0	0
83		0	0	0	0	1	1	1	0		0	0	0	0	1.11	1.06	1.21	0
84	as specified in cell B7	0	0	0	0	.452	.452	1	0		0	0	0	0	.578	.442	1.42	0
85		.667	1	.667	.667	.333	.667	0	.667		.481	1.17	.969	.498	.487	.443	0	.689
86		.696	.913	.739	1	.217	.435	0	.652		.854	1.29	.616	1.23	.161	.333	0	.874
87		.893	.893	.915	.915	.976	.976	1	.924		.466	1.02	1.12	.677	.489	.577	.871	.860
88	Right Table:	.793	.788	.743	.736	.882	1	.680	.730		1.08	1.16	.527	.477	.961	1.23	.843	.675
89		.917	.917	.833	.833	1	1	.771	.833		1.32	1.31	.966	.732	.834	1.31	1.07	1.14
90	Randomized, Standardized	1	.987	.961	.948	.819	.954	.769	.902		1.30	1.34	.684	.873	.927	1.37	.601	.610
91	Component Scores	1	1	.874	.874	.612	.612	.621	.709		.501	.574	.673	1.13	.763	.529	.658	.908
92		.347	.347	.526	.526	.884	.884	1	.695		.420	.448	.614	.438	.887	1.19	1.03	.512
93	Computed by adjusting the	1	.935	.954	.880	.981	.907	.824	.759		1.06	1.30	1.11	.987	1.22	.692	1.07	.703
94	Standardized Component Score	1	.954	.977	.897	.989	.920	.828	.782		.788	1.41	1.31	.473	.989	1.00	1.16	.449
95	(in the Left Table) by a random	.992	.980	1	.992	.808	.788	.820	1		.759	.664	.714	1.28	1.02	1.17	.866	1.09
96	factor drawn from the range,	.876	.886	.868	.878	.853	.859	.908	1		.605	1.18	.992	.800	.743	.680	.844	1.18
97	unity plus and minus the value																	
98	in cell E9	.850	.850	.850	.850	.768	.768	.859	1		.755	.912	1.02	.750	1.00	.689	.922	1.24
99		.990	1	.935	.935	.761	.772	.640	.703		1.02	1.31	1.01	1.17	.644	.684	.532	.614
100		.603	.491	.499	.509	1	.843	.377	.229		.704	.440	.498	.451	.759	.753	.260	.309
101		1	.559	1	.548	.825	.373	.571	.299		1.26	.341	1.17	.447	.771	.336	.446	.222
102		.895	.878	.939	1	.453	.486	.570	.735		.670	.482	.485	.535	.421	.598	.485	1.02
103		.378	.382	1	.910	.655	.515	.478	.467		.493	.398	1.28	.844	.550	.482	.381	.280
104		1	.749	.828	.577	.669	.419	.445	.418		1.00	.723	.496	.541	.809	.457	.268	.527
105		.339	.419	.916	1	.513	.596	.376	.522		.248	.233	1.34	.754	.627	.550	.472	.444
106		1	.8	.6	.7	.6	.7	.5	.7		1.22	.592	.337	.892	.779	.852	.577	.882
107		.5	.5	.5	.75	1	1	.75	.75		.729	.645	.655	.564	1.07	1.26	.720	.795
108		.667	.333	0	0	0	0	0	1		.979	.206	0	0	0	0	0	1.04
109		0	0	.167	1	.167	.167	.167	0		0	0	.141	.818	.200	.117	.155	0
110		.195	.195	.512	.512	1	1	.061	.061		.176	.199	.696	.387	1.40	.654	.042	.073
111		.969	.969	1	1	.990	.990	.774	.784		.877	1.03	.952	.946	1.18	.702	.471	.854
112		1	1	.875	.875	.813	.813	.625	.625		1.08	1.46	.867	.759	1.15	.537	.414	.516
113		1	1	1	1	.8	.8	.4	.6		1.28	.758	.779	1.32	.660	.905	.592	.607
114		.889	1	.889	1	-.78	-.67	-1	-.56		1.30	1.05	.753	.838	-.55	-.42	-1.5	-.59
115		1	2	1	2	2	3	2	2		.657	1.30	.959	1.78	1.10	4.39	2.17	1.74
116		1	1.5	1	1.5	2.5	3	2.5	1		.782	1.08	1.30	.913	2.72	2.86	2.74	1.14
117																		
118		1	2	1	2	4	5	3	2		1.29	1.68	1.08	2.33	2.17	4.19	2.12	2.25
119		.857	1	.857	1	.857	1	.571	.286		.547	.791	.740	.860	.457	.590	.423	.148
120		.5	.75	.5	.75	.75	1	-.5	-.75		.447	.488	.631	.475	.871	1.21	-.44	-.85
121		.333	.444	.333	.444	.889	1	.889	.778		.279	.307	.448	.364	1.27	.560	.836	.484
122		.494	.494	.539	.539	1	1	.573	.258		.428	.558	.611	.335	.675	1.40	.366	.197
123		.460	.460	1	1	.299	.299	.195	.103		.561	.664	1.45	.892	.431	.343	.114	.063
124		3	4	1	2	6	7	4	3		4.17	3.13	.887	1.73	6.35	9.36	4.78	4.21
125		1	2	1	2	4	5	5	3		.518	1.46	.539	2.78	2.45	3.32	4.96	3.16

Row	A	D	E	G	H	I	J	K	L	M	N	P	Q	R	S	T	U	V	W
127	Index of Relative Attractiveness											G	G1	T	T1	F	F1	P	O
128												-.19	-.23	-.28	-.20	-.25	-.26	-.10	-.13
129	Computed by multiplying the											-.20	-.25	-.17	-.25	-.24	-.17	-.14	-.33
130	Randomized, Standardized Component Scores (in the Right Table) by the											-.07	-.16	-.19	-.22	-.16	-.24	-.08	-.13
131	Randomized, Standardized Composite Weights (in cells W12:W67)											-.06	-.05	-.13	-.08	-.10	-.07	-.07	-.05
132												.154	.169	.081	.080	.066	.123	.056	.027
133	Summary of			G	G1	T	T1	F	F1	P	O	-.02	-.03	-.05	-.07	-.03	-.06	-.09	-.02
134	Results	Current		.550	.260	-.64	.411	-.34	-.41	-1.1	-.55	0	0	0	0	0	0	0	-.05
135	for 20 Runs	Run	1	.147	1.24	.614	.801	-.89	1.22	-.36	-.07	.288	.311	.112	.209	0	0	0	0
136		Run	2	.628	1.03	.311	.830	-.70	-1.2	-1.8	-.86	.126	.182	.091	.148	.034	.028	.014	.026
137		Run	3	-.39	.591	.342	-.04	-1.5	-1.4	-1.6	-1.5	>>>>	-.01	-.01	-.01	0	0	0	0
138		Run	4	.709	.497	-.14	.837	-1.0	-.42	-.93	-1.2	0	0	0	0	0	0	-.05	-.04
139		Run	5	-.11	.804	.655	-.66	-.69	-.82	-1.5	-1.4	-.03	-.01	-.02	-.02	-.02	-.03	-.02	-.01
140		Run	6	.297	1.29	.860	.767	.401	-.57	-.49	-.65	-.02	-.01	-.02	-.01	-.01	0	0	0
141		Run	7	.626	.861	.223	.308	.582	-.36	-.58	-.78	0	0	0	0	-.10	-.09	-.11	0
142		Run	8	.145	1.10	-.32	-.34	-1.1	.115	-1.5	-1.3	0	0	0	0	-.05	-.04	-.12	0
143		Run	9	-.14	-.05	-.06	-.05	-1.3	-.13	-1.2	-1.1	-.02	-.05	-.04	-.02	-.02	-.02	0	-.03
144		Run	10	.190	.602	.590	1.67	.112	.647	-1.1	-.92	-.05	-.08	-.04	-.07	-.01	-.02	0	-.05
145		Run	11	1.18	1.30	.993	.621	.530	.448	-1.2	-.61	-.03	-.07	-.07	-.04	-.03	-.04	-.06	-.06
146		Run	12	-.13	.411	.007	-.25	-.73	-.97	-.97	-1.3	-.10	-.11	-.05	-.05	-.09	-.12	-.08	-.06
147		Run	13	-.49	.149	-.34	-.46	-1.2	-.88	-1.5	-1.1	-.07	-.07	-.05	-.04	-.05	-.07	-.06	-.06
148		Run	14	.356	1.10	-.12	.959	-.13	-.57	-1.4	-.56	-.05	-.05	-.02	-.03	-.03	-.05	-.02	-.02
149		Run	15	.910	1.28	-.18	.835	-.88	.591	-.51	-.66	-.03	-.03	-.04	-.07	-.04	-.03	-.04	-.05
150		Run	16	.106	-.06	-.51	-.63	-1.4	-.78	-1.5	-1.3	-.06	-.06	-.09	-.06	-.13	-.17	-.15	-.07
151		Run	17	.993	.642	.076	.965	-.14	.400	-.43	-.96	-.02	-.02	-.02	-.02	-.02	-.01	-.02	-.01
152		Run	18	.374	1.03	-.02	.154	-.10	.368	-1.0	-.63	-.30	-.54	-.50	-.18	-.38	-.39	-.44	-.17
153		Run	19	.462	.671	-.10	-.01	-1.2	-.53	-1.3	-.97	>>>>	>>>>	>>>>	-.01	-.01	-.01	>>>>	-.01
154		Run	20	.017	.129	-.61	.221	-.84	-.67	-1.6	-1.2	-.10	-.19	-.16	-.13	-.12	-.11	-.13	-.19
155																			
156		Mean		.284	.686	.075	.272	-.59	-.31	-1.1	-1.0	-.03	-.04	-.04	-.03	-.04	-.03	-.04	-.05
157		StdDev		.455	.436	.441	.628	.675	.592	.390	.292	-.25	-.31	-.24	-.28	-.15	-.16	-.13	-.15
158		Std Err		.107	.103	.104	.148	.159	.139	.092	.069	-.08	-.05	-.06	-.05	-.09	-.09	-.03	-.04
159		Hi Conf		.508	.901	.292	.581	-.26	-.02	-.93	-.86	-.04	-.01	-.04	-.02	-.03	-.01	-.02	-.01
160		Lo Conf		.060	.471	-.14	-.04	-.92	-.60	-1.3	-1.1	-.02	-.01	-.01	-.02	-.01	-.02	-.01	-.03
161												-.05	-.04	-.12	-.08	-.05	-.04	-.04	-.03
162	Macro to Run the Model 20 Times and Graph the Results											-.08	-.06	-.04	-.04	-.07	-.04	-.02	-.04
163												-.05	-.05	-.27	-.15	-.12	-.11	-.09	-.09
164		\S	{home}{paneloff}					home cursor				-.07	-.04	-.02	-.05	-.05	-.05	-.03	-.05
165			/b G135:N154 ~					blank previous run				-.07	-.07	-.07	-.06	-.11	-.13	-.07	-.08
166			{let B9,0}					zero counter				-.04	-.01	0	0	0	0	0	-.04
167			/c [5;B9+135],B9,v					[E167] increment				0	0	-.03	-.14	-.04	-.02	-.03	0
168			/c G134:N134,[7;B9+135],v					copy current results				.010	.011	.040	.022	.080	.037	.002	.004
169			!					recalc				.963	1.13	1.05	1.04	1.29	.771	.517	.938
170			{if B9<20}{branch E167}					if counter<20, loop				.082	.111	.066	.058	.088	.041	.031	.039
171			{panelon}					dialog on				.424	.252	.258	.438	.219	.301	.197	.201
172			{beep}{view}					signal and graph				.381	.306	.220	.245	-.16	-.12	-.43	-.17
173												.031	.064	.047	.088	.054	.216	.107	.086
174												.040	.056	.067	.047	.140	.147	.141	.058
175																			
176												.130	.169	.109	.234	.218	.420	.212	.226
177												.086	.124	.116	.135	.072	.092	.066	.023
178												.025	.028	.036	.027	.050	.069	-.03	-.05
179												.006	.006	.009	.008	.027	.012	.017	.010
180												-.03	-.04	-.04	-.02	-.05	-.10	-.03	-.01
181												-.01	-.02	-.03	-.02	-.01	-.01	>>>>	>>>>
182												.021	.016	.004	.009	.032	.047	.024	.021
183												.030	.086	.032	.163	.144	.195	.292	.186
184																			
185	Source: Adapted from a model in Joseph Zieman et al., "Optimum Pathway Matrix Analysis Approach to the Environmental																		
186	Decision Making Process" (Athens, GA: Institute of Ecology, University of Georgia, 1971).																		

and just calculates a weighted average of the present and future weights. In order to standardize this average, the formula in cell V12—

```
IF($B$5=1,U12/V$7,U12/W$7)
```

—checks to see which technique is specified in cell B5 and then makes the appropriate calculation. Note that only techniques 1 and 2 are available for standardizing component weights. The third formula randomizes the standardized result obtained in cell V12 to the extent specified in cell E9. The following formula is used:

```
$V12*(1-$E$9+2*$E$9*RANDOM)
```

The next part of the worksheet (rows 69-125) deal with component scores rather than weights. In the left-hand table, the component scores are standardized; in the right-hand table, the standardized component scores are also randomized. Thus, the formulas in cells G70 and P70 (for example) are, respectively, as follows:

```
IF(ISERROR(1/$P12/$R12/($P12-$Q12)/$S12)=1," ",
    IF($B$7=1,G12/$P12,IF($B$7=2,G12/$R12,
    IF($B$7=3,(G12-$Q12)/($P12-$Q12),
    (G12-AV($G12:$N12))/$S12))))

IF(G70=" "," ",G70*(1-$E$9+2*$E$9*RANDOM))
```

The first formula (for standardization) checks for zeros in the statistical values computed in cells P12:S12; then it checks to see which method of standardization is specified in cell B7 (any one of the four techniques may be selected) and performs the appropriate operation. The second formula randomizes the results obtained in the first formula to the extent specified in cell E9.

The third part of the model—rows 127-83 on the right-hand side—computes what is called an "index of relative attractiveness" for each component under each option. This index is calculated by multiplying the component score (standardized and randomized) by the component weight (also standardized and randomized). As before, note that provision is made for blanking out any error messages that may have arisen from earlier calculations. Thus, the formula in cell P128 (for example) is as follows:

```
IF(P70=" "," ",P70*$W12)
```

The fourth part of the model—rows 133-60 on the left-hand side—is for recording the results of up to 20 runs of the model (prior to copying them up to the summary at the top of the worksheet). Thus, in column G, for example, the following formula is found in cell G134:

```
SUM(P128:P183)
```

Similar formulas are found in cells H134:N134. Further down in the same part of the model, certain descriptive statistics are computed for all of the runs. The formulas in cells G156:G160, for example, are as follows:

```
Mean                        AV(G137:G154)
Standard deviation          STD(G137:G154)
Standard error              G157/SQRT(COUNT(G137:G154))
Upper confidence limit      G156+2.093*G158
Lower confidence limit      G156-2.093*G158
```

Finally, the fifth part of the model, in the bottom left-hand corner, contains a short macro to run the model 20 times, store and summarize the results of each run, and then graph them. The macro works by first blanking out the results of any previous run and zeroing a counter; then the macro increments the counter, stores the results of the run, and loops back to do the same thing again (unless the counter has reached 20). After the macro has run 20 times, it displays a preprogrammed graph to illustrate the results.

Operation

The model in Listing 11.1 contains all the original data on the alignment of Route I-75 near Marietta (Georgia). So let us use the model to illustrate what happened in that case.[6] Begin by adjusting the parameters of the model to suppress any randomization effect (i.e., set the value in cell E9 to a small number, such as .1, 10%) and setting to zero the importance attached to future interests. Then invoke the macro to run the model (Alt-S).

---

6.   It proved impossible to exactly reproduce the results shown in the original study (Zieman et al. 1972). Even allowing for the inevitable differences caused by randomization, it is not clear how the results shown in the original study were obtained with the program shown there.

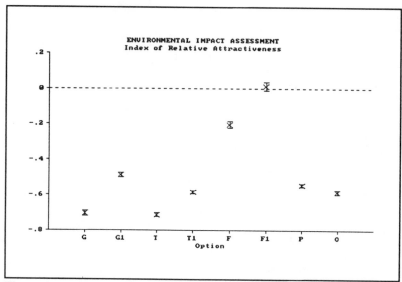

Figure 11.1: Low Randomization; High Present Weight

You should get a result similar to that shown in Figure 11.2, which favors option F or F1. This is exactly what the Georgia Department of Transportation concluded, before Zieman and his colleagues arrived on the scene. Try adjusting the randomization factor in cell E9 upwards, to see if this result (in favor of option F or F1) is sensitive to possible errors in the input data. You should discover that it is not, at least not up to randomization levels of .5, or 50%.

Next, set the randomization factor back to, say, .1, and reverse the relative importance of present and future weights: i.e., set the first to zero and the second to one. In this case, you should find (just as Zieman and his colleagues did) that, while option F or F1 is still the preferred easterly alignment, *all* of the westerly routes (options G, G1, T, and T1) are preferable to *any* of the easterly ones. (See the map again, in Figure 11.1, above.) This result is illustrated in Figure 11.3. You can vary the randomization factor again, but the results appear to be relatively insensitive to randomization levels up to .5. In the end, Zieman and his colleagues used a randomization factor of .5 and present/future weights ratio of 1:10 to argue for a westerly alignment—specifically, option G1 or option T1. This result is easily duplicated with the model presented here: see Figure 11.4.

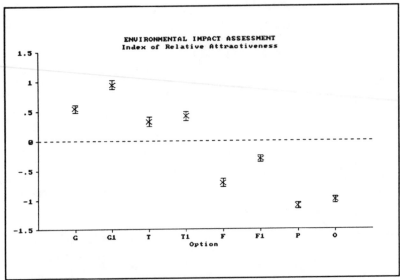

Fig. 11.2. Low Randomization; High Future Weight

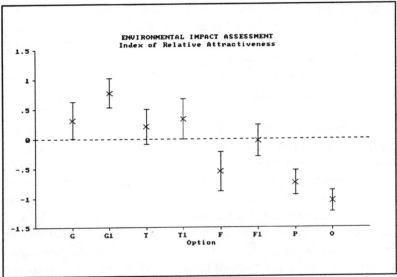

Fig. 11.3. High Randomization (.5); High Future Weight

Interpretation and Use

With the version of the model presented here, we can go one step further than Zieman et al., since we can explore the effect of varying the standardization techniques used.

For example, suppose we change the technique for standardizing component *weights* from technique 2 to technique 1 (i.e., from "raw/sum" to "raw/maximum").[7] The results should show that option G1 is still the preferred alignment—but that rejection of options T and T1, particularly relative to option F1, is more problematical than Zieman and his colleagues were willing to recognize. This is because (as explained in the technical note, below) one of the effects of using the raw/sum technique is to attenuate the influence of higher values relative to that of lower ones. In the I-75 case, as it turned out, that worked out to the relative disadvantage of options T and T1.

Continue to use technique 1 (raw/maximum) for standardizing the component weights, and change the method applied to component *scores* from technique 1 to technique 3 (raw/range). This time you should get a very clear division between the westerly alignments (all superior) and the easterly alignments (all inferior). Moreover, with this technique, you may notice that *lower* scores tend to enjoy narrower confidence bands than high scores. However, purists may argue that technique 3 is more appropriate for pair-wise comparisons than for cases (like the I-75) where there are multiple options.

Next, you can try using technique 4 (the normalized raw value) for the component scores. Once again, you should see a very clear division between the superior westerly and the inferior easterly alignments. You may also notice that, with this technique, *higher* scores tend to have narrower confidence bands. Again, however, purists may argue that use of this technique is not justified here, because the shape of the underlying distribution cannot really be assumed in this case.

In short, the model provides a great deal of flexibility for exploring the sensitivity of an impact assessment to some of the parameters that define it. But

---

7. See the technical note beginning on the next page for more discussion of these formulas.

perhaps the most striking feature of the model, at least for those who have worked with other kinds of impact assessment tools, is the "free form" allowed for the input data. You can enter as many components as you like, and you can score them in terms of whatever units or scales you like. There are no restrictions. This is not to suggest that data inputs will not affect the results produced by model: of course, they will. But the model itself does not require or prohibit the use of any particular kinds of data. The model can cope with almost any data imaginable, from precise technical units (such as hectares or decibels) to highly subjective "coefficients" or scales of approval or acceptability.

Evaluation and Extension

The model presented here is not intended to serve as the ultimate tool for environmental impact assessment. Probably no single technique will ever prove suitable for every occasion (Faludi and Voogd 1985). But this model does serve at least three useful purposes. First, it illustrates how computerization can enhance a fairly simple, well known technique for environmental impact assessment. Second, the model clarifies some of the assumptions entailed in making decisions about complex, multi-criteria problems like impact assessment. Third, the model demonstrates how simple statistical techniques (such as standardization and randomization) can, at least in some cases, reduce the ambiguity and uncertainty surrounding policy evaluation. That may not be everything, but it is a help.

Technical Note

Standardization is a mathematical technique for transforming values on two or more different scales into values on one common scale (usually the interval from zero to one). For example, suppose we have three options (options A, B, and C), whose impacts include the loss of forest resources (measured in hectares) and the gain of economic benefits (measured in the number of new jobs created). Such a situation is presented in Table 11.1. The object of standardizing the measurements of the two impacts of each option is to be able to compare them in some reasonably objective way.

There are several recognized techniques for standardization. Four of them are used in the model discussed here. These four are known as: raw/maximum, raw/sum, raw/range, and normalized raw.

1. Raw/Maximum:

This technique works by setting the maximum value to unity and scaling all the other values accordingly. That is, you divide each raw value by the maximum value. Thus, the standardized ith value, $V_i{}'$, is given by the formula

$$V_i' = \frac{V_i}{V_{max}}$$

where V_i is the ith raw value and V_{max} is the maximum raw value. Use of this technique means that there will always be at least one standardized value equal to unity (i.e., the maximum). Moreover, any raw values of zero will also have standardized values of zero; but if there are no zero values in the raw data, there will be none in the standardized data.

2. Raw/Sum

This technique works by dividing each value by the sum of all the values. Thus, the standardized ith value, $V_i{}'$, is given by the formula

$$V_i' = \frac{V_i}{V_{sum}}$$

where V_i is the ith raw value and V_{sum} is the sum all the values. The effect of this technique is to make all the standardized values lower than they would be, had the previous technique (raw/maximum) been used. Use of the raw/sum technique means that there cannot be a standardized value of unity, since even the maximum value will be less than the sum of all values. As before, if there are zero values in the raw data, their standardized values will also be zero, but if there are no zero values in the raw data, there will be none in the standardized data.

3. Raw/Range

This technique works by dividing the difference between each raw value and the minimum raw value by the range between the maximum and minimum values. In other words, instead of "mapping" the interval from zero to the maximum value to the interval from zero to unity (as in the raw/maximum technique), we take the range from the minimum value to the maximum value and map that to the interval from zero to unity. Thus, the standardized ith value, $V_i{}'$, is given by the formula

Table 11.1:   Hypothetical Component Scores for Three Options, Showing the Effect of Four Different Transformations

	Option A	Option B	Option C
Forest lost			
Raw data (ha)	1,000	2,000	2,500
Raw/maximum	.40	.80	1.00
Raw/sum	.18	.36	.45
Raw/range	0.00	.67	1.00
Normalized raw	-1.34	.27	1.07
Economic benefits			
Raw data (new jobs)	600	400	500
Raw/maximum	1.00	.67	.83
Raw/sum	.40	.27	.33
Raw/range	1.00	0.00	.50
Normalized raw	1.22	-1.22	0.00

$$V_i' = \frac{V_i - V_{min}}{V_{max} - V_{min}}$$

where V_i is the ith value and V_{max} and V_{min} are, respectively, the maximum and minimum values of the range. Use of this technique means there is always at least one raw value (i.e., the minimum) that is standardized into a value of zero and at least one raw value (i.e., the maximum) that is standardized into a value of unity.

4. Normalized Raw

This technique works by *normalizing* the raw values, instead of just standardizing them. Normalization is done by dividing the deviation of each value from the mean by the standard deviation from the mean. Thus, the normalized ith value, V_i', is given by the formula

$$V_i' = \frac{V_i - V_{mean}}{V_{std}}$$

where V_i is the ith value, V_{mean} is the mean value, and V_{std} is the standard deviation of all the raw values. Use of this technique means that the original values are not mapped to the interval from zero to unity but, instead, are distributed about a mean of zero and with a standard deviation of unity. Consequently, normalization (unlike the other three transformations) may yield negative as well as positive values and values that are greater than plus or minus unity.

None of these standardization techniques is intrinsically superior to the others, and the model in this chapter allows the user to select the one that seems appropriate. Professor Henk Voogd has made a detailed study of these techniques in relation to urban and regional planning, and he makes the following points (1983, pp. 77-78 and 87-89; see also Nijkamp 1986):

- Standardization in relation to the maximum (i.e., technique 1) has the advantage of preserving the ratio-scale character of the original data. Thus, this technique is "very useful in standardizing an evaluation matrix that will be analyzed by a weighted summation technique or any other technique which utilizes the magnitude of the individual scores."

- Standardization in relation to the sum of all the values (i.e., technique 2) has the advantage that the standardized values will always add up to unity, which may be "especially appropriate" in the case of weights, coefficients, and similar applications. The disadvantage of this technique is that it "results in a skew[ed] distribution with emphasis on the lower scores."

- Standardization in relation to the range from minimum to maximum (i.e., technique 3) is "especially appropriate where a technique is used which performs a pairwise comparison" or where it is felt useful to add a hypothetical maximum value to the original data, in order to represent some kind of goal or level of achievement. The disadvantage of this technique is that it produces only an interval scale.

- Normalization, although popular in the statistical literature, is unlikely to be appropriate in multi-criteria situations. Normalization is useful when (for example) a comparison of the shape of the distribution of the values is important, but this is rarely the case in multi-criteria evaluation. "What counts are the differences between scores, which for that reason should be influenced in the standardization process by as few as possible other criterion scores. . . . [Normalization] introduces an additional weight based on the numerical properties of the data, [and this] is not necessary from a multicriteria evaluation point of view."

The model in this chapter allows users to experiment with these techniques to see how far the choice of technique affects the outcome in any particular case of impact assessment. For standardizing weights, the user may select either of the first two techniques; for standardizing component scores, the user may select any one of the four techniques.

Bibliography and References

Biswas, A.K., and Q. Geping, eds. 1987. *Environmental Impact Assessment for Developing Countries.* London: Tycooly.

Brown, C.A., D.P. Stinson, and R.W. Grant. 1986. *MATS-PC: Multi-Attribute Tradeoff System: Personal Computer Version.* Bureau of Land Reclamation, Department of the Interior. Denver: mimeo.

Cartwright, T.J. 1988. "Using Microcomputers for Managing Land Information in Developing Countries." *Plan Canada*, June, pp. 16-20.

———. 1987. "Information Systems for Planning in Developing Countries." *Habitat International*, 11.3 (July), pp. 190-208.

———. 1975. *Royal Commissions and Departmental Committees in Britain: Adaptiveness and Participation in Government.* London: Hodder and Stoughton.

Clark, Brian D., Ronald Bisset, and Peter Wathern. 1980. *Environmental Impact Assessment: a Bibliography with Abstracts.* London: Mansell.

Cox, Louis Anthony, Jr. 1982. "Artifactual Uncertainty in Risk Analysis." *Risk Analysis*, 2.3 (September), pp. 121-35.

Dee, N., J.K. Baker, N.L. Drobny, I.L. Witman, and D.C. Fahringer. 1973. "Environmental Evaluation System for Water Resource Planning." *Water Resources Research*, 9, pp. 523-35.

Draggan, Sidney, J.J. Cohrssen, and R.E. Morrison, eds. 1984. *Environmental Monitoring, Assessment, and Management: The Agenda for Long-Term Research and Development.* New York: Praeger for the U.S. Council on Environmental Quality.

Environment Canada. 1974. *An Environmental Assessment of Nanaimo Port Alternatives.* Ottawa: Environment Canada.

Faludi, Andreas, and Henk Voogd, eds. 1985. *Evaluation of Complex Policy Problems.* Delft: Delftsche Uitgevers Maatschappij B.V.

Fortlage, C.A. 1990. *Environmental Assessment: A Practical Guide.* Aldershot, England: Gower.

Hobbs, B.F., M.D. Rowe, B.L. Pierce, and P.M. Meier. 1984. "Comparisons of Methods for Evaluating Multiattributed Alternatives in Environmental Assessments: Results of the BNL-NRC Siting Method Project." In *Improving Impact Assessment: Increasing the Relevance and Utilization of Scientific and Technical Information*, edited by S.L. Hart, G.A. Enk, and W.F. Hornick. Boulder, Colo.: Westview.

Holling, C.S., ed. 1978. *Adaptive Environmental Assessment and Management*. For the International Institute of Applied Systems Analysis. Chichester, U.K.: Wiley.

Kneese, Allen V. 1984. *Measuring the Benefits of Clean Air and Water*. Baltimore: Johns Hopkins University Press for Resources for the Future.

Leopold, L., F.E. Clarke, B.B. Hanshaw, and J.R. Balsley. 1971. *A Procedure for Evaluating Environmental Impact*. U.S.G.S. Circular no. 645. Washington, D.C.: U.S. Geological Survey.

Luhar, Ashok K., and P. Khanna. 1988. "Computer-Aided Rapid Environmental Impact Assessment." *Environmental Impact Assessment Review*, 8, pp. 9-25.

Massam, Bryan H., and Ian Skelton. 1986. "Application of Three Plan Evaluation Procedures to a Highway Alignment Problem." *Transportation Research Record*, 1076, pp. 54-58.

McHarg, Ian L. 1968. *A Comprehensive Highway Route-Selection Method*. Highway Research Record, no. 246. Washington, D.C.: Highway Research Board.

———. *Design with Nature*. Garden City, N.Y.: Natural History Press.

Midgley, J., and D. Piachaud, eds. 1984. *The Fields and Methods of Social Planning*. London: Heinemann.

Munn, R.E., ed. 1975. *Environmental Impact Assessment: Principles and Procedures*. For the Scientific Committee on Problems of the Environment (SCOPE). Chichester, U.K.: Wiley.

Nijkamp, Peter. 1986. "Multiple Criteria Analysis and Integrated Impact Analysis." In *Methods and Experiences in Impact Assessment*, edited by H.A. Becker and A.L. Porter, pp. 226-61. Dordrecht: Leidel.

Odum, E.P., J.C. Zieman, et al. 1975. "Optimum Pathway Analysis and Highway Route Selection." In *Environmental Impact Assessment*, edited by M. Blisset. Austin: University of Texas at Austin.

———. 1976. "Totality Indexes for Evaluating Environmental Impacts of Highway Alternatives." *Transportation Research Record*, 561, pp. 57-67.

Organization for Economic Cooperation and Development (OECD). 1989. *Environmental Policy Benefits: Monetary Valuation*. Paris: OECD.

Roberts, R.D., and T.M. Roberts, eds. 1984. *Planning and Ecology*. London: Chapman and Hall.

Sorensen, J.C. 1971. *A Framework for the Identification and Control of Resource Degradation and Conflict in Multiple Use of the Coastal Zone*. Master's thesis, University of California at Berkeley.

Stover, L.V. 1972. *Environmental Impact Assessment: A Procedure*. Pottstown, Pa.: Sanders and Thomas.

Thor, E.C., et al. 1978. "Forest Environmental Impact Assessment—A New Approach." *Journal of Forestry*, November, pp. 723-25.

Voogd, Henk. 1983. *Multicriteria Evaluation for Urban and Regional Planning*. London: Pion.

Zieman, Joseph, E.P. Odum, et al. 1971. "Optimum Pathway Matrix: Analysis Approach to the Environmental Decision Making Process. Testcase: Relative Impacts of Proposed Highway Alternatives." Institute of Ecology. Athens, Ga.: University of Georgia.

Part Three

MODELING
ARTIFICIAL SYSTEMS

Chapter 12

LIFE IN A SPREADSHEET: JOHN CONWAY'S GAME

> *"If I were a pattern, I'd be very careful where*
> *I fired my gliders! That game plays a rough game!"*
>
> *"It does," replied Cal, "as does all nature."*
>
> *Piers Anthony, Ox*

The flexibility of spreadsheets is truly formidable. It is no exaggeration to say that, with a spreadsheet, we can explore the meaning of life and the nature of the universe. Of course, we pay a price for this flexibility, a price in terms of speed of execution and elegance of output. But what we gain with a spreadsheet is accessibility. Spreadsheets do more than almost any other programming environment to demystify computing for the ordinary user.

The purpose of this chapter is to describe and discuss a spreadsheet version of John H. Conway's famous "Game of Life". In doing so, we have two objectives. First, we will use Life to illustrate some of the less frequently exploited features of spreadsheets and to reveal some of their weaknesses and limitations. In particular, we will examine conditional branching, both in the worksheet itself and in macros, and use several index and statistical functions, notably LOOKUP() and COUNT(). At the same time, it has to be admitted that there are faster and more elegant ways of programming Life than in a spreadsheet. For example, Niemiec 1979 gives an APL version of Life that uses only a single line of code! But the spreadsheet version seems somehow more open and immediate, even if it is a bit slow and awkward.

The second objective of this chapter is to make use of Life to explore the idea of chaos and the complex relationship between order and predictability in the environment. Thus, the implications of this chapter go far beyond computers and spreadsheets, right to the roots of our assumptions about human behavior. What Life illustrates is that local order does not necessarily mean global predictability (Cartwright 1991a). Life and other examples of what are called "cellular automata" are fundamentally models in which each cell behaves entirely and exclusively in accordance with (usually quite simple) rules. Yet the global results of such behavior can still be unpredictable. If humans (whom John von Neumann [1966] liked to refer to as "natural automata", in contrast to machines, or "artificial automata") are more complex than the forms used in such models, then social behavior can hardly fail to be less predictable than that of artificial life. Even when we think we know the rules!

Purpose of the Model

The Game of Life was invented in 1970 by John Conway, a mathematician at Cambridge University (Berlekamp et al. 1982). Life occurs in an infinite, two-dimensional space in cyclical time. Each cell in this spatial matrix can have one of only two states—alive or dead. The state of a cell in any particular cycle is determined by the state of its neighbors in the preceding cycle. As Life evolves, therefore, new cells are born, some cells live on, and some cells die.

The rules governing cellular birth, life, and death are called transition rules. In Life, they are as follows:

- Birth occurs in an empty cell surrounded by three live neighbors.

- Life survives in a cell surrounded by 2 or 3 live neighbors.

- Death occurs in all other cases, due either to "exposure" (too few live neighbors) or to "overcrowding" (too many live neighbors).

As simple as these local rules may seem, the global behavior of Life is astonishingly rich and complex. One outcome is that all the cells die off and Life disappears. Another possibility is that Life stabilizes in a fixed configuration and goes on apparently forever. Or Life may oscillate over varying periods between two or more forms. Life can also evolve in such a way that it appears to generate various kinds of "travelers" that migrate across space at different speeds. Life can also produce "machines" that appear to create and destroy other Life forms. One of these is the "glider gun", a kind of factory whose output (gliders) is described in the novel by Piers Anthony cited in the epigraph above. Perhaps most astonishing of all, however, is the fact that, no matter how complex it becomes, the evolution of Life depends entirely on the configuration of live cells from which it began.

In fact, Conway has demonstrated (Berlekamp et al. 1982, p. 830ff.) that Life can be regarded as a kind of universal computer, similar to what computer scientists call a Turing machine. Indeed, interest in cellular-automata models such as Life was originally inspired (von Neumann 1966; Burks 1970) by just such motives (Levy 1992). In the last decade, research on cellular automata has been spurred by developments in VLSI (very large-scale integration) technology, parallel computers, systolic (or rhythmically pulsed) algorithms, and RISC (reduced instruction set computer) architecture (Fogelman Soulie et al. 1987, pp. xx-xxi). For environmentalists, too, Life poses some fundamental questions: if social behavior is orderly yet fundamentally unpredictable, how then can we plan?

Conceptual Basis

The model is created by turning the cellular structure of the spreadsheet into a two-dimensional cellular universe for Life. Into this universe, an initial distribution of "live" cells is inserted. Then the model is set in motion, one cycle or generation succeeds another, and Life unfolds as it will.

Clearly, populations of only one or two live cells disappear immediately, as any cell requires at least two live neighbors to survive. Similarly, of the five different triplet forms—

—the first goes into immediate oscillation between a horizontal and vertical alignment, the second immediately forms a block of four (which is perfectly stable), and the other three die out in two cycles.

Asymmetrical or rook-wise connected sets of four or five live cells are called "tetrominoes" and "pentominoes" respectively. Most of them either disappear or reach a stable state (which Conway calls "still life") within at most nine cycles. One very interesting exception is the "r-pentomino" (so named by Conway because it seemed to him to resemble a lower-case "r"):

```
  . .
 .  .
    .
```

The r-pentomino also reaches a stable state—but not until after 1,104 cycles! Conway coined the term "Methuselah" for patterns of ten or fewer live cells that do not stabilize within 50 generations. Other "methuselahs" include the pi-heptomino and the remarkable "acorn" discovered by Charles Corderman. The acorn looks like this:

```
    .
 .
. .  . . .
```

According to Gardner 1983, the acorn does not stabilize until cycle 5,206. In the process of stabilizing, moreover, some of these patterns (including the r-pentomino and the acorn) seem to spawn or throw off various other patterns that then take on a life of their own.

One of the most remarkable of all patterns is the five-cell "glider", which looks like this:

```
  .
   .
. . .
```

The glider travels across the Life universe forever, moving down one cell and right two cells every four cycles. This behavior led Conway to suggest that the glider travels at "half the speed of light", since the maximum speed of cellular movement across the grid is obviously one cell per cycle and the glider travels at half that rate.

Conway also discovered larger patterns which he called "spaceships". Spaceships travel in the same way and at the same speed as gliders, but they have other intriguing characteristics as well:

> As they move, [spaceships] throw off sparks that vanish immediately as the ships continue on their way. Unescorted spaceships cannot have bodies longer than six [cells] without giving birth to objects that later block their motion. . . [but] longer spaceships . . . can be escorted by two or more smaller ships that prevent the formation of blocking [objects]. . . . A spaceship with a body

of 100 [cells], Conway finds, can be escorted safely by a flotilla of 33 smaller ships [Gardner 1983, pp. 222-23].

An even more remarkable discovery was made in 1970 by a group at MIT led by William Gosper. This was a pattern they christened a "glider gun" because it throws off or "fires" a new glider into space every 30 cycles. Among other things, this discovery seems to prove that a finite population of live cells can beget an infinitely large one. In addition, the MIT group discovered a way to position a pentadecathlon (a pattern with a 15-cycle period of oscillation) so that it can "catch" and "eat" any glider that bumps into it. They also discovered how to position two pentadecathlons so that they can "bounce" a glider back and forth between them forever. Gardner continues:

> Streams of intersecting gliders produce fantastic results. Strange patterns can be created that in turn emit gliders. . . . In other cases the collision mass destroys one or more guns by shooting back. The [MIT] group's latest burst of virtuosity is [to find] a way of placing eight guns so that the intersecting streams of gliders build a factory that assembles and fires a middleweight spaceship about every 300 ticks [ibid., p. 233].

Still another area of interest in Life is what are called "Garden of Eden" patterns. These are configurations of live cells that cannot arise naturally, because there exists no precedent pattern that could possibly have created them. It has been proved that Garden of Eden patterns exist in all cellular-automata systems (Moore 1962) and some have been found in Life (Berlekamp et al. 1982, pp. 828-9; Gardner 1983, p. 248). But it is not known how many such patterns there are in Life nor, in general, how to find them, except by "brute force", or trial and error.

Furthermore, while a Garden of Eden pattern has (by definition) no "parent", it can perfectly well have "children" (i.e., a sequence of patterns that evolve from it). It is also the case that a child can be the "offspring" of many different parents. All of which gives rise to the intriguing question, is there a class of patterns that have multiple parents but no grandparents? In other words (to continue the analogy), are there Adams and Eves as well as Gardens of Eden? According to Gardner (ibid., p. 249), this is still an open question.

Conway's Game of Life is defined primarily by its three rules of birth, life, and death. But there are two other aspects of cellular-automata systems that need to be discussed before we can describe this spreadsheet version. First, cellular automata depend on the concept of a neighborhood—that is, the set of cells surrounding a given cell and on which the latter depends. Originally, von Neumann (1966) worked with a neighborhood of four cells—those orthogonally adjacent (i.e., directly above, below, left, and right). Subsequently, Edward F. Moore (1962) proposed an expanded neighborhood, which included the four diagonally adjacent cells as well. Conway's Game of Life illustrates the use of the larger Moore neighborhood.

Second, cellular automata operate in a grid that is (at least potentially) infinite in size. The topology of an infinite rectangular surface is simulated by means of a toroid (or doughnut), rather than a sphere. On a computer, this effect can be achieved by "wrapping around" the opposite edges of the grid and connecting them to each other—i.e., the left edge to the right edge and the top edge to the bottom edge. Nevertheless, small toroids differ from large ones, because "wrap-around" (and thus the possibility of feedback or interference) occurs sooner on small toroids than on large ones. In Life, this feedback can have the effect of either prematurely ending or unduly extending the life of a population.

Data Requirements

The only "data" required for Life is the initial configuration of live cells. In this version of Life, live cells are indicated by a "1" and dead cells by a dot (".").[1] On initialization, each cell in the universe is set to a dot. This means there is no life in any of the cells. Before you can start the Game, therefore, you must create life somewhere in the universe. You do this by moving the cursor to any cell or cells in the display area and entering the number "1".[2] There is also a macro to provide a random population, in case that is what you would like to try. However it is produced, this initial pattern of live cells is the only input data the model requires.

---

1.   This choice of symbols is dictated partly by the need to make live cells stand out from dead cells (without the latter being completely invisible) and by the need to be able to apply the COUNT() function (which recognizes numbers and ignores other symbols). Strictly speaking, you should enter a space and then a dot (rather than just a dot) for a dead cell. This will ensure that the dots, like the ones, are right-justified in their cells and produce a visually more pleasing pattern on the screen.

2.   In fact, cells can be reset to life or death at any point in the Game, just by entering (over-writing) the appropriate symbol into the cell: "1" for a live cell and " ." for a dead one.

Overview

Life in a spreadsheet (see Listing 12.1) is divided into three sections arranged one above the other. The top part of the model provides the user interface: namely, a grid of 20 cells by 20 cells. Here the starting population is entered for each run of the model, and here the computer displays the new pattern following each cycle. On the left, there is a running record of the status of the model.

The core of the spreadsheet is in the middle section, which contains a grid identical to the one in the display area above it. The model is fundamentally cyclical. First, it "reads" the population displayed in the upper section. Then it computes and displays in the middle section the revised population for the next cycle, as dictated by the transition rules. Then it copies the new population data from the middle section back into the upper one. This process is repeated for the next cycle, and the next, and so on.

The third section of the model consists of three macros to execute and control the model. There is a macro to start and run the model, until either the user stops it or the population dies out. There is another macro to reinitialize the model by restoring all the cells to a dead state. And there is a third macro that fills up the display with a randomly distributed population of live cells, in case you want to start with such a pattern.

Operation

To get a taste of Life, you can experiment by entering some of the patterns described above into the display area.[3] The display area (F1:Y20) is just a matrix of cells containing dots: it contains no formulas of any kind. Once the initial pattern of Life has been entered, the display area is renewed by copying from the grid below.

---

3. There are also books with detailed examples of initial populations of live cells: e.g. Berlekamp et al. 1982; Gardner 1983; and Poundstone 1985. Naturally, more complicated patterns are sometimes difficult to follow on the relatively small grid (20 cells by 20 cells) used here. Life in a conventional programming language (e.g. Acero 1989) is typically plotted at the scale of the pixel, thereby permitting much larger grids (e.g., 320 x 200 or 640 x 350).

Listing 12.1

```
      |   A    ||  B ||  C ||  D |
 1    JOHN CONWAY'S GAME OF LIFE      . . . . . . . . . . . . . . . .
 2    © 1993 by T.J. Cartwright        . . . . . . . . . . . . . . . .
 3    ─────────────────────────        . . . . . . . . . . . . . . . .
 4                                     . . . . . . . . . . . . . . . .
 5    Game cycle ----->       1        . . . . . . . . . . . . . . . .
 6                                     . . . . . . . . . . . . . . . .
 7                                     . . . . . . . . . . . . . . . .
 8    Population              0        . . . . . . . . . . . . . . . .
 9                                     . . . . . . . . . . . . . . . .
10                                     . . . . . . . . . . . . . . . .
11    Maximum population      0        . . . . . . . . . . . . . . . .
12      in cycle             0        . . . . . . . . . . . . . . . .
13                                     . . . . . . . . . . . . . . . .
14                                     . . . . . . . . . . . . . . . .
15    ─────────────────────────        . . . . . . . . . . . . . . . .
16    Alt-I  =  initialize game        . . . . . . . . . . . . . . . .
17    Alt-S  =  start game             . . . . . . . . . . . . . . . .
18    Alt-Z  =  randomize              . . . . . . . . . . . . . . . .
19    Ctrl-Break  =  abort             . . . . . . . . . . . . . . . .
20    ─────────────────────────        . . . . . . . . . . . . . . . .
21
22
23    ─────────────────────────────────────────────────────────
24
25
26
27    WORKSHEET
28
29    Calculations for Next Cycle
30
31    New game cycle -->      2        . . . . . . . . . . . . . . . .
32                                     . . . . . . . . . . . . . . . .
33                                     . . . . . . . . . . . . . . . .
34                                     . . . . . . . . . . . . . . . .
35                                     . . . . . . . . . . . . . . . .
36                                     . . . . . . . . . . . . . . . .
37                                     . . . . . . . . . . . . . . . .
38                                     . . . . . . . . . . . . . . . .
39                                     . . . . . . . . . . . . . . . .
40                                     . . . . . . . . . . . . . . . .
41                                     . . . . . . . . . . . . . . . .
42                                     . . . . . . . . . . . . . . . .
43                                     . . . . . . . . . . . . . . . .
44                                     . . . . . . . . . . . . . . . .
45                                     . . . . . . . . . . . . . . . .
46                                     . . . . . . . . . . . . . . . .
47                                     . . . . . . . . . . . . . . . .
48                                     . . . . . . . . . . . . . . . .
49                                     . . . . . . . . . . . . . . . .
50                                     . . . . . . . . . . . . . . . .
51
52    ─────────────────────────────────────────────────────────
```

```
        |    A    || B || C || D |
53
54
55
56
57
58
59   MACROS        Commands              Comments
60
61   {\S}          {home}                home cursor
62                 {paneloff}            turn dialog line off
63                 !                     [B63] manual recalc
64                 {if C8=0}{branch B68} exit if population reaches zero
65                 /c C31,C5,v           update number of cycles
66                 /c F31:Y50,F1,v       copy results of calculation to display
67                 {branch B63}          loop until stopped
68                 {beep} {panelon}      [B68] turn dialog line on again
69
70   {\I}          {home}                home cursor
71                 {let C5,1}            set number of cycles to one
72                 {let F1," ."}         assign value " ." to f1
73                 /c F1,F1:Y20,v        copy " ." throughout display
74                 !                     manual recalc
75                 =010 ~ {beep}         go to middle of display
76
77   {\Z}          {home}                home cursor
78                 {let C5,1}            set number of cycles to one
79                 {let F1," ."}         assign value " ." to f1
80                 /c F1,F1:Y20,v        copy " ." throughout display
81                 =F1 ~                 go to upper left corner
82                 {if ran>.5} 1         [b82] replace with "1" at random
83                 {if currow<20} {down} {branch B82}
84                 {if curcol<25} {up 20} {right} {branch B82}
85                 ~{home}
86                 ! {beep}              manual recalc
87
88   ─────────────────────────────────────────────────────
```

The menu area on the left shows the four commands available for operating the model. In addition, cell C5 displays the current cycle, and cell C8 displays the current population. The first is incremented through the regular recalculation process, while the second uses the COUNT() function to compute how many cells in the display area contain numerical characters (a 1 as opposed to a dot). Cells C11 and C12, respectively, record the maximum population reached during the current run and the cycle in which it occurred.

In contrast to the display area, the lower grid (cells F31:Y50) is a matrix of complicated, but ultimately symmetrical, formulas. The formulas are all built around conditional branching using the IF() and COUNT() functions. The formula in cell G32, for example, is as follows:

```
IF(OR(AND(COUNT(F1:H3)=4,G2=1),COUNT(F1:H3)=3),1," .")
```

There are two steps to understanding how this formula works. First, note the syntax of the IF() and COUNT() functions. The IF() function—for which the syntax is IF(a,b,c)—means "If a, then b; otherwise c." The COUNT() function—for which the syntax is COUNT(range)—returns the number of cells within the specified range of cells that have a numerical value (in this case, a one). Second, recall that the state of any cell in Life depends on the state of its "Moore neighbors" (that is, its eight surrounding cells) *in the previous cycle*. In this version of Life, the *previous* state of the neighbors of any cell in the lower area is found in the upper (display) area—because the latter contains the previous cycle for the lower area. Thus, the state of any cell in the lower area depends on the state of its neighbors in the upper area; so, for example, the state of cell G32 depends on the state of cells F1:H3.

Now, according to the rules of Life, there are three conditions in which there can be life in a cell at the end of a cycle. These can be summarized as follows:

New Cell State	Cell State in Previous Cycle	Neighbors in Previous Cycle
Birth	No life	Life in 3 cells
Survival	Life	Life in 2 cells
Survival	Life	Life in 3 cells

In all other cases, there is no life, only death.

In spreadsheet terms, we can summarize the first two conditions by saying that life exists if the total number of ones in the range (i.e., in the neighborhood) is three. This occurs either when there is no life in the central cell and life in three neighboring cells, or life in the central cell and life in two neighboring cells. Similarly, we can restate the third condition by saying that life exists if the total number of ones in the range (neighborhood) is four, and one of them is in the central cell. Thus, we arrive at the formula in cell G32 (and all the other cells in the lower matrix). Life will exist in cell G32 in the current cycle if either of the following conditions is true:

    COUNT(F1:H3)=3

or

    AND(COUNT(F1:H3)=4,G2=1)

These are the rules of Life transformed into spreadsheet terms.

The macros in the third section of the model should be more or less self-explanatory by now. The most important is Alt-S, which starts and runs the model. The macro begins by removing the cursor from the display area, turning off the dialog line, and doing a manual recalculation. Then it checks to see if the entire population of Life is dead. If so, the macro branches to the last line, turns the dialog line back on, and exits. If not, the macro updates the number of cycles the model has run and copies the contents of the lower worksheet area to the upper display area. Then it loops back to do another manual recalculation (to update the lower worksheet area to a new cycle) and continues as before.

The other two macros initialize the model for a new run by resetting the counters and clearing the display area. One of the macros (Alt-I) zeros the display area—i.e., sets all the cells to a no-life state (" .")—while the other macro (Alt-Z) fills the display area with a random distribution of cells with life ("1") and cells with no life (" .").[4]

---

4.   Programmers will see from these macros that spreadsheets do not have commands for looping. In most conventional programming languages, there are commands to establish a loop on entry and monitor the number of times that, or the conditions under which, the loop is to be repeated. In BASIC, for example, there are FOR...NEXT loops, WHILE...WEND

Interpretation and Use

Once Life is running, there is nothing to do but sit back and watch it unfold! The model records the current cycle and current population, as well as the maximum population reached and the cycle in which that occurred. Life goes on as long as life exists, which may be forever.

Enthusiasts may want to strive for those initial Life forms that last longest (without either dying out or stabilizing); for those forms that produce the highest "peak" populations (a population of 200 is the theoretical maximum in a 400-cell square grid); for those forms that yield the highest stabilized level of population; and so on. Even in the simple version shown here, the Game of Life can be an endless source of fascination.

Evaluation and Extension

Obviously we should not read too much into Life. Its transition rules are very simple, and its grid has only two dimensions. It may be a metaphor, but it is hardly an analogy. On the other hand, the underlying principle of Life is very provocative. What if Life were generalized to a much larger grid and a more complex set of rules? What if we postulated a grid of many more dimensions? Some scientists (e.g., Kemeny 1955; Gardner 1983) have speculated that the universe might be some kind of gigantic and highly complex system of cellular automata, just like Life. In fact, Professor Ed Fredkin of the Artificial Intelligence Laboratory at MIT has been quoted (Levy 1992, p. 63) as saying:

Living things may be soft and squishy. But the basis of life is clearly digital. We don't know how it works exactly, but instead of computer bits, there's a four-state code [i.e., the four base chemicals that make up DNA and whose sequencing forms the genetic code]. And there's some kind of process that

---

loops, etc. In spreadsheets, however, there is no similar command. So, if you want to have a loop in a spreadsheet, you have to make it yourself, using conditional branching. You can do this in three steps. First, you establish a counter or a condition to be met; then you proceed with the commands that are to be repeated; finally, you determine whether to repeat the loop by using an IF() function to check on the state of the counter or condition you set up earlier. This may not be elegant, but it works.

interprets it. [Life is] obviously some sort of program, running on a digital computer. It's just that the messages don't come in from a model; they come from chemicals. . . . Put it another way—nothing is done in nature that can't be done by a computer. If a computer can't do it, nature can't.

After all, the simple decision-making entities involved in Life do not have to "know" that they are cellular automata, in order for Life to unfold as it does. If real life does imitate the Game of Life, would we humans know it? Would we need to know it? The implication of Conway's game is that we would not.

In his novel *Ox*, Piers Anthony adds a wry twist to the conversation cited in the epigraph to this chapter on page 277. The conversation, which is about patterns, is between two characters (Cal and Aquilon) who are themselves "pattern entities." Although they are clearly intelligent and sentient beings, Cal and Aquilon have reached this state through a process of adaptation in a world of dimensions higher than those of merely human space and time. Thus, there is a clear implication in Anthony's novel that those who play the Game of Life may be participating in it in a deeper sense than they realize.[5]

In nature, there is certainly evidence of Life-like processes at work. For example, there are studies suggesting that the complex ways in which birds flock, fish school, and bees swarm may be the result of quite simple local rules.[6] Of course, these rules are applied thousands and even millions of times over relatively short periods of time, and that may give the illusion of complexity. But the rules themselves are simple and knowable, even if their overall effect is not.

The significance of Life for environmental choice is potentially profound. If Life does indeed reflect the ad hoc, incremental way in which rational (i.e., rule-governed) decisions are made in real life, then we may have to rethink a lot of our ideas about decision making and planning. We are so conditioned to thinking in terms of global trends and societal change that we often overlook the effects of purely local behavior. We explain population growth (for example) in terms of social and

---

5.   The title of the novel, *Ox*, refers not to the animal but to the combination of two binary states, life (O) and death (X).

6.   On flocking birds, see (for example) Reynolds 1987 and Levy 1992, pp. 76-82; Cartwright 1991b cites numerous other cases of chaos in life.

cultural trends, national economic policies, and political choices—all of them exogenous to the process we want to explain. But suppose that population growth is not really affected by exogenous factors at all, but is inherent in the development process. Suppose that population growth is merely the inevitable result of thousands or millions of local decisions. Of course, that kind of assumption implies that we may need to rethink our entire approach to population policy. On the other hand, such an assumption may provide a more accurate basis for understanding what is going on. Similarly, overall patterns of land use, residential choice, traffic flows, and other phenomena that characterize a modern, industrial society may likewise be the result not of broad social causes so much as of a lot of individual decisions based on essentially local and limited evidence.

Such a world may not seem altogether desirable but it is realistic. It may just be what life is like.

Bibliography and References

Acero, A.A. 1989. *AAALIFE*. Ver. 1.0 (May 11). Microcomputer program.

Anthony, Piers. 1976. *Ox*. London: Avon.

Aspray, William. 1990. *John von Neumann and the Origins of Modern Computing*. Cambridge: MIT Press.

Banks, Edwin. 1971. "Information Processing and Transmission in Cellular Automata." Project MAC Technical Report TR-81. Massachusetts Institute of Technology. Mimeo.

Berlekamp, E.R., J.H. Conway, and R.K. Guy. 1982. "What Is Life?" In *Winning Ways for Your Mathematical Plays*. Vol. 2, *Games in Particular*, chapter 25. London: Academic Press.

Buckingham, D.J. 1978. "Some Facts of Life." *BYTE*, 3.12 (December), pp. 54-67.

Burks, Arthur W., ed. 1970. *Essays on Cellular Automata*. Urbana: University of Illinois Press.

Cartwright, T.J. 1991a. "Experimental Systems Research: Towards a Laboratory for the General Theory." *Cybernetics and Systems: An International Joournal*, 22.1 (March), pp. 135-49.

———. 1991b. "Planning and Chaos Theory." *Journal of the American Planning Association*, 44 (Winter), pp. 44-56.

Dewdney, A.K. 1989. "A Cellular Universe of Debris, Droplets, Defects and Demons." *Scientific American*, 261.2 (August), pp. 102-5.

Englander, William. 1978. "Life." *BYTE*, 3.12 (December), pp. 76-82.

Fogelman Soulie, Francoise, Yves Robert, and Maurice Tschuente, eds. 1987. *Automata Networks in Computer Science: Theory and Applications*. Manchester: Manchester University Press.

Gardner, Martin. 1983. *Wheels, Life and Other Mathematical Amusements*. New York: Freeman.

Helmers, Carl. 1975-76. "Lifeline," parts 1-4. *BYTE*, 1 (September), pp. 72-80; 1 (October), pp. 34-42; 1 (December), pp. 48-55; and 2 (January), pp. 32-41.

Kemeny, John G. 1955. "Man Viewed as Machine." *Scientific American* (April), pp. 58-68.

Levy, Steven. 1992. *Artificial Life: the Quest for a New Creation*. London: Jonathan Cape.

Macaluso, Pat. 1981. "Life after Death." *BYTE*, 6 (July), pp. 326-33.

Millen, J.K. 1978. "One-Dimensional Life." *BYTE*, 3.12 (December), pp. 68-74.

Millium, J., J. Reardon, and P. Smart. 1978. "Life with your Computer." *BYTE*, 3.12 (December), pp. 45-50.

Moore, Edward F. 1962. "Machine Models of Self-Reproduction." In Burks 1970, Essay 6. First published in *Proceedings of Symposia in Applied Mathematics*, 14, pp. 17-33.

Niemiec, Mark D. 1979. "Life Algorithms." *BYTE*, 4.1 (January), pp. 90-97.

Poundstone, William. 1985. *The Recursive Universe*. New York: Morrow.

Reynolds, Craig. 1987. "Flocks, Herds, and Schools: a Distributed, Behavioral Model." *Computer Graphics*, 21 (July), p. 25ff.

Soderstrom, Randy. 1979. "Life Can Be Easy." *BYTE*, 4 (January), pp. 166-69.

Toffoli, Tommaso, and Norman Margolus. 1987. *Cellular Automata Machines: a New Environment for Modeling*. Cambridge: MIT Press.

Von Neumann, John. 1966. *Theory of Self-Reproducing Automata*, edited by Arthur W. Burks. Urbana: University of Illinois Press.

Wolfram, Stephen. 1984. "Computer Software in Science and Mathematics." *Scientific American*, 251.3 (September), pp. 188-203.

Chapter 13

THE GAME OF CHOICE: NEIGHBORHOOD TOLERANCE

Ich bin ein Berliner.

President John F. Kennedy,
Berlin City Hall, June 26,1963

Like John Conway's Game of Life (discussed in the previous chapter), the model in this chapter is also a model of local decision making. But the transition rules in this model are probabilistic rather than deterministic (Dawkins 1986). In this model, the future state of each cell is not entirely predetermined (as in Life) by the state of its neighborhood. Rather, the fate of each cell is influenced in proportion to the balance, or mix, of cell-states in its neighborhood.

In other words, as long as a cell is surrounded by cells of its own kind, its state is unlikely to change. But, the more the *diversity* of a cell's neighborhood exceeds what that cell is prepared to accept, or *tolerate*, the more *probable* it is that the state of that cell will change.

Thus, this model is not so much a Game of Life as it is a Game of Choice—the choice to tolerate those who are different from oneself.

Purpose of the Model

The Game of Choice is meant to simulate how communities may evolve over time, how coexistence occurs, and how ghettos form.[1] In form, the Game of Choice

---

1.   In fact, I think I recall seeing a similar program some ten years ago written in BASIC. It may have been in an early issue of *80Micro* (published for users of Tandy TRS-80 microcomputers). The program was called GHETTOS.

is very similar to the Game of Life; but the way they function is fundamentally different. The purpose of Life is to show how individual and global behavior are related. The purpose of Choice, on the other hand, is to show how individual *attitudes* affect both individual and global behavior. Where Life simulates decision making, Choice simulates tolerance.

Thus, a typical run of Life begins with a deliberate pattern of cells of one particular kind (i.e., live cells denoted in the model by ones). Choice, on the other hand, usually begins with a random distribution of both types of cells (i.e., ones and dots) sperad over the entire grid. Furthermore, in Choice (but not in Life), we have to define how *tolerant* each community is of the other. We can make both types of cells relatively tolerant of each other, both types of cells relatively intolerant of each other, or one type of cell more tolerant in varying degrees than the other.

Conceptual Basis

In contrast to Life, Choice highlights not the birth and death of cells but rather the interactions between two different kinds of cells. The cells may represent different communities, populations, species, etc. Thus, there are no empty or dead cells. Instead, each cell is recognized as belonging to one of two communities, either the ones or the dots.[2]

Like Life, Choice uses a what is called a Moore neighborhood. A Moore neighborhood (named after the mathematician, Edward Moore) consists of a cell and its eight orthogonal and diagonal neighbors. The cells in the neighborhood can be referred to as North, South, East, West, Northwest, Southwest, Northeast, and Southeast. Choice also uses the same, quasi-infinite grid as Life (Kenner 1989). That is, opposite sides of the rectangular grid are wrapped around and joined to each other, to create what is topologically equivalent to the surface of a toroid (or doughnut).

As in Life, the state of any cell in any cycle of the Game is a function of what the state of its neighborhood was in the previous cycle (Poundstone 1985).   In

---

2.   The reason for using "ones" and "dots" is that they can be recognized efficiently in a spreadsheet environment, both visually on the screen by the user and operationally by means of the COUNT() function. See also chapter 12, above.

Choice, however, the user has an additional parameter to indicate how tolerant each type of cell is of the other type. For example, if ones are said to have a tolerance level of 2 and dots a tolerance level of 6, this would indicate that:

- a one will tolerate up to 2 dots in its neighborhood (but not 3 or more), and

- a dot will tolerate up to 6 ones in its neighborhood (but not 7 or 8).

Once the tolerance level of a cell is exceeded, there develops a likelihood, or probability, that the cell will be replaced by a cell of the opposite type.

Moreover, the probability of a cell changing its state is a direct function of the extent to which its tolerance level has been exceeded. Thus, for example, a cell that can tolerate up to 4 cells of the opposite type in its neighborhood (but not 5, 6, 7, or 8) is 25% likely to be replaced if its neighborhood contains 5 foreign cells, 50% likely to be replaced if it contains 6 foreign cells, 75% likely to be replaced if there are 7, and certain to be replaced if there are 8. Thus, where Life is deterministic, Choice is probabilistic (Condon 1989).

Data Requirements

The only input data that Choice requires is the tolerance levels of the two communities. As for the initial pattern of ones and dots, simulations in Choice typically begin with a random distribution. However, if desired, you can enter specific, user-designed patterns, such as solid blocks (ghettos) or mixed neighborhoods, and see what happens to those under different conditions of tolerance.

Overview

The structure of Choice (see Listing 13.1) is very similar to that of Life. The model consists of three sections arranged one above the other. The top part of the model provides the user interface: a square display area, 20 cells by 20 cells, where the computer shows the results of each cycle. The display area (cells F1:Y20) is a matrix of dots and ones; it contains no formulas. This is because data appearing in this area are copied there, by means of a macro, from a working area below. To the left of the display area, there is a running record of the status of the model and a list of the four commands used for operating it.

Listing 13.1

```
    |    A    || B || C || D |
 1  THE GAME OF CHOICE              . . . . . . . . . . . . . . . . .
 2  © 1993 by T.J. Cartwright        . . . . . . . . . . . . . . . . .
 3  ─────────────────────           . . . . . . . . . . . . . . . . .
 4                                   . . . . . . . . . . . . . . . . .
 5  Game cycle ------>    25         . . . . . . . . . . . . . . . . .
 6                                   . . . . . . . . . . . . . . . . .
 7              Ones  Dots           . . . . . . . . . . . . . . . . .
 8              ----  ----           . . . . . . . . . . . . . . . . .
 9  Current pop    0   400           . . . . . . . . . . . . . . . . .
10  Maximum pop  208   400           . . . . . . . . . . . . . . . . .
11                                   . . . . . . . . . . . . . . . . .
12  Level of                         . . . . . . . . . . . . . . . . .
13  Tolerance      1     2           . . . . . . . . . . . . . . . . .
14                                   . . . . . . . . . . . . . . . . .
15  ─────────────────────           . . . . . . . . . . . . . . . . .
16  Alt-I  =  initialize game        . . . . . . . . . . . . . . . . .
17  Alt-S  =  start game             . . . . . . . . . . . . . . . . .
18  Alt-Z  =  randomize              . . . . . . . . . . . . . . . . .
19  Ctrl-Break  =  abort             . . . . . . . . . . . . . . . . .
20  ─────────────────────           . . . . . . . . . . . . . . . . .
21
22
23  ──────────────────────────────────────────────────────────
24
25
26
27  WORKSHEET
28
29  Calculations for Next Cycle
30
31  New game cycle -->    26         . . . . . . . . . . . . . . . . . .
32                                   . . . . . . . . . . . . . . . . . .
33                                   . . . . . . . . . . . . . . . . . .
34                                   . . . . . . . . . . . . . . . . . .
35                                   . . . . . . . . . . . . . . . . . .
36                                   . . . . . . . . . . . . . . . . . .
37                                   . . . . . . . . . . . . . . . . . .
38                                   . . . . . . . . . . . . . . . . . .
39                                   . . . . . . . . . . . . . . . . . .
40                                   . . . . . . . . . . . . . . . . . .
41                                   . . . . . . . . . . . . . . . . . .
42                                   . . . . . . . . . . . . . . . . . .
43                                   . . . . . . . . . . . . . . . . . .
44                                   . . . . . . . . . . . . . . . . . .
45                                   . . . . . . . . . . . . . . . . . .
46                                   . . . . . . . . . . . . . . . . . .
47                                   . . . . . . . . . . . . . . . . . .
48                                   . . . . . . . . . . . . . . . . . .
49                                   . . . . . . . . . . . . . . . . . .
50                                   . . . . . . . . . . . . . . . . . .
51
52  ──────────────────────────────────────────────────────────
```

```
         |   A    || B || C || D |
53
54
55
56
57
58
59   MACROS        Commands                 Comments
60
61   {\S}          {home}                   home cursor
62                 {paneloff}               turn dialogue line off
63                 !                        [B63] manual recalc
64                 {if B9=0} {branch B69}   exit if ones population reaches zero
65                 {if C9=0} {branch B69}   exit if dots population reaches zero
66                 /c C31,C5,v              update number of cycles
67                 /c F31:Y50,F1,v          copy results of calculation to display
68                 {branch B63}             loop until stopped
69                 {beep} {panelon}         [B69] turn dialogue line on again
70
71   {\I}          {home}                   home cursor
72                 {let C5,1}               set number of cycles to one
73                 {let F1," ."}            assign value " ." to F1
74                 /c F1,F1:Y20,v           copy " ." throughout display
75                 !                        manual recalc
76                 =O10 ~ {beep}            go to middle of display
77
78   {\Z}          {home}                   home cursor
79                 {let C5,1}               set number of cycles to one
80                 {let F1," ."}            assign value " ." to F1
81                 /c F1,F1:Y20,v           copy " ." throughout display
82                 =F1 ~                    go to upper left corner
83                 {if ran>.5} 1            [B83] replace with "1" at random
84                 {if currow<20} {down} {branch B83}
85                 {if curcol<25} {up 20} {right} {branch B83}
86                 ~{home}
87                 !                        manual recalc
88                 =B13 ~ {beep}            go to tolerance settings
89
90   ─────────────────────────────────────────────────────────────
```

As in Life, the important part of the model is the middle part, which contains a grid similar in form to the display area above it—but containing a matrix of formulas. The model is basically cyclical in nature. It works by reading the population displayed in the upper section; computing and displaying in the middle section the revised population for the next cycle, as dictated by the transition rules; and then writing (copying) the new population data from the middle section back to the upper one. This process is repeated for the next cycle, and the next, and so on.

The third section of the model consists of three macros that initiate and control the simulation. The most important is Alt-S, which starts and runs the whole model. The macro begins by removing the cursor from the display area, turning off the dialogue line, and doing a manual recalculation. Then it checks to see if either population has completely died off; if so, the macro branches to the last line, where it turns the dialogue line back on and exits. If not, the macro updates the number of cycles the model has run and copies the contents of the lower working area into the upper display area. Then it loops back to do another manual recalculation (in order to update the lower working area to a new cycle) and continues as before. The other two macros initialize the model for a new run by resetting the counters and clearing the display area. Then one macro (Alt-I) zeros the display area—i.e., sets all the cells to dots (" .")—while the other (Alt-Z) fills the display area with a random mixture of ones ("1") and dots (" .").

Operation

To run Choice, an initial population distribution is entered. Normally, this is done with the randomizing macro (Alt-Z), although ones and dots can be keyed in one at a time, if desired. In addition, the tolerance level of each community (or type of cell) for the other has to be defined. These values are entered and displayed in cells B13 and C13. When this is done, the start macro (Alt-I) is invoked. As the model progresses, cell C5 displays the current cycle, cells B9 and C9 record the current population of each community, and cells B10 and C10 record the maximum population achieved by each community.

The important part of the model is the grid in the middle (cells F31:Y50), which is a matrix of complicated but ultimately symmetrical formulas. The formulas are all built around the COUNT() function and conditional branching using the IF() function. For example, the formula in cell G32 is as follows:

```
IF(G2=1,IF(((8-$B$13-COUNT(F1,F2,F3,G1,G3,H1,H2,H3))/
    (8-$B$13))>RAN," .",1),
    IF(((COUNT(F1,F2,F3,G1,G3,H1,H2,H3)-$C$13)/
    (8-$C$13))>RAN,1," ."))
```

As in the previous chapter, there are two steps to understanding how this formula works. First, recall the syntax of the IF() and COUNT() functions—IF(a,b,c) and COUNT(range). Second, recall that the *current* state of any cell depends on the state of its Moore neighbors (that is, its eight surrounding cells) *in the previous* cycle. In this model, the previous state of the neighbors of any cell in the lower area is found in the upper (display) area—because the latter contains the previous cycle for the lower area. Thus, the state of any cell in the lower area depends on the state of its neighborhood in the upper area: e.g., the current state of cell K40 depends on the state of cells F1, F2, F3, G1, G3, H1, H2, and H3.

In order to simplify the formula in cell G32, let us (just for a moment) use the expression "..." to represent the Moore neighborhood of a cell. Then the formula in cell G32 could be written like this:

```
IF(G2=1,IF(((8-$B$13-COUNT(...))/(8-$B$13))>RAN," .",1),
    IF(((COUNT(...)-$C$13)/(8-$C$13))>RAN,1," ."))
```

In plain language, this means that (a) if cell G2 is a one (i.e., if cell G32 was a one in the previous cycle), then if the relative tolerance of ones for dots is greater than a random number, then display a dot (otherwise display a one); but if cell G2 is not a one (i.e., if it is a dot), then if the relative tolerance of dots for ones is greater than a random number, display a one (otherwise display a dot). These (succinctly!) are the rules of Choice.

Interpretation and Use

Once begun the Game proceeds automatically until one community disappears or the model is stopped by the user. For any pair of tolerance levels, we can run the model starting from randomized, integrated patterns of ones and dots. Alternatively, we can create deliberate initial configurations, such as a solid block, a hollow square, an unbroken line from edge to edge, a checkerboard pattern, etc. In each case, the the model continues until one community fills the entire grid or the user intervenes to stop the run.

Evaluation and Extension

The implications of Choice are probably less profound than those of Life, but they are nonetheless interesting and instructive. The basic lesson of Choice is that different communities—or, at least, communities that perceive themselves as being different—tend to segregate themselves from each other automatically, as long as there exists any level of intolerance at all. Even if members of both communities are so tolerant of each other that they do not change their state until they are entirely surrounded by neighbors of the opposite kind, the two communities will still (given enough time) separate into ghettos.

In short, there seems to be a kind of "iron law" of segregation at work. Given any level of preference for their own kind, communities will form ghettos. The stronger the preferences (or the higher the levels of intolerance), the faster the segregation occurs. But as long as there is any kind of communitarian identification or preference influencing behavior, segregation is an inevitable byproduct. That is a sobering thought for people predisposed towards both community integration and the maintenance of cultural identity: we may not be able to have our cake and eat it too.

The iron law has a corollary. As long as the two communities are *equally intolerant* of each other, both are likely to survive. It does not appear to matter whether the communities are highly tolerant or highly intolerant of each other. As long as the strength of their feelings towards each other is equally matched, neither community is likely to become dominant.[3] But as soon as the two communities develop different levels of tolerance for each other, one will inevitably start to overwhelm the other. The striking—and perhaps encouraging—fact is that it is the *more* tolerant community that is likely to replace the *less* tolerant community. Indeed, the greater the disparity in levels of tolerance, the faster the less tolerant

---

3.   In practice, one community may overwhelm the other, especially on a relatively small grid like the one used here. For example, if one community is significantly larger than the other at the outset, that community is more likely to become dominant. Similarly, for any particular run of the model, the sequence of random numbers actually generated will in fact favor one community at the expense of the other. But, in principle, communities that are equally tolerant or intolerant of each other are likely to co-exist forever.

community is likely to decline in numbers and eventually to disappear altogether. In short, it is not necessarily the strong who survive; rather, it is the more tolerant.

By the same token, Choice implies that the intolerant may become victims of their own preferences and prejudices. Take, for example, the tactic of residential "block-busting", in which representatives of a minority community deliberately move into an area inhabited exclusively by members of the other community for the purpose of forcing it to become integrated. The implication of Choice is that the tactic works only if, and to the extent that, the second community is more intolerant of the first than vice versa. Thus, the best defence against block-busting is to remove its cause: that is, to reduce the level of intolerance towards the immigrants.

As with Life, we should not read too much into Choice. The rules are simple and the grid has only two dimensions. Nevertheless, Choice provides a dramatic and compelling perspective of the relationship between individual and group behavior, between private values and public policies. As President Kennedy declared in Berlin, we all belong to our respective communities; but we are all free to choose whom we will identify with.

Bibliography and References

Aspray, William. 1990. *John von Neumann and the Origins of Modern Computing.* Cambridge: MIT Press.

Berlekamp, E.R., J.H. Conway and R.K. Guy. 1982. "What Is Life?" In *Winning Ways for Your Mathematical Plays.* Vol. 2, *Games in Particular*, chapter 25. London: Academic Press.

Burks, Arthur, ed. 1970. *Essays on Cellular Automata.* Urbana: University of Illinois Press.

Cartwright, T.J. 1991a. "Experimental Systems Research: Towards a Laboratory for the General Theory." *Cybernetics and Systems: An International Joournal*, 22.1 (March), pp. 135-49.

———. 1991b. "Planning and Chaos Theory." *Journal of the American Planning Association*, 57.1 (Winter), pp. 44-56.

Condon, Anne. 1989. *Computational Models of Games*. Cambridge: MIT Press.

Dawkins, Richard. 1986. *The Blind Watchmaker*. London: Longman.

Dewdney, A.K. 1989. *The Turing Omnibus: 61 Excursions in Computer Science*. New York: Computer Science Press.

Gardner, Martin. 1983. *Wheels, Life and Other Mathematical Amusements*. New York: Freeman.

Kenner, Hugh. 1989. Review of Dewdney 1989. *BYTE*, 14 (December), pp. 444-45.

Levy, Steven. 1992. *Artificial Life: The Quest for a New Creation*. London: Jonathan Cape.

Moore, Edward F. 1962. "Machine Models of Self-Reproduction." In Burks 1970, essay 6. First published in *Proceedings of Symposia in Applied Mathematics*, 14, pp. 17-33.

Poundstone, William. 1985. *The Recursive Universe*. New York: Morrow.

Chapter 14

THE GAME MACHINE: CELLULAR AUTOMATA, CHAOS, AND FRACTALS

Global order can be generated by forces that act solely between neighbors [and] the output of such automata . . . can, in fact, be fractal.

Benoit Mandelbrot, The Fractal Geometry of Nature

The model in this chapter is derived from one that was developed in a project at MIT a few years ago (Toffoli and Margolus 1987). The original version uses customized computer hardware and software, whereas the model described here is similar to the versions of Life and Choice described in chapters 12 and 13, above, except that the model in this chapter is confined to a von Neumann (orthogonal) neighborhood rather than a Moore (orthogonal and diagonal) neighborhood.

The aim of the model is to illustrate how environmental patterns on various scales—neighborhoods, squatter settlements, urban form, land-use patterns, forest boundaries, soil erosion, sand dunes, air currents, etc.—may be susceptible to modeling as fractal forms. We usually think of physical objects, like a pencil or a piece of paper, as having either two or three dimensions. (The "fourth dimension" sounds like something out of Hollywood!) On reflection, we might be persuaded to agree that "more complex" objects (such as cities or databases) could be said to have more than just two or three dimensions. It all depends on what is meant by "dimensions", we would say. Fractal objects, however, pose an even more profound conceptual challenge for most of us.

Fractals are objects that have non-integer, or fractional, dimensions—that is, a fractal is neither two-dimensional nor three-dimensional but might be 2.35-

dimensional, for example.   The mechanics of calculating the dimensionality of fractals is fairly complicated and is not pursued here.[1] The key point is that there is this "nether world" (mathematically speaking) between objects with integer dimensions, and it appears to be inhabited by real-world objects. Contemporary interest in such forms is barely a decade old; yet already chaos and fractals are challenging some widely held assumptions. For example, it has long been assumed that order and predictability were two sides of the same coin. Now, it appears that there are phenomena that are perfectly orderly but, for all practical purposes, unpredictable. (The weather is a favorite example.) Similarly, we normally assume that locations, boundaries, and shapes of objects have to be smoothed into idealized points, lines, and curves in order for us to be able to study them. Now, it seems they can be modeled with infinite detail by conceiving of them as fractals—like the coastline in Mandelbrot's celebrated essay, "How Long Is the Coast of Britain?" (Mandelbrot 1983) or the shape of the planet Genesis in the movie *Star Trek II: The Wrath of Khan* (Barnsley et al. 1988, p. 13).

The purpose of this chapter, therefore, is to explore some of the implications of fractals for environmental modeling. The conventional approach to explaining behavior is to assume that it is determined by exogenous forces—by access to raw materials; by physical barriers, such as rivers or mountains; by social factors, such as race or class; by power relationships, laws, and sanctions; and so on. But one of the most important implications of chaos theory is that, while exogenous factors may play a part in explaining behavior, there are also endogenous factors at work. In other words, the dynamics of interaction may themselves shape the patterns in which we live. In fact, even when the effect of forces like those just described seems clear and undisputed (in the sense that we know the underlying, scientific laws that are at work), behavior may still be unpredictable. This may seem like an obvious point; yet it is frequently omitted from any serious discussion of environmental issues. In the end, fractal dynamics may provide the indispensable framework of understanding, the essential stage on which all the other factors play their part. This is the point we shall try to illustrate here with a simple, artificial model (Simon 1981).

---

1.   Among the standard references are Mandelbrot 1983; Peitgen and Richter 1986; and Barnsley et al 1988. Gleick 1987 and Levy 1992 provide more popular, general introductions to chaos theory; see also Cartwright 1991 for some implications for environmental planning. My favorite exposition of the Mandelbrot set (one of the stars of fractal theory) is Dewdney 1989, in which the set is explained in terms of taking a ride on a "Mandelbus".

Purpose of the Model

To illustrate how the dynamics of interaction at the local level can affect behavior at the global level, the model in this chapter is as simple as we could make it. The model consists of two idealized populations in a two-dimensional grid, just like the dots and ones of Life and Choice in chapters 12 and 13. The object of the the model here is to examine how the spatial distribution of these populations varies over time, subject not to random influences but to variation in its transition rules. If these rules play little or no part in the evolution of the model's results, we should expect to see smooth, uniform patterns of change. If the transition rules play an important role, we should expect to see chaotic or fractal behavior in the spatial distribution of the two populations.

As noted, the model contains the simplest possible kind of nonhomogeneous population: cellular automata that can have one of two possible states. The meaning of these two states can be interpreted in any way you like—as long as each is independent of the other, except to the extent specified in the transition rules. Similarly, the transition rules are made as simple as possible. The present state of any cell is determined entirely by its own past state and that of its four orthogonal neighbors. This set of cells is sometimes called a "von Neumann neighborhood" (von Neumann 1966)—in contrast to a Moore neighborhood (Moore 1962), which includes the four diagonal neighbors as well. Since there are five cells in a von Neumann neighborhood and each cell can have one of two possible states (here, a dot or a one), the neighborhood as a whole has 32 (or 2^5) possible configurations. By contrast, a Moore neighborhood of nine cells has 512 (or 2^9) possible configurations.

The purpose of the model presented here is to simulate the behavior of two different communities of cellular automata in a von Neumann neighborhood under different sets of transition rules. In Life and Choice, the transition rules were fixed and we concentrated on the effect of different initial states. Here it is the transition rules we want to be able to vary. In any model based on a von Neumann neighborhood, we need (as noted) transition rules covering a total of 32 possible configurations.[2] Even in such a simple model, however, the simulation possibilities are truly

---

2.   In practice, of course, a single rule may apply to several configurations. In Life, for example, the birth rule covers all 56 cases in which an empty cell is surrounded by three (but only three) live cells. But, in principle, there must be a rule for every possible configuration.

mind-boggling. With the model shown here, you can create more than 4 billion (2^{32}) different sets of transition rules.[3] Moreover, each of these games can be run with what is, for all practical purposes, an infinite number of possible initial states. Many, if not most, of these variations will prove uninteresting but the possibilities are there. Needless to say, we will not be able to consider all of them here.

Conceptual Basis

To create all these different simulations, we need a "Game Machine" that can be programmed with any desired arrangement of the 32 transition rules. Because it deals with automata in grid-like neighborhoods, such a machine is sometimes called a cellular automaton machine. Spreadsheets also have a grid-like structure; so they provide a natural environment for programming cellular automata.

Like Life and Choice, the Game Machine presented here consists of a grid, 20 cells by 20 cells. The individual cells can be thought of as plots, blocks, housing units, trees, grains of sand, airborne particulates, or whatever makes sense in terms of the phenomenon you want to model. Each cell can have one of two values (we use dots and ones again, as in previous chapters), and these can be interpreted in any way that makes sense in the context of the simulation. Here, we shall take a dot to mean that a cell is empty or "off" and a one to mean that a cell is occupied or "on". Each cell in the model is regarded as forming the center of a neighborhood consisting of five members: the central cell itself and its orthogonal neighbors to the West, North, South, and East.

As noted, such a neighborhood can have any one of 32 possible configurations. The state of any cell in the current cycle of the game is a function of the state of its neighborhood in the previous cycle. In order to run the model, the user must provide both an initial distribution of dots and ones (which may be done either manually or by means of a random procedure) and a set of 32 transition rules.

---

3.   If we had a model of a Moore neighborhood with 512 transition rules, there would be 2^{512} different games we could play. According to Toffoli and Margolus (1987, p. 30), the number 2^{512} is so large that it is roughly the square of the estimated number of elementary particles in the entire universe!

Once the model is initialized and running, there are various global patterns that we can look for. Among these are: parity, or balance; lack of parity, or distinctiveness; consistency, such as uniform rows and columns; symmetry; diversity; and so on. These are discussed further below, after we have discussed how the model works.

Data Requirements

As in Life and Choice, the user of the Game Machine must provide an initial population distribution. The Game Machine also requires that the user specify the effect on the central cell that will result from each of the 32 possible combinations of dots and ones in the five-cell neighborhood.

Overview

The Game Machine model is shown in Listing 14.1. It consists of four sections arranged one above the other: a display and data-entry area, a working area, a lookup table, and three macros.

The top part of the model provides the user interface: a square, 20-cell-by-20-cell, display area in which the user enters an initial configuration for each run of the model and where the computer displays the results of each cycle. To the left of the grid is a running record of the state of the model, the transition rules currently in effect, and a list of the commands available for operating the Game Machine.

The core of the model is below this, in a grid similar to the display area above it. The Game Machine is basically cyclical in nature. It works by reading the population displayed in the upper grid; calculating and displaying, in the lower grid, a revised configuration for the next cycle, as dictated by the transition rules; and then writing (copying) the new configuration from the lower grid back up to the display grid. This process is repeated, over and over, for each succeeding cycle.

The third part of the model contains a lookup table that performs two functions. First, it translates each of the 32 possible neighborhood configurations into a unique numerical value. Second, it links that value to whatever transition the user has stipulated in the rules. Let us look at these two steps in more detail.

Listing 14.1

```
    | A |    | D |    | G |    | J |
 1  THE GAME MACHINE
 2  © 1993 by T.J. Cartwright
 3
 4  Cycle-->    1   Pop---->    0
 5
 6  CWNSE     CWNSE     CWNSE     CWNSE
 7  00000 0   01000 0   10000 1   11000 1
 8  00001 0   01001 0   10001 1   11001 1
 9  00010 0   01010 0   10010 1   11010 0
10  00011 0   01011 1   10011 0   11011 1
11  00100 0   01100 0   10100 1   11100 0
12  00101 0   01101 1   10101 0   11101 1
13  00110 0   01110 1   10110 1   11110 1
14  00111 1   01111 1   10111 1   11111 1
15
16  Alt-I  =  initialize game
17  Alt-S  =  start game
18  Alt-Z  =  randomize
19  Ctrl-Break  =  abort
20
21
22
23
24
25
26
27  WORKSHEET
28
29  Calculations for Next Cycle
30
31  New game cycle ----->      2
32
...
58
```

```
     | A |    | D |    | G |   | J |
59   LOOKUP TABLE             Count  Value
60
61                            0   0
62                            2   0
63                            4   0
64                            6   0
65                            8   0
66                           10   0
67                           12   0
68                           14   1
69                           16   0
70                           18   0
71                           20   0
72                           22   1
73                           24   0
74                           26   1
75                           28   1
76                           30   1
77                           32   1
78                           34   1
79                           36   1
80                           38   0
81                           40   1
82                           42   0
83                           44   1
84                           46   1
85                           48   1
86                           50   1
87                           52   0
88                           54   1
89                           56   0
90                           58   1
91                           60   1
92                           62   1
93
94
95
96   MACROS          Commands                 Comments
97
98   {\S}            {home}                   home cursor
99                   {paneloff}               turn dialog line off
100                  !                        [G100] manual recalc
101                  {if J4=0} {branch G105}  exit if population reaches zero
102                  /c J31,D4,v              update number of cycles
103                  /c M31:AF50,M1,v         copy results to display area
104                  {branch G100}            loop until stopped
105                  {beep} {panelon}         [G105] turn dialog line on again
106
107  {\I}            {home}                   home cursor
108                  {let D4,1}               set number of cycles to one
109                  {let M1," ."}            assign value " ." to f1
110                  /c M1,M1:AF20,v          copy " ." throughout display
111                  !                        manual recalc
112                  =v10 ~ {beep}            go to middle of display
113
114  {\Z}            {home}                   home cursor
115                  {let D4,1}               set number of cycles to one
116                  {let M1," ."}            assign value " ." to f1
117                  /c M1,M1:AF20,v          copy " ." throughout display
118                  =M1 ~                    go to upper left corner
119                  {if ran>.5} 1            [G119] replace with "1" at random
120                  {if currow<20} {down} {branch G119}
121                  {if curcol<32} {up 20} {right} {branch G119}
122                  ~{home}
123                  ! {beep}                 manual recalc
```

Consider the pattern 10110, which can be found in cell G13. This cell is near the bottom of the third of the columns labeled "CWNSE" (Center-West-North-South-East) in the upper left part of the worksheet. This particular pattern (10110) signifies a one in the Center cell, a dot (or zero) in the West cell, a one in the North cell, a one in the South cell, and a dot (or zero) in the East cell. This can be visualized like this:

```
        1
      • 1 •
        1
```

In the model, this pattern is converted to a single numerical value by multiplying each digit (each 0 or 1) in the pattern by the power of two corresponding to the position of the digit in the pattern; then the results are added together. Thus, the pattern 10110 is translated into the number 44 like this:

$$1 * 2^5 + 0 * 2^4 + 1 * 2^3 + 1 * 2^2 + 0 * 2^1 = 44$$

There is no particular significance to this result (44). The important thing is that no other five-digit pattern of ones and zeros except 10110 generates the same result. In other words, this method produces a unique numerical code for each of the 32 transition rules.[4]

The second step is to link each rule, through its code, to a particular consequence for the Center cell. Consider the pattern 10110 again in cell G13. Suppose the user wants to stipulate that such a neighborhood causes the Center cell (now a one, as indicated by the first digit in the pattern) to remain a one. This rule is programmed into the model by entering a one in the cell beside the code (i.e., by entering a one in cell H13). As soon as this is done, this value (a one) is copied down to the lookup table in cells A59:K92. Specifically, the value (one) is copied into the cell beside the number 44 (i.e., the cell beside cell J83). The result of all this is that whenever a cell in the grid corresponds to a pattern that is in turn converted (using the process described above) into a value of 44 and that value (44) is looked up in the lookup table, a one is found. In this way, the computer "knows"

---

4.   It does not matter what coding sequence is used for the cells in the neighborhood (e.g., Center first, West before North, etc.), as long as the same sequence is used in all of the transition rules.

that the Center cell in such a neighborhood is supposed to remain a one in the current cycle.

The bottom part of the model consists of three macros that initialize and run the model. There is a macro (Alt-S) to start the model and run it, until the user stops it or until one population disappears altogether. There is another macro (Alt-I) to initialize the model, restoring it to a clean slate after it has been run. And there is a third macro (Alt-Z) that fills up the display with a random pattern of dots and ones, if you prefer to start with that. The macros are almost identical to those used in the Life and Choice models in chapters 12 and 13.

Operation

Much of the structure of the Game Machine is similar to that of the two previous models, but there are also some important differences. The most obvious is in the top part of the model, where there is provision for specification of the transition rules. These are arranged in four columns (columns A, D, G, and J) each headed "CWNSE" (for Center, West, North, South, East). Beside each of the columns is another column (columns B, E, H, and K) in which the user enters the desired transition state—either a one, or (in lieu of a dot) a zero.

As in Life and Choice, the lower grid is matrix of formulas, this time using the LOOKUP() function to access the table lower down. For example, the formula in cell R40 is as follows:

```
IF(LU((R10=1)*2^5+(Q10=1)*2^4+(R9=1)*2^3+
   (R11=1)*2^2+(S10=1)*2,$J$61:$J$92)>0,1," .")
```

What this formula means is this: if the LOOKUP value of the neighborhood of cell R40 (defined as 32 times cell R10 plus 16 times cell Q10 plus 8 times cell R9 plus 4 times cell R11 plus twice cell S10) in the table in cells J61:J92 is greater than zero, then display a one; otherwise display a dot.[5]

---

5.   Note that individual cells in the neighborhood are evaluated by means of a Boolean operation (e.g., R10=1). Spreadsheets evaluate equalities inside formulas like this as either true (1) or false (0); so this provides an efficient way of evaluating each pattern. See also appendix 1, below, on Boolean operations.

In a nutshell, this is exactly how the Game Machine works. Calculate the pattern value for a neighborhood and look up its transition value in the table below. This table is derived from instructions entered in the upper left-hand corner of the model. In this way, the user can specify whatever transition rules he or she may desire.

Interpretation and Use

As noted previously, the simulation possibilities in the Game Machine are potentially enormous. Cellular automata have been studied for several decades now; so quite a lot is known about them (Codd 1968; Preston and Duff 1984). Inspired by the model (Toffoli and Margolus 1987) on which this one is based, let us consider some of the possible scenarios.

One of the simplest we can simulate is that of unconstrained growth. This is the case where one of the populations (say, the ones) never dies and all that is required to convert a dot into a one is for there to be another one somewhere nearby (i.e., in the neighborhood). To simulate this kind of situation, the following procedure can be followed (ibid., pp. 37-38):

- First, edit each transition rule to match the state of the Center cell. In other words, if the Center cell is a one, enter a one beside the rule; if the Center cell is a zero, enter a zero beside the rule.

- Second, edit each rule containing a one in any other cell (i.e. in the West, North, South, or East cell) to set the Center cell to a one (even if you set it to zero in the previous step).

The result is that all but one of the 32 rules stipulate a one for the Center cell in the next cycle. The exception is the very first rule (00000), which causes the Center cell to remain a zero or dot. As might be expected, as soon as the simulation is initialized and run, ones spread rapidly outward from any cells where they existed originally, until there are no dots left and the grid is entirely filled with ones.

Obviously, this is a rather extreme case. If we alter the transition rules to require at least two neighboring ones (instead of just one) as a condition for growth, we get a more stable world. This is the case even if we retain the previous assumption that, once they exist, ones never die.

To create this scenario, proceed as follows:

- First, edit each transition rule to match the state of the Center cell.

- Second, edit every rule containing a one in at least two other cells to set the Center cell to a one.

This time, 27 of the rules stipulate a one in the Center cell, and 5 stipulate a dot or dot (or zero).[6] When the simulation begins, the number of ones immediately starts to grow, but only up to the point where the rectangle surrounding each original seed is filled. Then growth stops and the world stabilizes, since none of the ones ever dies and none is capable of breaking into a new neighborhood on its own.

More interesting are cases of constrained growth. In one such scenario, growth can be defined to occur when there is one (and only one) one in a cell's neighborhood. You can set up this scenario as follows:

- First, edit the transition rules to match the state of the Center cell.

- Second, edit only those rules containing a single one in the other cells to set the Center cell to a one.

The result in this case is that 20 of the rules stipulate a one in the Center cell and 12 of the rules stipulate a dot (or zero). This time, when the Game Machine goes to work, growth proceeds outward from any initial seeds in a pattern that is fractal in nature (ibid., pp. 39-40). Thus, growth continues until ones nearly, but never completely, fill the grid.

So far we have focused on growth. In all the scenarios up to this point, none of the ones disappears once it has come into existence. So let us see if we can introduce the notion of death as well as birth. Suppose we begin by trying to imitate Life in the Game Machine. Since Life is based on a von Neumann neighborhood and the Game Machine on a Moore neighborhood, we cannot exactly simulate Life with

---

6. Of course, the initial distribution for this scenario must include at least one neighborhood containing two or more ones; otherwise, nothing happens.

the Game Machine. But we can see what happens if we scale down Life's criteria of birth and survival to reflect the smaller neighborhood of the Game Machine.

The key feature of Life (in contrast to the growth models discussed up to now) is that the default transition rule for the Center cell is not maintenance of the status quo; it is death. In other words, to set up a Life-like scenario in the Game Machine, the procedure is as follows:

- First, edit all the transition rules to set the Center cell to a dot (or zero).

- Second, edit only those rules where birth or survival in the Center cell is justified to set the Center cell to a one.

In real Life, three ones in the neighborhood are required for birth, and two or three ones for survival (or continued life). In the Game Machine, however, this would mean only 14 out of the 32 rules would create or maintain Life: 4 rules governing birth, where the Center cell is a dot and there are 3 ones in the rest of the neighborhood; and 10 rules governing survival, where the Center cell is a one and there are 2 or 3 ones in the rest of the neighborhood. But in the smaller neighborhood of the Game Machine, these standards are too high. At best, the ones end up reduced to a minimal rectangular core; at worst, they are wiped out.

If we lower the standards to, say, two neighboring ones for birth and one or two for survival, some Life-like behavior appears. Basic Life forms of two or three cells are stable, as in real Life, and some larger patterns also stabilize. But it is more difficult to find forms that grow, travel, and throw off other forms. In short, Life in a von Neumann neighborhood is a lot bleaker than in a Moore neighborhood.

The advantage of the Game Machine, however, is that it allows us to experiment with some more anomalous conditions for Life. For example, imagine a version of (von Neumann) Life in which birth occurs when and only when there is one additional one in the neighborhood, and that survival occurs when there is either one or four (but not two or three) additional ones. In this case, the result is a strange world of far-flung communities moving dramatically about the space. Life forms that flourish in the Moore version (such as the "r-pentomino", the "pi-pentomino", and the "acorn") exhibit some of their old character. At the same time, quite complex Life forms can suddenly disappear in this version, as though without warning. In this harsher, more polarized version of Life, survival is the name of the Game.

Another possible scenario is decision-making by voting. In this case, the transition rules are set to simulate majority voting: i.e., if there were three, four, or five ones in the neighborhood in the previous cycle, the Center cell becomes a one; if not, the Center cell becomes a dot. The effect is that 16 of the 32 transition rules lead to a one in the Center cell, 16 lead to a dot. Even if this simulation run is seeded with a random distribution, the two populations quickly consolidate into a few homogeneous domains or blocks. Throughout the simulation, the combined size of the domains of each population conforms more or less to its initial populations.

An interesting variation on this voting principle has been proposed by Gérard Vichniac (cited in Toffoli and Margolus 1987, pp. 41-42). In what we might call the "Vichniac variation", the principle of majority rule is modified by reversing the outcome for the two "closest" votes. In other words, in the two cases where the "vote" is 3-2 (3 ones and 2 dots, or 2 ones and 3 dots), the "losers" are made into the winners. In practice, this means that neighborhoods of 2, 4, or 5 ones ones generate a one in the Center cell, whereas neighborhoods of 0, 1, or 3 ones generate a dot in the Center cell. The effect of these variations is to bring about a kind of

> reshuffling at the boundary . . . where the majority is [by definition] marginal. The net effect is one of gradual *annealing* of domains: in the long term, each cell behaves as if the vote reflected not only the state of the immediate neighbors, but also, with decreasing weights, that of cells that are further and further away from it. Domains form as before, but now the boundaries are in continual ferment; each cell can "feel", so to speak, the curvature in its general vicinity, and will dynamically adjust its state so as to make the boundary straighter [Toffoli and Margolus 1987, pp. 41-42]. . . .

As far as is known, the fate of a Vichniac simulation is unpredictable. In some cases, even when the two populations start out on a more or less equal footing, one of them eventually dominates the other, almost to the point of annihilation. In other cases, the populations seem to work out a modus vivendi and neither predominates. But no one yet knows why or how this occurs.

Voting paradigms like these focus attention on the quantitative character of neighborhoods. But we can alter some of their qualitative features as well. For example, we can try simulating *parity* rule instead of *majority* rule. This idea was first mooted by Ed Fredkin as early as 1971 (ibid., p. 30; see also Levy 1992, p. 60ff). In this scenario, each transition rule is set so that the Center cell reflects the parity (oddness or evenness) of its neighborhood in the previous cycle. Thus, if the number of ones in a neighborhood is an odd number (i.e., one, three, or five), then the Center cell becomes a one. If the number of ones in a neighborhood is an even number (i.e., zero, two, or four), then the Center cell becomes a dot. If this kind of scenario is seeded with a small square block of ones, the result is a pulsating "Persian rug" type of pattern that stabilizes with a fairly long periodicity. The same scenario can be run in such a way that the Center cell reflects not the parity of its neighborhood but the opposite.

One interesting feature of parity simulations is that they are essentially *linear*, according to Toffoli and Margolus (1987, p. 32). This makes it possible to provide formal proofs of certain kinds of behavior. For example, Toffoli and Margolus have found that, in parity simulations, "waves emerge unaffected after going through one another." Among other things, this means that parity simulations are unaffected by the toroidal topology of the grid used in this model to simulate infinite space. "Another property that can be proved . . . is that, for any initial figure on a uniform background, this figure will be found exactly reproduced in five copies after a suitable lapse of time (and later on in twenty-five copies, etc.)" Such self-similarity at varying scales is, incidentally, one of the hallmarks of fractal forms.

Another more complex scenario has been proposed by Edwin Banks (Banks 1971; cited in Toffoli and Margolus, pp. 42-44). In this case, the idea is to organize the transition rules so as to "fill in the gaps" and "round off the corners" in the spatial distribution of the two populations. To set up this kind of scenario, the procedure is as follows:

- First, edit all the transition rules to match the state of the Center cell.

- Second, for any rule in which the Center cell is a dot, set the rule to a one (i.e., "fill in the gap") if either one or both pairs of neighborhood opposites (i.e., North and South, and/or West and East) are also ones.

- Third, for any rule in which the Center cell is a one, set the rule to a dot (i.e., "round off the corner") if one (and only one) pair of adjacent cells (i.e., North and East, or North and West, or South and East, or South and West) are ones.

In this case, the model begins to display some quite different properties.

If we run this [scenario] starting from random initial conditions, after a few steps we obtain [a] pleasant but undistinguished texture. . . ; in a few places we can make out little pockets of activity, with signals shuttling back and forth. Can we "tame" this activity, and turn it to more purposeful tasks?

It turns out that signals can be made to run on "wires" [in this scenario]. If . . . you cut a slanted notch on the edge of a solid black area [i.e., by removing some ones from a solid block of them], the notch will move one position at every step [cycle]; with the opposite slant, the notch will move in the opposite direction. To support [such] a signal, the black area [i.e., the block of ones] need only be three cells deep [Toffoli and Margolus 1987, p. 44]. . . .

Note that a one-cell "serif" has to be added to the ends of the wire to prevent them from unraveling.

Table 14.1 illustrates how the Game Machine can be set up for this scenario. The minimum thickness of the wire is three cells, but it can be wider than that. When the simulation is run, the notch in cells W17 and V18—it could equally well be anywhere else on the wire—travels up the wire until it reaches the end; then it disappears and the wire remains stable and whole.[7]

You can also try cutting a notch in the ends of the wire instead of the side (i.e., changing the ones in cells V2 and V19 to dots). This time, a lot of "noise" develops along the wire, until eventually the wire is reduced to a single strand. Make the wire one row shorter and it recovers completely from the noise and regains its full thickness. Make it one row longer and the noise oscillates back and forth along the wire. With patience, Banks was able to demonstrate that signals on wires could be made to turn corners, cross each other, and fan out into space. The M.I.T. group reported (ibid., p. 44) that "it is possible to build a 'clock' that generates a stream of pulses at regular intervals, much like the 'glider gun'" in Life.

Another way of using the Game Machine is just to experiment with different sets of transition rules. While many of these produce results that are really quite uninteresting, some are quite intriguing. For example, you can try setting the 32

---

7.   Alternatively, you can remove the serifs and extend the wire to the top and bottom of the grid. Then, because the top and bottom of the grid are connected, the notch will travel forever.

Table 14.1: Game Machine Initial Configuration for Sending a Signal along a Wire

```
     | A |    | D |   | G |   | J |
 1   THE GAME MACHINE                       . . . . . . . . . . . . . . . . . . . .
 2   © 1993 by T.J. Cartwright             . . . . . . . . 1 1 1 1 . . . . . . . .
 3   ─────────────────────────────         . . . . . . . . 1 1 1 . . . . . . . . .
 4   Cycle-->    1   Pop---->    0         . . . . . . . . 1 1 1 . . . . . . . . .
 5   ─────────────────────────────         . . . . . . . . 1 1 1 . . . . . . . . .
 6   CWNSE    CWNSE    CWNSE    CWNSE       . . . . . . . . 1 1 1 . . . . . . . . .
 7   00000 0|01000 0|10000 1|11000 1       . . . . . . . . 1 1 1 . . . . . . . . .
 8   00001 0|01001 1|10001 1|11001 1       . . . . . . . . 1 1 1 . . . . . . . . .
 9   00010 0|01010 0|10010 1|11010 0       . . . . . . . . 1 1 1 . . . . . . . . .
10   00011 0|01011 1|10011 0|11011 1       . . . . . . . . 1 1 1 . . . . . . . . .
11   00100 0|01100 0|10100 1|11100 0       . . . . . . . . 1 1 1 . . . . . . . . .
12   00101 0|01101 1|10101 0|11101 1       . . . . . . . . 1 1 1 . . . . . . . . .
13   00110 1|01110 1|10110 1|11110 1       . . . . . . . . 1 1 1 . . . . . . . . .
14   00111 1|01111 1|10111 1|11111 1       . . . . . . . . 1 1 1 . . . . . . . . .
15   ─────────────────────────────         . . . . . . . . 1 1 1 . . . . . . . . .
16   Alt-I  =  initialize game             . . . . . . . . 1 1 1 . . . . . . . . .
17   Alt-S  =  start game                  . . . . . . . . 1 1 . . . . . . . . . .
18   Alt-Z  =  randomize                   . . . . . . . . 1 . 1 . . . . . . . . .
19   Ctrl-Break  =  abort                  . . . . . . . . 1 1 1 1 . . . . . . . .
20   ─────────────────────────────         . . . . . . . . . . . . . . . . . . . .
```

Table 14.2: Randomly Selected Transition Rules for Use with Listing 14.1

	Column B	Column E	Column H	Column K
Row 7	0	0	1	0
Row 8	0	0	0	1
Row 9	0	0	0	0
Row 10	0	0	1	1
Row 11	1	0	1	1
Row 12	0	0	0	0
Row 13	0	0	0	0
Row 14	0	1	0	1

Source: Adapted from Tommaso Toffoli and Norman Margolus, <u>Cellular Automata Machines: A New Environment for Modeling</u> (Cambridge, Mass.: M.I.T. Press), p. 45.

(in the order shown in the model) to the sequence of values shown in Table 14.2. In this case, the result is a profusion of fractal patterns (ibid., pp. 29, 45) that vary according to the initial distribution or seed that is used. There are, of course, many other possibilities besides this one—something like 4 billion of them, in fact!—which we cannot explore here.

Evaluation and Extension

There are several ways in which the Game Machine can be extended. For example, the present model is based on two matrices or grids: the display area, which effectively "remembers" the current cycle, and the working area, which calculates the next cycle based on the current one. We could create a third grid (although performance might become painfully slow), so that the model could remember not just the current cycle but also the previous cycle. Then we could create decision rules in which future actions were based on past as well as present experience.

What is even more intriguing about a three-stage model is that you could swap "past" and "present". This would create the effect of traveling backward instead of forward in time (ibid., chapters 6 and 14). In this way, we could not only simulate a future world; we could also "reverse engineer" a past that had never existed!

Alternatively, the third grid could be used to expand the binary state of each cell—now, either a 1 or a 0 (or a dot)—to four different states: namely, 00, 01, 10, and 11. In this way, we could add a little color or shading to our simulations (ibid., chapter 7).

Another possibility is not to expand the Game Machine at all but to interpret it in a different way. Instead of thinking of it as a spatial model in two dimensions, we could think of it as having one spatial dimension (in, say, the horizontal) and one chronological dimension (in, say, the vertical). In its present form, this would mean the model could have a 20-cycle history built in. Thus, for example, the top row could be defined as the world in the current cycle and then be pushed down one row in each succeeding cycle, as a new world took its place at the top. In such a case, of course, the spatial neighborhood would be limited to the Center cell and its Western and Eastern neighbors; there would be no North cell or South cell.

Such a neighborhood might seem a lot more limited than a von Neumann one. But it has at least two advantages. For one thing, we could create a kind of one-dimensional version of the von Neumann neighborhood by including not just adjacent cells but adjacent pairs of cells. Such a neighborhood would consist of five cells: namely, Center, West, East, West of West, and East of East. Stephen Wolfram (1984), Charles Bennett (cited in Toffoli and Margolus 1987, p. 97), and others have worked with neighborhoods like these. Such neighborhoods are topologically identical to von Neumann ones and require 32 transition rules. Yet they invite different scenarios and their evolutionary behavior is not at all the same.

The second advantage of modeling one-dimensional neighborhoods is that they suggest more extensive vertical, or chronological, linkages than does a von Neumann neighborhood. In a "chronological neighborhoods, one does not feel som impelled to link influence to spatial contiguity. That is, we could modify the Game Machine to include transition rules so that they link the current state of a cell to the state of cells not only in the previous cycle but over several previous cycles—up to 20 previous cycles in the model shown here. Compared to the Game Machine that is described earlier in this chapter, this new version effectively plays down the influence of "environment" and plays up the influence of "heredity". Furthermore, it is easier in such a scenario to rationalize neighborhoods of six or seven cells: for example,

Other configurations are also possible, of course. The key point is that neighborhoods of six or seven cells are more sophisticated than a von Neumann neighborhood (which contains only five cells). Yet they are more attractive from a computational standpoint than a Moore neighborhood (which contains nine cells), because they require only 64 (2^6) or 128 (2^7) transition rules instead of the 512 (2^9) required for a Moore neighborhood.

Norman Margolus (Toffoli and Margolus 1987, chapter 12) has been able to study spatial neighborhoods of six, seven, and eight cells—that is, neighborhoods that are larger than von Neumann ones but smaller than Moore ones. Some of these are based on alternating blocks of two cells by two cells. Using such models, Margolus found he could generate patterns that are fractal in nature. He has also discovered rules that produce "critters" which display "a variety of structure and a liveliness reminiscent of LIFE":

If we start from a very nonuniform initial state (such as a blob of randomness . . .) we'll see a rich evolution. Small "critters" race horizontally and vertically; when they collide they may bounce back or execute a right-angle turn; occasionally they stick together—at least until freed by being hit again—or even pile up to form complicated little pieces of circuitry [ibid., p. 133].

Similar results have been obtained with other models in other contexts.[8]

In summary, the possibilities inherent in the Game Machine are staggering. There seems to be no shortage of ingenious and thought-provoking variations and extensions that can be made to the basic model. Perhaps the most sobering lesson of all is the overwhelming variety inherent in even the simplest of models. Here we have a world of only 400 individuals (in a grid of 20 cells by 20 cells) who can behave in one of only two different ways (a one or a dot, life or death, on or off, etc.). Yet, if the behavior of each individual is influenced by that of only eight others, not all the computers in the world with all the time in the world could simulate all possible outcomes. The aims of science, it seems, are nothing if not ambitious.

Bibliography and References

Banks, Edwin. 1971. "Information Processing and Transmission in Cellular Automata." Project MAC Technical Report TR-81. Massachusetts Institute of Technology, Cambridge, Mass. Mimeo.

Barnsley, Michael, et al. 1988. *The Science of Fractal Images.* With a foreword by Benoit Mandelbrot. Berlin: Springer-Verlag.

Berlekamp, E.R., J.H. Conway, and R.K. Guy. 1982. "What Is Life?" In *Winning Ways for Your Mathematical Plays.* Vol. 2, *Games in Particular*, chapter 25. London: Academic Press.

---

8.   See Cartwright 1991a for discussion of the behavior of "Tur-mites" in what is called a two-dimensional Turing-machine environment.

Burks, Arthur W., ed. 1970. *Essays on Cellular Automata*. Urbana: University of Illinois Press.

Cartwright, T.J. 1991a. "Experimental Systems Research: Towards a Laboratory for the General Theory." *Cybernetics and Systems: An International Journal*, 22.1 (March), pp. 135-49.

———. 1991b. "Planning and Chaos Theory." *Journal of the American Planning Association*, 57.1 (Winter), pp. 44-56.

Choffrut, C., ed. 1988. *Automata Networks*. Lecture Notes in Computer Science, no. 316. Berlin: Springer Verlag.

Codd, E.F. 1968. *Cellular Automata*. New York, Londong: Academic Press.

Dewdney, A.K. 1989. "A Cellular Universe of Debris, Droplets, Defects and Demons." *Scientific American*, 261.2 (August), pp. 102-5.

———. 1989. "A Tour of the Mandelbrot Set aboard the Mandelbus." *Scientific American*, 260.2 (February), pp. 108-11.

Fogelman Soulie, Francoise, Yves Robert, and Maurice Tschuente, eds. 1987. *Automata Networks in Computer Science: Theory and Applications*. Manchester: Manchester University Press.

Gardner, Martin. 1983. *Wheels, Life and Other Mathematical Amusements*, especially chapters 20-22. New York: Freeman.

Gleick, James. 1987. *Chaos: Making a New Science*. New York: Viking Penguin.

Helmers, Carl. 1975-76. "Lifeline," parts 1-4. *BYTE*, 1 (September), pp. 72-80; 1 (October), pp. 34-42; 1 (December), pp. 48-55; and 2 (January), pp. 32-41.

Levy, Steven. 1992. *Artificial Life: The Quest for a New Creation*. London: Jonathan Cape.

Macaluso, Pat. 1981. "Life after Death." *BYTE*, 6 (July), pp. 326-33.

Mandelbrot, Benoit. 1983. *The Fractal Geometry of Nature*. New York: Freeman.

Moore, Edward F. 1962. "Machine Models of Self-Reproduction." In Burks 1970, essay 6. First published in *Proceedings of Symposia in Applied Mathematics*, 14, pp. 17-33.

Preston, Kendall, Jr., and M.J.B. Duff. *Modern Cellular Automata*. New York, London: Plenum.

Peitgen, Hans-Otto, and P.H. Richter. 1986. *The Beauty of Fractals*. Berlin: Springer-Verlag.

Poundstone, William. 1985. *The Recursive Universe*. New York: Morrow.

Simon, Herbert. 1981. *The Sciences of the Artificial*. Cambridge, Mass.: M.I.T. Press.

Toffoli, Tommaso, and Norman Margolus. 1987. *Cellular Automata Machines: a New Environment for Modeling*. Cambridge, Mass.: M.I.T. Press.

Von Neumann, John. 1966. *Theory of Self-Reproducing Automata*, edited by Arthur W. Burks. Urbana: University of Illinois Press.

Wolfram, Stephen. 1984. "Computer Software in Science and Mathematics." *Scientific American*, 251.3 (September), pp. 188-203.

Part Four

THE WORLD IN A
SPREADSHEET

MODELING WITH MICROCOMPUTERS

I can get along very well without such a program.
After all, it's clear to anyone who's ever thought about it
that—I mean, it's not a very difficult matter to resolve,
once you understand how—or rather, conceptually,
one can clear up the whole thing by thinking of,
or at least imagining a situation where . . . .
Hmmm . . . I thought it was quite clear in my mind.

Achilles to the Tortoise,
in D.R. Hofstadter, Gödel, Escher, Bach: An Eternal Golden Braid

The theme of this book is that simulation modeling is a useful way of learning about the world and our place in it, and that microcomputers and spreadsheets provide a convenient way of doing it. Perhaps readers may also have concluded that modeling on microcomputers is also good fun!

Simulation modeling can help us learn how the world works and how better to manage it. Modeling does not provide *direct* evidence of this, of course. There are no fundamental truths or "right" answers to be found in modeling. If life itself does not reveal its secrets, we can hardly expect modeling to provide the key. History does not repeat itself, at least not exactly, so there are no scientific certainties to be revealed, either by modeling or in any other way. Few scientists believe, at least since Gödel, that we shall ever achieve complete understanding of the world (Hofstadter 1980). Human capacities are finite; the world is not. It is as simple as that.

But simulation modeling can nonetheless play a useful *indirect* role in helping us to appreciate what is going on and providing a means of deciding what if anything we should do to try and influence it. Models provide an analogy to, or a surrogate for, the real world. They are not copies of the real world, so much as artificial realities in their own right. Models are finite, calculable, and predictable. So we can understand them. We can experiment with them—and with our own reactions to them. We can copy them, rerun them, speed them up, slow them down, and otherwise alter and adapt them. Models are essentially a cheap and harmless laboratory (Cartwright 1991a) for testing our understandings of the world and our place in it.

The lessons we draw from our modeling—as indeed from any experience—is, and will forever remain, a matter of very human judgment. The British politician Edmund Burke complained more than two centuries ago that the "age of chivalry" was dead and had been replaced by "that of sophisters, economists, and calculators." But it is not so much our perceptions of the world that are right or wrong, whether they are fundamentally romantic like chivalry or mechanistic like economics. The mistake is to take either on faith. The mistake is to suspend the quintessentially human capacities for intuition, judgment, and equity, and replace them with blind reliance on either ideology or technology.

For many of the world's problems, there just are no simple answers or quick fixes. As much as we might not like it, and as much as we might find such a view excessively humbling, understanding and change do not come easily and do not even come in ways that are expected. This is not a plea for passivity or fatalism. Quite the contrary. It is a plea for the patience, the perseverance, and the flexibility that are necessary in both research and intervention.

The models in this book, therefore, are opportunities, not answers. They are artifacts for examining, tools for learning, laboratories for experimenting, case studies for testing—even toys for playing with. Neither models nor history can teach; but there is much that a wise person can learn from both.

Good environmental simulation models are not exactly plentiful. Kurt Fedra of the International Institute of Applied Systems Analysis (IIASA) in Vienna recently completed a comprehensive survey of environmental impact assessment (EIA) models for the United Nations Industrial Development Organization (UNIDO). He concluded that the record was rather disappointing. According to Fedra:

> A very rough, random and preliminary screening has resulted in more than 200 models [for environmental impact assessment]. . . . However, only a few of

these models can be regarded as potentially and generally useful. . . . Most of them were never built to be generally useful. Many serve a very special, well-defined purpose. Others have been constructed at [only] a prototype level, to demonstrate a principle rather than to develop a generally useful tool for widespread application. Also, by and large, these models are developed by scientists, not by software professionals. They [the models] tend to be idiosyncratic, and most of them are not sufficiently documented . . . [and] were never designed for non-expert users [Fedra 1989, p. 36]. . . .

To illustrate his point, Fedra quotes a study of groundwater transport models (Bachmat et al 1980) which found that, out of 29 models examined, only 3 were "rated usable"! Other studies of computer-aided methods of environmental impact assessment (e.g., Luhar and Khanna 1988) tend to confirm this view.

Nevertheless, at least two basic kinds of criticisms will doubtless be made of this book. On the one hand, modeling enthusiasts will argue that the models presented here fall short of the current state of the art in environmental modeling. On the other hand, non-modelers will argue that models are by definition simplifications of a complex reality and are, therefore, inevitably misleading. There is some justice to both views, but neither is as critical as its proponents tend to assume.

Certainly, there exist more recently developed and more sophisticated models than some of those presented here. The science and art of modeling are constantly advancing in many areas. But the purpose of this book is not to present the most advanced models. Rather, this book is intended to provide an introduction to the design and use of environmental simulation models generally. Its purpose is to show how much can be achieved with no more than a microcomputer and some off-the-shelf software. Besides, new models do not always replace old ones. Nor do complexity and sophistication necessarily make for better models—or, at least, for more useful ones. So the models in this book should be judged more for their operational usefulness than for their technical sophistication. This is not to disparage science or to suggest that the models presented here are somehow less scientific than others. My point is that there is a place for simple, easy-to-understand models, even when more complex models may be available. In short, the aim of this book is to contribute more to the diffusion of modeling than to its advancement. This book is aimed at people who have not done much modeling up to now—because they think it is too difficult or too time-consuming, or because they never dreamed it might be useful or even possible. My aim is to spread the idea that modeling is something that anyone can do, and that modeling is often a worthwhile exercise.

By contrast, there will be some readers who mistrust the very idea of modeling. Models are *by definition* something less than the reality they purport to represent. So, the argument goes, how can we possibly rely on them? The point is well taken—but it is not confined to models! In a sense, *all* our perceptions of the "real world" are incomplete, since they are (in the end) finite representations, or images, of an infinitely complex world (Boulding 1969). Thus, we should always be cautious about relying on what we think we know, regardless of what the source of that knowledge might be. There is nothing inherently good or bad, right or wrong, about choosing to program, formally and explicitly in a model, what we think we know about the world. The risks arise not in creating the model but in deciding how far we are prepared to trust it.

Judging the limits of modeling is often exceedingly difficult. On the one hand, a clearly articulated model is in some respects more honest than an understanding or appreciation whose nature and assumptions are less explicit. Certainly, the assumptions and rules of a model are obvious, but I am not convinced that that makes them *a priori* more or less significant than those of other techniques. On the other hand, it may be that models are somehow psychologically more compelling to people nowadays than arguments based on intangible factors like experience and skill. If so, it may be that the use of formal models demands greater care and responsibility than the use of other kinds of techniques. But I am not really convinced of that either. In the end, what strikes me as curious is that some people regard modeling is either a virtue or a vice in itself.

Modeling, like any other research or planning technique, is a means to an end. The value of modeling lies in the hand that holds it. Certainly, there are disciplines and professions which have suffered from exaggerated claims on behalf of "quantitative revolutions" or "large-scale modeling". But that is our mistake; it is not the techniques themselves that are at fault. The best defence against this kind of mistake, it seems to me, is demystification. The more familiar we become with models and with modeling, the less likely we are to be fooled by excessively optimistic claims for their results.

Even so, there is a hidden assumption in all of this. The assumption is that quantitative models tend to bias our outlook. It could be equally be argued that our understanding of the world has up to now been biased by our primitive ability to compute. In reflecting on their research with cellular automata, for example, Toffoli and Margolus write:

> Science is concerned with explaining things. . . . We say we "understand" a complex system when we can build, out of simple components that we already understand well, a model that behaves in a similar way.

> If the assortment of components at our disposal is too lavish, it is often too easy to arrive at models that display the expected phenomenology just because the outward symptoms themselves . . . have been directly programmed in. . . [1987, p. 142].

By contrast,

> The simpler the primitives used to describe a complex system the greater is the computational burden. . . . For this reason, the development of mathematics . . . reflects to a much greater extent than many would suspect the nature of the computational resources available. . . . In the past three centuries, enormous emphasis has been given to (1) models that are defined and well-behaved in a *continuum*, (2) models that are *linear*, and (3) models entailing a small number of *lumped* variables. This emphasis does not reflect a preference of nature, but rather the fact that the human brain, aided only by pencil and paper, performs best when it handles a small number of symbolic tokens having substantial conceptual depth. . . .

> The advent of digital computers has shifted the region of optimum performance. While much progress can still be made in the more traditional areas, the horizon has dramatically expanded in the complementary areas, namely (1') *discrete* models, (2') *nonlinear* models, and (3') models entailing a large number of *distributed* variables. Such models give more emphasis to the handling of a large number of tokens of a simple nature (e.g., Boolean variables and logic functions)—a task at which computers are particularly efficient.

What is true of mathematics may also be true of science in general. While they may not all be able to claim three centuries of experience for their disciplines, most have indeed emulated mathematicians in their preference for models that are continuous, linear, and parsimonious in their use of variables. The inevitable effect has been to focus attention on macro models, macro explanations, and macro solutions. Now that we have access to microcomputers and user-friendly software like spreadsheets, we can pay more attention to research and intervention on more micro scales.

Finally, I hope that readers have had fun with some of the models in this book. Personally, I still get a kick out of watching this little machine on my desk doing all that tedious calculation, while I sit and ponder the meaning and implications of the results. From some of the models, I even get a sense of wonderment, as they impel me to think about the extraordinary complexity of life on both a microscopic and macroscopic scale. Think of the countless twists and turns of a particle of smoke as it drifts through the air; of a molecule of pollution as it travels through an aquifer; of a tree as it bends in the wind and searches for the best supply of food, water, and light; or of the grizzly bear as it struggles for survival in an ever-shrinking habitat. Think of the countless incremental adjustments that occur as individual people buy houses, choose jobs, go on trips, dispose of wastes, and so on. We are brave indeed to want to understand and manage these processes.

Suppose in any case that we are all just intelligent patterns in some massively parallel computer model; suppose (as chaos theory suggests) that free will and determinism are not fundamentally incompatible with each other. If that were the case, is it likely that we would know it was true? If we were to find out, would that threaten the functioning of the model? If we are only patterns, are we even capable of knowing that we are?[1] I do not expect my spreadsheet and computer to answer these questions for me, but they have certainly stimulated my thinking.

The future of modeling is highly promising, as microcomputers and their software become steadily more powerful. Spreadsheets, in particular, are acquiring more built-in functions and commands, as well as more elaborate graphing capabilities (Stinson 1991). Perhaps most intriguing of all, some spreadsheets now can operate with a grid, or cellular matrix, of three dimensions instead of two. None of the models in this book requires a three-dimensional grid, but several of them could be

---

1.   There are already models that appear to simulate the behaviour of genes, including natural selection, symbiosis, parasitism, "arms races", mimicry, niche formation, and speciation. See, for example, Dawkins 1986 and Holland 1992. There is a simulation of Dawkins' creatures (Biomurffs) by Damian Murphy in Prata 1993, and Oehler et al 1990 provides a more global evolutionary framework.

redesigned to take advantage of it. In addition, formal simulation frameworks ranging from the esoteric to the popular are gaining more widespread use all the time. Expert-systems software is also gaining ground—there is a neural-network program called Braincel that runs inside Lotus 1-2-3 and Excel—and that adds an exciting new dimension to simulation modeling. There are also reports that "virtual reality" simulations are finding diverse practical applications (e.g., in the stock markets) beyond the flight simulator and similar devices. In short, the potential for "desktop modeling" is likely to explode in the years to come.

So the theme of this book can be summarized as follows. Thanks to microcomputers and spreadsheets, simulation modeling is no longer an arcane activity reserved for mathematicians and computer programmers. Anyone can do it. This book contains some models to illustrate what you can do with a modest effort to simulate processes and activities of an environmental nature. Naturally, you can do more with more powerful computers and more specialized software. But the models here provide a good introduction to the capabilities of microcomputers and off-the-shelf software like spreadsheets.

These models may not explain the world; but they can help us to deepen our understanding of it. It is to be hoped that this will encourage us to look after the planet a bit better than we have done up to now.

Bibliography and References

Bachmat, Y., J. Bredehoeft, B. Andrews, D. Holtz, and S. Sebastian. 1980. *Groundwater Management: The Use of Numerical Models*. AGU Water Resources Monograph no. 5. Washington, D.C.: American Geophysical Union.

Boulding, Kenneth. 1959. *The Image: Knowledge in Life and Society*. Ann Arbor: University of Michigan Press.

Cartwright, T.J. 1991a. "Experimental Systems Research: Towards a Laboratory for the General Theory." *Cybernetics and Systems: An International Journal*, 22, pp. 135-49.

———. 1991b. "Planning and Chaos Theory." *Journal of the American Planning Association*, 57.1 (Winter), pp. 44-56.

———. 1987. "The Lost Art of Planning." *Long-Range Planning*, 20.2 (April), pp. 92-99.

———. 1973. "Problems, Solutions, and Strategies: A Contribution to the Theory and Practice of Planning." *Journal of the American Institute of Planning*. 39.3 (May).

Dawkins, Richard. 1986. *The Blind Watchmaker*. London: Longman.

Fedra, Kurt. 1989. "Environmental Impact Assessment of Industrial Development. A State-of-the-Art Review." Final Report by the International Institute for Applied Systems Analysis (IIASA) for the United Nations Industrial Development Organization (UNIDO). Mimeo.

Hofstadter, Douglas R. 1980. *Gödel, Escher, Bach: An Eternal Golden Braid*. New York: Vintage.

Holland, John H. 1992. "Genetic Algorithms". *Scientific American*, 267.1 (July), pp. 66-72.

Luhar, Ashok K., and P. Khanna. 1988. "Computer-Aided Rapid Environmental Impact Assessment." *Environmental Impact Assessment Review*, 8, pp. 9-25.

Oehler, Peter, and the AXP Programming Cooperative. 1990. *BUGS*. Version 0.6 (October). Microcomputer program.

Prata, Stephen. 1993. *Artificial Life Playhouse: Evolution at Your Fingertips*. Corte Madera, California: The Waite Group.

Stinson, Craig. 1991. "Spreadsheets Begin to Put the User First." *PC Magazine*, 10.24 (December), pp. 241-310.

Appendices

A. CREATING THE MODELS
B. THE RECIPES

CREATING THE MODELS

*. . . it vanished quite slowly, beginning with the
end of the tail, and ending with the grin,
which remained some time after the rest of it had gone.*

Lewis Carroll, Alice's Adventures in Wonderland

With the appropriate instructions, spreadsheet models are easy to create, even from scratch. At first, the sheer size of some of the models (over 100 Kb) may serve to discourage people from trying to build the model themselves. But spreadsheet models tend to be quite repetitive (not least because of the lack of a looping function in most spreadsheets, as noted at various points in the chapters above). Thus, the same formula may be repeated in many different cells, with only minor variations in cell references. Of all the models in this book, for example, only two contain more than 30 formulas. However, these formulas are repeated over and over again in hundreds and hundreds of different cells.

Of course, it would be very tedious to have to enter the contents of each of these cells one at a time, copying them from an itemized listing of every cell—all the more so, given that differences between neighboring cells are often very slight. Fortunately, however, while they may not be as elegant as some other programming languages, spreadsheets have a unique ability to help program themselves. That is, most spreadsheets have a "copy" command that can be used to copy the contents of one cell (or of a range of cells) into another cell or range of cells. What makes this command particularly useful is that, in so doing, the spreadsheet can be instructed to vary any cell references in the original cell or cells to make them appropriate to the location of the new cell or cells.[1] For example, if you copy a cell that contains a reference to cell A1 into a cell one row down and one column to the right, that reference will change to B2. At the same time, if you want to hold some cell

---

1.   Some spreadsheets originally called it the "replicate" instead of the "copy" command.

references constant (i.e., keep them as "absolute" cell references and prevent them from being changed), it is usually enough to insert a dollar sign in front of the column or row reference. For example, if you copy a cell that contains a reference to cell A$1 into a cell one row down and one column to the right, that reference will change to B$1. These are powerful tools for the programmer.

Thus, the approach taken here and in appendix B, below, is to present not a cell-by-cell listing for each model but rather, what is called a "recipe". A recipe is not a printout of the entire contents of every cell in the worksheet, since this is virtually indigestible. Instead, a recipe is a set of instructions for (a) formating the worksheet as required and entering appropriate labels and constants and (b) entering the key formulas and then copying them into all the other cells where (subject to appropriate changes in cell references) they are meant to be. In both cases, the instructions provided in the recipes make use of the spreadsheet's own abilities. The result is that recipes are typically no more than three or four pages long. By contrast, detailed printouts can run to 30 or 40 pages each, and frequently give little clue to the structure and organization of the model.

The models in this book were all written in SuperCalc4 and are presented here in that form. This is not because I think SuperCalc is necessarily the best of all spreadsheets. It has both weaknesses and strengths compared to other spreadsheet programs. It is just that SuperCalc was the spreadsheet I happened to start out with (on a 1970 Tandy Radio Shack TRS-80 Model I, for readers old enough to remember those days!), and SuperCalc still seems to me sufficiently powerful and user friendly for me to be able to do the things I want to do with spreadsheets. Other people who have different needs or who have already invested in learning to use another spread-sheet, may prefer to use that one. That is fine. Fortunately, there is so much similarity among the various brands of spreadsheets—perhaps more than in any other kind of software—that there should be little difficulty in converting the recipes provided here to whatever spreadsheet you may want to use. With that in mind, the rest of this appendix provides some general guidelines on following the instructions in the recipes presented in appendix B, below, and in adapting the instructions to spreadsheets other than SuperCalc.

System Requirements

The first step in getting a model to run on your system is to check that your hardware and software are capable of doing so.

As far as hardware is concerned, all of the models should be able to run on any computer that has sufficient memory (RAM) to contain both the spreadsheet program and the model itself. (This is because spreadsheet programs, unlike word processors or database management programs, usually require the entire worksheet to be in memory. Spreadsheets do not normally like to swap parts of a worksheet to and from disk.) Exactly how much RAM memory is required depends, of course, on the size of the model and of the software in use. In most cases, 640 Kb RAM will be enough for any of the models in this book. In some cases, less will suffice. In other cases (where network drivers, terminate-and-stay-resident (TSR) programs, graphic user interfaces, etc. are in use), 640 Kb may not be enough—unless you can unload some of this overhead. But if you have a stand-alone computer with 640 Kb RAM and a typical spreadsheet program, you should be able to run any of the models in this book. It does not matter what kind of disk drives or capacity you have, as long as you can run your spreadsheet program. It does not matter what video standard you use, as long as your spreadsheet can address it. Nor is there any special need for mouses, printers, or any other peripherals, although you can use them as appropriate, if you like. In short, these models are all meant to run on "plain vanilla" hardware.

Of course, the speed and power of your hardware may also be a consideration in the case of some models. Even though computers work blindingly fast (at the speed of light), they do so in painfully small steps. Some of the models in this book, particularly those that run through a number of cycles, involve very large numbers of operations (e.g., more than 250,000 discrete arithmetic or logical operations). So most of these models do take a discernible amount of time to run their course. For example, on my laptop computer (which has a '386 microprocessor and a mathematics coprocessor), some models can take several minutes to complete their programmed runs. On slower machines, therefore, the models should all work perfectly well; but the time they need to run may make it impractical to use them for what-if modeling and other applications. Since this is at least in part a matter of perception, the only thing to do is try the model and see if performance is adequate for your purposes, whatever they may be.

As far as software is concerned, there are at least eight key points that are worth checking before trying to run any of the models in this book on your system. These are:

1.   Is there enough free memory after your spreadsheet program is loaded for it to be able to load and run the model you want to run? Table A.1 shows the approximate size of each model and, therefore, the amount of free memory each model will require.[2] If you do not know how much free memory is available with your spreadsheet, consult the user's manual.

2.   What is the maximum number of characters that can be entered in a single cell? In some of the models, there are cells containing well over 100 and, in some cases, 150 characters (see Table A.1). Naturally, if your spreadsheet does not permit as many characters as that in a single cell, you will have to redesign the model to suit your spreadsheet.[3]

3.   Does your spreadsheet support all of the functions used in the model? The functions used in each model are listed in Table A.1, below. Functions are shown in their SuperCalc form; your spreadsheet may express them in a somewhat different way.   If your spreadsheet does not have all the required functions, it is sometimes feasible to compute them directly. For example, if your spreadsheet has no AVERAGE() function, it can readily be computed by means of the SUM() and COUNT() functions.

4.   Does your spreadsheet provide a means of generating random numbers?[4] If the model requires random numbers (see Table A.1), then you cannot run the model without them (except for the first and last of the three "game" models, as they use the random function not for running the model itself but only for an optional initialization macro).

_____

2.   Model sizes shown in Table A.1 are those of the SuperCalc versions. Other versions may require slightly more or slightly less space than this. You can add 10% to these figures to be on the safe side.

3.   This can usually done by performing the calculation in two or more stages, and providing for storage of the intermediate results in some appropriate area of the worksheet. Of course, this usually means increasing the size of the model and reducing its speed.

4.   Strictly speaking, we are talking about pseudorandom, not random, numbers. See the Technical Note at the end of this appendix.

Table A.1: Summary of System Requirements for Running the Models

Model	Approx. Size (Kb)	Max. Cell Size	Required Functions	Uses Random Numbers	Circular Links	Uses Macros	Uses Error Check
SMOKE	29	110	IF, MIN, SQRT, EXP, LN, VLOOKUP	No	No	No	No
WATER	65	149	IF, ABS, SQRT, LOG, EXP, ITER	No	Yes	No	Yes
BEARS	42	92	SUM, ROUND, VLOOKUP	Yes	No	Yes	No
TREES	122	168	IF, SUM, AV, MAX, AND, OR, LN	Yes	No	Yes	Yes
SOLAR	78	185	SUM, AV, STD, MAX, SQRT, EXP, SIN, COS	No	No	No	No
KLEIN	10	71	None	No	Yes	No	No
LOWRY	11	59	SUM	No	Yes	No	No
BERTAUD	16	81	IF, SUM	No	No	No	No
TRAFFIC	46	98	SUM	No	Yes	No	No
WASTE	19	21	SUM, AV, MAX, MIN, EXP, VLOOKUP	Yes	No	No	No
EIA	172	146	IF, SUM, AV, STD, ABS, MAX, MIN	Yes	No	Yes	Yes
LIFE	46	109	IF, COUNT, MAX, AND, OR	Yes*	No	Yes	No
CHOICE	79	153	IF, COUNT, MAX, AND	Yes	No	Yes	No
GAMACH	55	89	IF, LOOKUP	Yes*	No	Yes	No

Notes:

Approximate size (in kilobytes) refers to the size of the SuperCalc versions of the models, as they are saved to disk. Model sizes in other spreadsheet programs may differ slightly from these figures.

The maximum cell size refers to the maximum number of characters per cell in the SuperCalc versions of the models; maximum cell sizes may vary in other spreadsheet versions.

Required functions are listed here in SuperCalc format and without parentheses. Other spreadsheet programs may have slightly different formats and syntax.

In the column referring to the use of random numbers, an asterisk indicates that macros are used, but they are used only in macros, not in the model itself.

Circular links occur when the formula in any cell refers, directly or indirectly, to itself.

Models that rely on error checking do so by making use of an ISERROR() function: if the argument (the expression inside the parentheses) contains an error, the function returns a value of 1; if not, it returns a value of 0.

5. Does your spreadsheet allow circular links and, if so, what does it do about them? (Circular links occur when the formula in a cell refers, directly or indirectly, to itself.) Most spreadsheets will tolerate such links, being content to flag them as potential sources of error. But not all spreadsheets will accept them at face value, so to speak, and keep on recalculating the value of the cell, over and over again, based each time on its own "new" value. This process is called iteration, and some of the models in this book (see Table A.1) will not work without iteration. If your spreadsheet does not interpret circular links as a case for iteration, it is often possible to write a macro to have the same effect.

6. Does your spreadsheet provide for macros? (Macros are sequences of pro-grammed instructions that the user can call at his or her discretion.) As shown in Table A.1, six of the models depend on macros. Most of these macros are designed primarily to run the model over and over again a given number of times; so it may be possible to manage without the macro, as long as you do not mind the tedium of repeating the same instructions again and again. It may also be possible to recast some models to substitute circular links (iteration) for the use of macros. However, neither of these options is discussed here in detail.

7. Does your spreadsheet provide the necessary capabilities for error-checking? If not, this may prevent models that use error-checking (see Table A.1 again) from working over certain ranges of input or parameter values.

8. Finally, all of the models except the last three "game" models make use of simple graphs (bar, stacked-bar, line, and pie graphs) to display the results of their calculations.[5] In most cases, this is highly desirable, since results are often difficult to absorb through numbers alone. However, even if you are using a spreadsheet that does not do graphs or does not do graphs as sophisti-cated as some of those shown here, you may still be able to run the models themselves. In no case are the graphs essential to the working of the model.

---

5. One model (the EIA model in chapter 11) uses a "high-low" graph to display a result and an associated confidence range. If this type of graph is not available on your spread-sheet, you could use a stacked-bar graph to achieve a similar effect.

In summary, the models in this book rely for the most part on fairly standard spreadsheet capabilities that have been part of most spreadsheets for many years. You are unlikely to need the latest version of your favorite software, in order to be able to run these models. In particular, you do not need a spreadsheet with a graphical user interface (or GUI) to run any of the models, although they will run perfectly well in such an environment, if that is what you prefer.

Programming Notes

The recipes all have a similar structure. First, there are instructions on how to format the worksheet appropriately for the model (i.e., suggested column widths, numerical and text formating, horizontal and vertical lines between different parts of the model, etc.). Second, text and numerical constants are specified and entered. Third, formulas are entered in specific cells and copied to other cells as appropriate. Finally, there are instructions for protecting the model (if desired) and defining suitable graphs. Here are some general notes on some of these points.

A.   Copying

As noted above, the recipes in appendix B make extensive use of the copy command. In most spreadsheets, it is not necessary to copy "like" into "like". That is, the contents of a single cell can be copied into another cell, into an entire row or column, or into any (rectangular) block of cells. Similarly, in cases where the contents of more than one cell (e.g., a row, a column, or a block) are to be copied to another location, most spreadsheets require you to specify only the "anchor" point (i.e., the upper left-hand cell) of the destination. For example, suppose that a block of 100 cells (e.g., cells A1:J10) is to be copied to the area immediately below (i.e., to cells A11:J20). In this case, only cell A11 need be specified as the destination, since the rest of the block will be assigned to the cells down from and to the right of the anchor cell, A11, in accordance with the structure of the source block. Similarly, to copy a column (e.g., cells A1:A10) into a series of columns to the right (e.g., columns B:J), only the topmost cells of each column need be specified. Thus, the destination need be only cells B1:J1, not cells B1:J10 (although both will work equally well). Nevertheless, if your spreadsheet does not conform to this practice, you may have to make appropriate adjustments to some of the copy instructions in the recipes below.

B.   Absolute Cell References

In spreadsheets, individual cells are identified by means of their column and row "coordinates". Thus, the cell in the upper left corner is referred to as A1; the cell beside it as B1 and the cell below it as A2.

Formulas refer to specific cells by means of these cell references—A1, A2, etc. In most cases, when a formula is copied from one cell to another, any cell references in the formula are adjusted accordingly. For example, suppose that a formula contains a reference to cell A1: if that formula is copied into the cell immediately to the right of its present location, any references in the formula to cell A1 will be changed to cell B1. Similarly, if the formula is copied into the cell immediately below its present location, any references to cell A1 will be changed to cell A2. This feature of spreadsheets can be exceedingly useful.

At other times, however, it may be desirable to suppress this kind of automatic adjustment of formulas when copying. Most spreadsheets allow the user to interrupt the copy process and override some or all of the automatic adjustments, but this is often tedious and error-prone. So there is another way to suppress the automatic adjustment of cell references when they are copied. This is by inserting a dollar sign ($) directly into the formula in front of each row or column reference that is *not* to be adjusted. This is called an absolute cell reference. Thus, in the example just given, references to cell A1 would be adjusted by the copy command, whereas references to cell $A$1 would not. Moreover, references to cell $A1 would be adjusted when they were copied down to another row but not when they were copied sideways to another column; and references to cell A$1 would be adjusted when they were copied sideways to another column but not when they were copied down to another row.

The key point to note is that adding dollar signs to a cell reference does not affect its value. In any cell, a reference to cell A1 is exactly the same as a reference to cells $A$1, $A1, or A$1. They all evaluate to exactly the same numerical value. Absolute cell references are used only to facilitate copying cell contents from one location to another. Absolute cell references serve no purpose when the model is running; they are useful only when the model is being built or modified. In fact, you can go through the model and delete all the dollar signs, if you like, once the model is running properly. That will reduce the size of the model and may even give noticeable improvements in speed on some machines.

C. Named Ranges

Many spreadsheets allow users to refer to cells or blocks of cells by names. These are usually referred to as "named ranges". The idea is that named ranges may make formulas easier to read and may reduce the risk of error, when formulas are being entered by hand.[6] Nevertheless, named ranges have not been used in any of the models in this book. Users are, of course, free to add this feature to the models wherever and whenever they feel it might be appropriate.

D. Repeating Text and Text Overflow

Most spreadsheets allow for entering repeating text. For example, the single (instead of the double) quotation mark or the backslash may be used to indicate that the character or characters that follow should be repeated (horizontally) across the screen, until a non-blank cell is encountered. If your spreadsheet has no such capability, then you will have to enter repeating text manually in every cell where it is required.

In particular, repeating text provides a convenient way to create horizontal dividing lines to separate one part of a worksheet from another part. Thus, to draw a horizontal line, you just enter the repeating-text symbol (the single quotation mark, the backslash, or whatever that symbol may be in your spreadsheet) followed by a hyphen or a dash; and the spreadsheet will repeat it right the way across the screen and beyond (until it comes to a non-blank cell). If your spreadsheet accepts character graphics, you can use ASCII code 196 (e.g., Alt-196 entered on the numeric keypad) instead of the hyphen or dash, and you should get a neater, smoother effect. Vertical dividing lines are more tedious, since there is no similar mechanism for repeating text in a vertical direction. To get a vertical line, you have to enter a vertical bar in one cell and then copy it up or down into all the other cells that are to contain the line. For this purpose, the vertical-bar key usually produces a satisfactory effect. But if your spreadsheet accepts character graphics, you can use ASCII code 179 (e.g., Alt-179 entered on the numeric keypad) instead of the vertical bar.

---

6. However, named ranges cannot be used as a mechanism for reducing the length of a formula that would otherwise be too long to fit within a cell. This is because most spreadsheets that allow named ranges still require that the full specification of the named range be able to fit within any cell where it is used.

In most spreadsheets, text that is too long to fit into the cell where it is entered is not lost, even if it is not displayed. (Typically, however, there is an upper limit on the number of characters that can be entered into a single cell: it is usually somewhat less than 256.) In fact, many spreadsheets allow text that exceeds the width of the cell where it is entered to overflow into the adjacent cell or cells to the right, providing they do not already contain text or formulas of their own. If your spreadsheet does not work like this, you may have to make appropriate adjustments.

E.   Numerical Overflows and Errors

An overflow occurs in cases where a cell is formated in such a way that there is not enough physical space for its numerical value to be displayed. (Note that it was *text* overflow that we discussed in the previous paragraph, whereas here we are talking about *numerical* overflow. The spreadsheet handles the first by displaying as much as it can and truncating the rest; but it cannot do the same with numerical values.) In this case, the value of the cell has been computed correctly and stored in the worksheet—but the cell is not wide enough for the value to be displayed, given the way the worksheet is currently formated. Since numerical results cannot just be truncated, a symbol for numerical overflow is required. The nature of this symbol varies from spreadsheet to spreadsheet: e.g. it may be a series of "greater-than" symbols (">>>>>") or a series of asterisks ("*****"). But whatever the symbol, such overflows have no effect on the functioning of a worksheet, beyond making it impossible for the user to see the result of that particular calculation (unless the worksheet is reformated somehow to make the cell wider). Thus, spreadsheet overflows are more of a nuisance than an error. In some of the models here, overflows are tolerated when they occur in nondisplay areas of the worksheet.[7]

An error message, on the other hand, is more serious than an overflow. An error means something is wrong and the spreadsheet has been unable to compute or store the result required. Naturally, any calculations that depend on a cell containing an error will also result in errors. Most spreadsheets display the word "ERROR" in any cell that contains an error. In general, if there are errors in your model, they must be tracked down and dealt with in some appropriate way.

---

7.   Overflows typically occur in vertically oriented worksheets like those advocated in this book, because column widths tend to be dictated by the requirements of the display area at the top of the worksheet and may not be as suitable for the rest of the model; see also p. 9 ff., above.

In models like those in this book, there are two common causes of errors. One is division by zero. For example, you may have a cell containing a formula which involves division by a second cell. Should that second cell ever evaluate to zero, then the first cell will display an error message. A second common source of errors is found in certain spreadsheet functions. For example, the EXP() function generates an error message whenever its argument (the part inside the parentheses) falls outside whatever range is allowed by the spreadsheet.[8] If the argument is a formula, then there is always the possibility that the formula may generate a value that is outside the range allowed by the function.

The easiest way to cope with the possibility of division by zero is to use an IF() function to avoid it. For example, suppose that cell A1 contains the formula,

```
B1/C1
```

and the formula in cell C1 (whatever it may be) sometimes evaluates to zero. In such a case, the formula in cell A1 can be altered to read

```
IF(C1=0,0,B1/C1)
```

—assuming (of course) that zero is an appropriate alternative value for cell A1.

For more sophisticated problems, most spreadsheets provide an error function of some kind—e.g., ISERROR() or ISERR()—that can be used for error trapping. In this case, the argument (the expression inside the parentheses) is a cell that may return an error message. Typically, the ISERROR() function is constructed so that it returns a one (for "true"), if the argument evaluates to an error, and a zero (or "false"), if it does not. For example, suppose that the formula in cell A1 contains the function EXP(B1) and you are worried that the value of cell B1 may sometimes fall outside the range allowed for the EXP() function. In that case, you can alter the formula in cell A1 to the following (assuming once again that zero is an appropriate alternative value for cell A1):

```
IF(ISERROR(EXP(B1))=1,0,EXP(B1))
```

---

8. In SuperCalc4, the argument in the EXP() function must fall within the range of approximately ± 128; in Lotus 3.1, the range is approximately $\pm 11,355$. For more discussion of the EXP() function (and its inverse, the LN() function, see p. 37, above.

F.   Lookup Functions

One of the unique features of spreadsheets is their built-in ability to look up values in a "table" and return another corresponding value. In most spreadsheets, lookup functions take one or more of the following three forms:

- LOOKUP(*value,range*), where *value* specifies the number that is to be looked up and *range* specifies the cell range where *value* is to be looked for; the function itself returns the value found in the cell adjacent to the cell within *range* where *value* is found, adjacent meaning immediately to the right if *range* is a column range or directly below if *range* is a row range.[9]

- VLOOKUP(*value,range,offset*), where *value* specifies the number that is to be looked up and *range* specifies the cell range where *value* is to be looked for and where the lookup values are found; the function itself returns the value found in the cell that is *offset* columns to the right of the cell within *range* where *value* is found.

- HLOOKUP(*value,range,offset*), where *value* specifies the number that is to be looked up and *range* specifies the cell range where *value* is to be looked for and where the lookup values are found; the function itself returns the value of the cell that is *offset* rows below the cell within *range* where *value* is found.[10]

G.   Iteration

Some spreadsheets have a built-in capacity to iterate, whenever there is a circular relationship anywhere in the worksheet—that is, whenever the value of one cell depends, directly or indirectly, on itself.

---

9.   If an exact match with *value* cannot be found within *range*, SuperCalc selects the next lower value. Thus *range* should always be sorted in ascending order.

10.   Note that *range* is just the source vector in the LOOKUP() function, whereas *range* is a matrix containing both the source vector and the corresponding *offset* columns and rows in the VLOOKUP() and HLOOKUP() functions. Other than that, LOOKUP() has the same result as the other two functions with their *offset* values set to unity.

The purpose of such repetitive calculations is to try to find a value or set of values that do not change or do not change significantly from one iteration to the next. If the formulas involved are such as to cause the value of the cell to converge on a specific value, iteration will eventually stop, as any changes in the value of the self-referential cell become smaller and smaller. If the formulas involved are not such as to cause the value of the cell to diverge, then iteration will continue until some other limit is reached. In some spreadsheets, the user is provided with mechanisms for controlling such iteration. This may be by specifying the number of times that iteration is to be done, a target range for a particular variable, or a "delta" value that defines the maximum amount of change in the value of a variable from one cycle to the next (e.g., 1%) that is consistent with stopping the iteration. In the absence of such control mechanisms, diverging iteration may continue until the numbers just get too big for the spreadsheet to handle.[11]

If your spreadsheet does not have a built-in capacity to iterate, it is not difficult to write a simple macro that will do it. In the example just given, such a macro might be:

```
{\I}   !
       {if iserror(A1) = 0} {branch XN}
```

The first line contains the name of the macro ("I") within braces on the left, and an exclamation mark to force a manual recalculation on the right. The second line checks to see if cell A1 has produced an error message and, if not, branches back to cell XN (which should refer to the cell containing the exclamation mark immediately above). If cell A1 has generated an error, then the macro continues to the next cell; this cell is blank and so the macro terminates. The macro language and syntax in your spreadsheet maybe slightly different from the SuperCalc style shown here, but

---

11.   Readers can readily experience iteration in a spreadsheet. Enter the formula $A2+1$ in cell A1 and the formula $A1$ in cell A2. Since cell A1 keeps evaluating to a larger and larger number (because the value of cell A2 keeps getting bigger and bigger), cell A1 does not converge on any particular value. So iteration will not stop until the value in cell A1 exceeds the capacity of the spreadsheet and generates an error messageWhat this maximum capacity is depends on the spreadsheet software you are using: seethe discussion in the Technical Note at the end of this appendix. If you want to see iteration continue until this limit is reached, format column A to the full width of the screen and change the formula in cell A1 to something like $A2+10^50$ or $A2+10^100$ to speed things up a bit.

the principle of writing an iteration macro will be the same. More complicated examples can be adapted from the macros used in chapters 3 and 4; see also Johansson 1985 (p. 405) for a Lotus macro to run the model in chapter 6.

H.   Boolean Logic

Some of the models in this book provide an opportunity to capitalize on the ability of spreadsheets to include Boolean operators in formulas. Typically this involves a conditional branch using an IF() function. Suppose we have a case where, if the value of an expression in (say) cell A1 is less than zero, we would like it to display a zero; otherwise, we would like it to display the actual value of cell A1. This can easily be achieved with the formula,

```
IF(A1<0,0,A1)
```

However, the same effect can be achieved in a slightly more elegant manner by writing:

```
(A1>0)*A1
```

If the value of cell A1 is greater than zero, then the Boolean expression A1 > 0 is true and it evaluates to 1. So the formula as a whole returns the value of cell A1 (whatever that may be) times unity. If, however, the value of cell A1 is zero or less, then the Boolean expression A1 > 0 is false and it evaluates to 0. So the formula as a whole returns a value of zero. In some cases (e.g., chapter 14), Boolean expressions can greatly simplify complex formulas.

I.   Macros

Macros are used sparingly in this book, essentially only when it is desirable to automate some sequence of commands that must be repeated over and over again. Although macros work in much the same way in most spreadsheets, there are rather more differences of detail in the language and syntax of macros than in most other areas of spreadsheet operation.

In most spreadsheets, writing macros involves two discrete steps. The first step is to write the macro. This usually involves listing the desired commands in an uninterrupted sequence of cells all in the same column, although some spreadsheets allow macros to branch to cells in other columns if that is desirable. Macro commands generally reflect what you can do manually, but may include some additional capabilities that are not appropriate when the program is being used in the normal way. Comments or other notes about each command can conveniently be written in an adjacent column. (This is good programming "form", but macros do not require comments.) The macros shown here are all written in SuperCalc format, of course, but they are heavily annotated; so users should be able to adapt them to whatever spreadsheet program they may be using.

The second step is to link the macro to a "name" that can then be used to invoke, or call, the macro. The mechanics for doing this vary from spreadsheet to spreadsheet; so users should check their software manual for details.

Notes for Lotus and Quattro Users

Subject to all of the general comments above, the recipes in appendix B can readily be adapted to creating models to run in Lotus, Quattro, and Quattro Pro. There are no fundamental differences between these programs and SuperCalc, and only a few differences in style and syntax. Some of the more important of these are summarized below:

- All formulas beginning with a cell reference must be preceded by a plus or minus sign; otherwise Lotus and Quattro will regard the formula as text.

- Ranges in Lotus and Quattro are indicated by two periods rather than a colon (as in SuperCalc): e.g., A1:J10 in SuperCalc must be written A1..J10 in Lotus and Quattro.

- All of the functions used in the models in this book are exactly the same in Lotus and Quattro as in SuperCalc, except that

  - all Lotus and Quattro functions begin with an @ symbol;
  - lookup functions in Lotus and Quattro cannot be shortened to LU(), VLU(), and HLU(), as they can in SuperCalc; and
  - the AND() and OR() functions in SuperCalc are rendered by #AND# and #OR# operators in Lotus and Quattro.

- SuperCalc allows up to 240 characters in a single cell; so does Lotus, while Quattro can take up to 254 characters. However, both Lotus and Quattro insist on a leading zero for all decimal values less unity, and will include the zero even if it is not entered by the user. (This and some of the other differences noted above may make Lotus and Quattro formulas longer than their SuperCalc equivalents.)

● SuperCalc can display up to 10 different variables in the same graph; in Lotus and Quattro, the maximum is only 6. Similarly, it may be necessary to shorten or otherwise adjust some of the titles and labels used in SuperCalc to suit the graphing parameters of Lotus and Quattro.

Notes for Excel Users

Subject to all of the general comments above, the recipes in appendix B can readily be adapted to creating models to run in Excel. Excel is, if anything, closer to SuperCalc than are Lotus and Quattro. All formulas in Excel must begin with an equals sign, but cell ranges are specified with a colon as in SuperCalc. As with SuperCalc, Excel functions do not require a leading "@" symbol, and all the functions used in the recipes in appendix B are exactly the same as in SuperCalc, except for the following minor differences:

● In Excel, RAND() and PI() require dummy arguments; in SuperCalc, they do not; in Excel 4, there is also a more sophisticated RANDOM() function that can generate normal, poisson, and other distributions.

● In Excel, the LOG10() function is written thus; in SuperCalc, it is written as just LOG().

● In Excel, the HLOOKUP() and VLOOKUP() functions cannot be shortened to VLU() and HLU(), as they can in SuperCalc.

Moreover, Excel allows up to 255 characters in a single cell. As with most other spreadsheets, the Excel programming language has some idiosyncrasies of its own; so users will have to convert the SuperCalc macros to run on Excel.[12]

---

12. Alternatively, Excel does have a macro translation facility for converting Lotus macros. So it may be easier for some users to convert the SuperCalc macros to Lotus or Quattro format, and then let Excel's translation facility convert them to Excel format.

Notes for Users of Integrated Programs

Integrated programs are programs that provide spreadsheet capabilities along with word processing, database management, and graphics. Generalization about their suitability for the models in this book is more difficult because of the variety of programs involved.[13] However, the following comments may be helpful:

- **Free memory**: Integrated programs may use more memory than dedicated spreadsheet programs; so you should be sure to check how much free memory you have after loading the spreadsheet portion of your program.

- **Maximum dimensions**: Integrated programs may have lower limits on the number of rows and/or columns allowed in a worksheet, compared to those allowed by dedicated spreadsheet programs; so you should consult table I.2 (in the Introduction) to see whether your program can accomodate the models you are interested in.

- **Maximum cell length**: Integrated programs may limit the number of characters that can be entered in a single cell to less than what is allowed by dedicated spreadsheet programs; so that too should be checked. In some cases, you can program your way around this problem using "scratch pads" for intermediate calculations (as we already have done to some extent, in chapters 5 and 11, above); but this is awkward and invariably slows down the model.

- **Functions**: The models presented here do not require any very unusual functions; so problems of missing functions are unlikely.

- **Graphs**: Most integrated programs provide some kinds of line, bar, and pie charts; but the extent of these capabilities can vary a lot. For example, some programs create bar charts but not stacked-bar charts, or pie charts but not exploded-segment pie charts, etc. Of course, none of these features is essential

---

13. In August 1991, *PC Magazine* reviewed nine integrated programs (Perratore 1991), including DeskMate, Eight-in-One, Enable/BP, Framework XE, LotusWorks, Microsoft Works, Personal Office, PFS:First Choice, and Venture. The following comments are based mainly on that review.

to getting any of the models to work. On the other hand, a good graph is always useful in helpful us to understand what the model is doing.

Notes for Users of Shareware Spreadsheets

There are several shareware spreadsheet programs capable of running many of the models in this book.[14] Indeed, some shareware spreadsheets include features still not found in all commercial spreadsheets: e.g., matrix multiplication; a weighted-average function; various string-conversion functions; and even the capability of creating customized, user-defined functions. For users of these programs, the following comments may be helpful:

- **ExpressCALC (ExpressWare):** Worksheets are limited to 64 columns and 256 rows, although the number of rows can be increased by reducing the number of columns. Cell contents are limited to 74 characters. There is no graphing capability, no macro language, no lookup functions, and no error-detection function.

- **EZ-SpreadSheet and TurboCALC (EZX Publishing/P&M Software Co.):** EZ-Spreadsheet worksheets are limited to 64 columns and 512 rows, while TurboCALC limits can vary from 256 columns and 90 rows to 2 columns and 8,192 rows. The first program has no graphing capabilities and the second program does only bar charts. Both programs have a macro language and an error-detection function.

- **PC-CALC+ (ButtonWare):** Worksheets are limited only by available memory. PC-CALC+ has good graphing capabilities, including the ability to switch from vertical to horizontal bar charts (which means you can draw a population pyramid), to "explode" not just one but several segments in a pie chart, and to sway time and point variables interactively in a graph. There is iteration control and a macro language.

---

14. Shareware is software in which the author retains copyright but allows it to be copied and distributed (but not altered or sold), on condition that people who use it register with the author and pay a (usually modest) fee. Shareware is typically available for downloading from computer services like Compuserve, from computer bulletin boards, or from Internet Internet sources such as WUARCHIVE.WUSTL.EDU or WSMR-SIMTEL20.ARMY.MIL.

• **AS-EASY-AS (Trius Inc.):** Worksheets are limited to 256 columns and 8,192 rows. The program has over 80 functions as well as the ability to create user-defined functions. It has extensive graphing capabilities, including log-log and semi-log plots. AS-EASY-AS has over 100 macro commands and, when it recalculates, it recalculates only those cells that need it (a feature that is sometimes called "intelligent recalculation"). The program also has built-in utilities for search-and-replace, for recording macros, for auditing, and for goal-seeking. A more advanced version of the program called PIVOT is reportedly available from the author.

• **Insta-Calc and Qube-CALC (FormalSoft):** Insta-Calc is a memory-resident spreadsheet in which worksheets are limited to 256 columns and 256 rows. Cells can contain up to 75 characters, except that formulas are limited to a maximum of 70 characters. There is a full range of macro commands, but no graphing capabilities and no lookup functions. Qube-CALC is a three-dimensional, non-resident version of Insta-Calc. WorkQUBEs are limited to a total of 262,144 cells (e.g., 64 columns, 64 rows, and 64 pages) but they can be structured in any regular cubic shape. Other than that, Qube-CALC has the same functions and capabilities as Insta-Calc.

Summary

The models described in appendix B range in size from 10 to 172 Kb and perform anywhere from a few hundred to over 10,000 arithmetic or logical operations. Yet most of the models consist of only a few (less than 30) different formulas that use only the four basic arithmetic operations plus exponentiation and a dozen or so logical, statistical, and trignometric functions. In short, the models do more to illustrate the basic adaptability of the spreadsheet than the diversity or specialization of its functions and commands. Thus, there should be little difficulty in adapting the recipes in the next appendix to a wide variety of computers and spreadsheets.

Technical Note

The results of all the models in this book are subject to the computational accuracy of the hardware and software used to achieve them. In the present case, there are at least four such issues that deserve consideration: namely, (a) maximum and minimum values; (b) significant figures; (c) rounding errors; and (d) random numbers.

Maximum and Minimum Values

Microcomputer spreadsheets are limited in terms of the biggest and smallest numbers they can handle. Try entering larger and larger numbers into a spreadsheet; for example, enter 10^{10} (entered as $10^{\wedge}10$ in most spreadsheets), then 10^{20}, then 10^{30}, and so on. At first, the spreadsheet displays the numbers you enter—in standard or exponential notation, depending on the width and formatting of the cell where the numbers are entered. After a certain point, however, the spreadsheet stops accepting the numbers you enter and starts giving an error message, because the numbers have simply become too big for the spreadsheet to handle. Similarly, try entering progressively smaller numbers (e.g., 10^{-10}, 10^{-20}, 10^{-30}, etc.). Again, the spreadsheet will accept the numbers you enter and display them (in standard or exponential notation, as the case may be) up to a certain point. Beyond that point, however, the numbers are just too small for the spreadsheet to handle and so it returns an error message.

The maximum and minimum numerical quantities that a spreadsheet can handle depend on how numbers are stored in a computer (Clarke 1987, p. B.3). Numbers are normally entered into a computer in decimal format. The computer, on the other hand, uses binary numbers—that is, numbers made up of zeros and ones. So it is ultimately the requirements of binary storage that determine the decimal limits on the numbers that can be handled.

Decimal numbers are often converted into pure binary form and stored as either "single precision" numbers (in four bytes) or "double precision" numbers (in eight bytes). In either case, the first byte is an exponent and the other three or seven bytes (depending on whether the number is single or double precision) form the mantissa. In this way, single-precision numbers can range from approximately to 10^{-38} to 10^{38}, while double-precision numbers can range from approximately 10^{-307} to 10^{307}.

Alternatively, decimal numbers are sometimes converted and stored in "binary coded decimal" (BCD) form. In this case, each decimal digit is converted independently to binary form and coded in four bits (0000 to 1001). In this way, numbers in the range of approximately 10^{-65} to 10^{62} can be accomodated in four bytes.

In practice, it is usually a good idea to be aware of the biggest and smallest numbers that your software can handle, in case these limits become critical. For example, SuperCalc4 accepts numbers in the range 10^{-65} to 10^{62}, whereas Lotus 3.1 accepts numbers in the range of 10^{-100} to 10^{100}.

Significant Figures

Because it has only a finite space in which to store any number, a computer is limited in not just the size but also the precision of the numbers it can handle. In pure binary form, numbers are accurate to about 7 significant figures when stored as single-precision numbers and about 17 significant figures when stored as double-precision numbers. This is because 3 bytes in the mantissa of a single-precision number amounts to 24 bits of binary precision. In decimal, 2^{24} is about 16 milliion, or $10^{7.22}$, which means about 7 significant (decimal) figures. In double-precision arithmetic, a mantissa of 7 bytes means 56 bits of binary precision, which translates into about 17 significant (decimal) figures. In BCD form, numbers are accurate to exactly six significant (decimal) figures. If you like, you can check to see which of these results you can verify with your own spreadsheet.

Once again, it is usually wise to understand the limitations of your software, in case they begin to affect your results. For example, SuperCalc4 provides 12 figures of accuracy, whereas SuperCalc5 and Excel provide 15 figures of accuracy, and Lotus 3.1 provides 18 figures of accuracy.

Errors Due to Rounding

Problems may also arise due to rounding. This occurs because most decimal fractions do not have exact binary equivalents. For example, 0.1 in decimal is equivalent to .00011001100 . . . . (repeating) in binary. Inevitably, such numbers have to be truncated or "rounded off". This may introduce inaccuracies into the results of any arithmetic operations in which fractions are involved. Even when they are relatively small, moreover, errors due to rounding can be significant if they are cumulative.

You can demonstrate this phenomenon with a simple worksheet such as the one shown in Listing A.1, above. Format the worksheet to make the columns fairly wide (e.g., 36 characters). Leave the default numerical format as "general". Then, in cell B9, enter the following formula:

```
B4+B6-B4-B6
```

Obviously, the formula should evaluate to zero. All you have done is added two numbers together and then subtracted each of them from the total.

Listing A.1

```
      |                    A              ||              B                  |
 1    ERRORS DUE TO ROUNDING IN A SPREADSHEET
 2    _____
 3
 4    Enter any number >=1:                                              1
 5
 6    Enter any fractional number:                             .123456789
 7
 8    Add the two numbers together . . .
 9    then subtract both from the result:   .0000000000000000138777878078
10
11    Of course, you should get an answer of zero;                     ^
12    in practice, you may get a small error. - - - - - - - - - - - - - ^
13    This error is caused by the need to round off the
14    binary value of the decimal fraction entered in cell B6.
15
16    _____
```

However, you may in practice get a non-zero answer! This is because the fractional number you enter in cell B6 may not have an exact binary equivalent within the limits of precision of the spreadsheet.[15]

Random Numbers

Random numbers are numbers generated by repeated, independent drawings from the population 0, 1, 2, . . . 9, where the probability of drawing any particular digit is equal to that of drawing any other digit: that is, the probability is 0.1.

This is equivalent to putting ten [identical] balls, numbered from 0 to 9, into an urn and drawing one ball at a time, replacing the ball after each drawing. . . . Any group of *n* successive random digits is known as a *random number*. [However,] the use of random numbers in electronic computers has resulted in a need for random numbers to be generated in a completely deterministic way. The numbers so generated are termed pseudo-random numbers [Abramowitz and Stegun 1965, p. 949].

_____

15.   Note that the formula in cell B9 must be entered exactly as shown. If the formula is altered to B4+B6−B6−B4, it will never produce an error.

Different spreadsheets use different procedures for dealing with this problem. Some just have built-in lists of random numbers which the random function draws from; some use an iterative procedure to compute a pseudo-random number; some do one or other of the above but then use a rapidly changing "seed" value (such as the tick of the computer clock) to incorporate an element of unpredictability into the calculation. The result is that some spreadsheets (including SuperCalc and Excel), generate exactly the same sequence of random numbers each time you start to use the RANDOM function. Other spreadsheets (such as Quattro Pro) seed their random number generator from the computer clock; so the sequence of random numbers is not identical.[16] Unfortunately, user manuals are rarely clear as to the exact procedures used for generating random numbers.

Bibliography and References

Abramowitz, Milton, and Irene A. Stegun, eds. 1965. *Handbook of Mathematical Functions, with Formulas, Graphs, and Mathematical Tables.* New York: Dover.

Clarke, David K. 1987. *Microcomputer Programs for Groundwater Studies.* Developments in Water Science, no. 30. Amsterdam: Elsevier.

Cranford, Richard. 1986. "Streamlining Formulas." *Lotus Magazine* (February), pp. 70-74.

Johansson, Jan-Henrik. 1985. "Simulataneous Equations with Lotus 1-2-3." *BYTE*, 10 (February, pp. 399-405.

Perratore, Ed. 1991. "Integrated Software under $200." *PC Magazine*, 10.14 (August), pp. 241-305.

16.   Note that Excel has a built-in function—NOW()—that effectively gives the user access to the tick of the computer clock. So Excel users can achieve the same effect as in Quattro Pro.

THE RECIPES

This appendix consists of 14 "recipes", one for each of the 14 models in this book. Each recipe consists of detailed instructions for creating the corresponding model. Each recipe consists of five sections: layout, text and numerical entries, formulas, macros, and graphs.

Note the following general considerations:

● Each recipe refers to the model Listing in the corresponding chapter.

● Recipes are for SuperCalc versions of the models; you may have to make some adjustments (see Appendix A) to create versions for use with other spreadsheet programs.

● All formulas should be entered as a continuous sequence of characters, without line breaks or spaces (unless enclosed in quotation marks or otherwise noted).

It should also be noted that all of the recipes make extensive use of the COPY command to replicate labels or formulas that have been entered in one cell or range of cells into another cell or range of cells. The syntax for the COPY command usually requires the user to define, first, the original or source cell or range of cells and, second, the destination cell or range of cells.

As for the destination, you need specify only an "anchor" cell or range of cells in most spreadsheets. In other words, you do not need to specify the entire range of cells that will be filled by the original range of cells. Thus, to copy a simple 2-by-2 range of cells (say, cells A1:B2) to another location (say, cells A11:B12), you need specify only cell A11 as the destination—because cell A11 serves to anchor the rest of the 2-by-2 range of cells. Similarly, to copy a row of five cells (say, cells A1:E1) into the next five rows, you need specify only cells A2:A5 as the destination (and not A2:E5). The recipes presented here follow this abbreviated practice for specifying the destination of the COPY command.

Recipe 1: BLOWING SMOKE

SMOKE uses 16 columns and 65 rows and is about 30 Kb in size. It contains a total of 21 different formulas, which together perform approximately 3,200 discrete arithmetic or logical operations (exclusive of those involved in graphing).

In addition to the four standard arithmetic operations, the model uses powers, the constant PI, and the following spreadsheet functions:

      MIN(), SQRT(), EXP(), LN(), IF(), and VLOOKUP().

Note that VLOOKUP() can be abbreviated to VLU() in SuperCalc.

Layout

Set the default column width to 5; then reset the width of columns A, B, C, and D to 4, 3, 4, and 4 respectively. If columns A to P do not fit within the width of your screen, you can reduce the width of columns D and P until they do. Set the default format to integer. Format row 13 as general. Set spreadsheet recalculation to manual.

In cell A2, enter a continuous horizontal line by using a repeating hyphen or, if your spreadsheet accepts character graphics, Alt-196. Copy the contents of cell A2 into cells A6, A10, A14, A25, and A63.

Text and Numerical Entries

Referring to the following rows in Listing 1.1 (page 26):

1	Enter the title and other information shown in row 1.
3-5	Enter the text shown in cells A3, A4, A5, I3, I4, and I5. Enter the values shown in cells F3, F4, F5, N3, N4, and N5.
7-9	Enter the text shown in cells A7:A9 (note the leading spaces in cells A8:A9), H7:H9, and M7:M9. Enter the value shown in cell E8.
11-13	Enter the text shown in cells A11:A13, C11:C13, and E11:E12. Enter the values shown in cells E13:O13.
15-24	Enter the values shown in cells A15:A24.

27   Enter the text shown in cell A27.

29-32   Enter the text shown in cells A29:A32 and D29:D32.

34-41   Enter the text shown in cell A34 and cells D36:D41.

43   Enter the text shown in cell A43.

45-52   Enter the text shown in cells A45:A46 and A52. Enter the values shown in
                cells B45:B50.

54-61   Enter the text shown in cells A54:A56 and A61. Enter the values shown in
                cells B54:B59.

64-65   Enter the source citation as shown in rows 64-65.

Formulas

Referring to the following rows in Listing 1.1 (page 26):

15-24   Enter the following formula in the cell shown:

```
C15    IF(E$8<5,F$3+B$40/A15,
          F$3+MIN(2.4*EXP(LN(B$38/(B$41*A15))/3),
          5*EXP(LN(B$38/(B$41*A15))/3)))
```

                Copy the formula in cell C15 into cells C16:C24.

                Enter the following formula in the cell shown:

```
E15    IF(OR(E$52<.01,E$61<.01),0,
          1000000*$F$5/(2*PI*E$52*E$61*$A15)*
          (EXP(-.5*($C15/E$61)^2)+
          EXP(-.5*($C15/E$61)^2)))
```

                Copy the formula in cell E15 into cells E15:O24.

36-41   Enter the following formulas in the cells shown:

```
B36    N4+273.15
B37    N5+273.15
B38    3.12*.785*N3*F4^2*(B36-B37)/B36
```

        B39   IF(B38>55,34*(EXP(.4*LN(B38))),
                     14*(EXP(.625*LN(B38)))))
        B40   1.6*EXP(LN(B38)/3)*EXP(LN(3.5*B39)*2/3)
        B41   IF(E8<6,9.806*.02/B37,9.806*.035/B37)

45-50   Enter the following formulas in the cells shown:

        E45   .22*1000*E13*1/SQRT(1+.1*E13)
        E46   .16*1000*E13*1/SQRT(1+.1*E13)
        E57   .11*1000*E13*1/SQRT(1+.1*E13)
        E48   .08*1000*E13*1/SQRT(1+.1*E13)
        E49   .06*1000*E13*1/SQRT(1+.1*E13)
        E50   .04*1000*E13*1/SQRT(1+.1*E13)

       Copy the formulas in cells E45:E50 into cells F45:O45.

52   Enter the following formula in the cell shown:

        E52   VLU($E$8,$B45:$O50,3)

       Copy the formula in cell E52 into cells F52:O52. Then edit each cell manually so that the offset value in the formula (that is, the number 3 in the example shown above) is incremented by 1 for each column. In other words, the formula in cell F52 should have an offset value of 4, the formula in cell G52 an offset value of 5, the formula in cell H52 an offset value of 6, and so on—up to cell O52, where the offset value is 13.

54-59   Enter the following formulas in the cells shown:

        E54   .2*1000*E13
        E55   .12*1000*E13
        E56   .08*1000*E13*1/SQRT(1+.2*E13)
        E57   .06*1000*E13*1/SQRT(1+1.5*E13)
        E58   .03*1000*E13*1/(1+.3*E13)
        E59   .016*1000*E13*1/(1+.3*E13)

       Copy the formulas in cells E54:E59 into cells F54:O54.

61   Copy the formulas in cells E52:O52 into cell E61.

Protection of the Model

To safeguard the model from inadvertent changes, you should "protect" the entire worksheet and then "unprotect" the following cells: F3:F5, N3:N5, E8, E13:O13, and A15:A24.

Graphs

To obtain the figures shown in chapter 1, above, select the line type of graph. Select the following data ranges as variables: E15:O15, E16:O16, E17:O17, E18:O18, E19:O19, E20:O20, E21:O21, E22:O22, E23:O23, and E24:O24. Label the graph with the text in cells D29 and D30 as title and subtitle respectively. Label the x-axis and y-axis with cells D31 and D32 respectively. Label the x-axis points with cells E13:O13. Label the variables with cells A15:A24.

Recipe 2: RUNNING WATER

WATER uses 11 columns and 121 rows and is about 65 Kb in size. It contains 23 different formulas, which together perform about 9,050 discrete arithmetic or logical operations (exclusive of those involved in graphing). Because the model is "circular", it is solved by iteration. With a delta value of .01 (the default in my spreadsheet), this model goes through about 12 iterations to reach a new solution.

In addition to the four standard arithmetic operations, the model uses powers, the constant PI, and the following spreadsheet functions:

```
ABS( ), SQRT( ), LOG( ), EXP( ), IF( ), and ISERROR( ).
```

WATER also uses a built-in variable, ITER, that keeps a running count of the number of iterations a model may have gone through.

Layout

Set the default column width to 8; then reset the width of columns A, B, and C to 6, 4, and 1 respectively. If columns A to K do not fit within the width of your screen, you can reduce the default column width to 7.

Set the default numerical format to general, but format row 12 as integer. Format cells D15:K20 to show commas and one decimal place.

In cell A2, enter a continuous horizontal line by using a repeating hyphen or, if your spreadsheet accepts character graphics, Alt-196. Copy the contents of cell A2 into cells A10, A13, H6, H8, D13, A21, and A119.

In cell G3, enter three spaces followed by a vertical line, with the last being either the vertical-bar key or, if your spreadsheet accepts character graphics, Alt-179. Copy the contents of cell G3 into cells G6:G9; leave cells G4 and G5 blank for the time being. In cell C11, enter a vertical line without the three preceding spaces and copy the contents of that cell into cells C12:C20.

Text and Numerical Entries

Referring to the following rows in Listing 2.1 (pages 49-50):

1	Enter the title and other information shown in row 1.
3-9	Enter the text shown in cells A3:A9, H3:H5, H7, and H9. Enter the values shown in cells F3:F9, K5, K7, K9.
11-12	Enter the text shown in cells A11:A12 and F11. Enter the values shown in cells D12:K12.
14-20	Enter the text shown in cells F14, A15, B16:B20 (with a single space before the letter "m"). Enter the values shown in cells A16:A20.
24-28	Enter the text shown in cells A24 and A28.
38-41	Enter the text shown in cells A38 and A40:A41.
43-61	Enter the text shown in cells A43, A45:A51, A53:A59, and A61.
63-111	Enter the text shown in cells A63, A64, A71, AA78, A91, A104, and A111.
120-21	Enter the source citation shown in rows 120-21.

Formulas

Referring to the following rows in Listing 2.1 (pages 49-50):

3-9	Enter the following formula in the cell shown, where the vertical line is either the vertical bar or Alt-179 (see above):

```
G4      IF(K4>0,"   |   *","   |   ")
```

Copy the contents of cell G4 into cell G5. Edit the contents of cell G5 to change the expression K4>0 to read K4=0.

14-20   Enter the following formulas in the cells shown:

```
D15    IF($K$4>0,IF($K$7=1,D26,D30),IF($K$7=1,D61,D112))
D16    IF($K$4>0,IF($K$7=1," ",D31),IF($K$7=1," ",D113))
```

Copy the formula in cell D15 into cells E15:K15. Copy the formula in cell D16 into cells D16:K20.

26   Enter the following formula in the cell shown:

```
D26    1000*$K$4/(2*$F$6*$F$7*$F$4*
       SQRT(PI*$F$8*$F$3*$K$9/$F$4))*
       EXP(-((D12-$F$3*$K$9/$F$4)^2/
       (4*$F$8*$F$3*$K$9/$F$4)))*
       EXP(-$F$5*$K$9)
```

Copy the formula in cell D26 into cells E26:K26.

30-35   Copy the contents of cells A15:A20 into cell A30. Enter the following formula in the cell shown:

```
D30    1000*$K$4/(4*PI*$F$7*$F$6*$F$3*
       SQRT($F$8*$F$9)*$K$9)*
       EXP(-((D$12-$F$3*$K$9/$F$4)^2/
       (4*$F$8*$F$3*$K$9/$F$4)*$A15^2/
       (4*$F$9*$F$3*$K$9/$F$4)+$F$5*$K$9))
```

Copy the formula in cell D30 into cells D30:K35.

40-41   Enter the following formulas in the cells shown:

```
D40    1000*K5/(F6*F7*F3)
D41    SQRT(1+4*F5*F8*F4/F3)
```

45-51   Enter the following formulas in the cells shown:

```
D45    (D$12-$F$3*$K$9*$D$41/$F$4)/2/
       SQRT($F$8*$F$3*$K$9/$F$4)
D46    1.061405429*1/(1+.327591117*ABS(D45))
D47    (D46-1.453152027)*1/(1+.3275911117*ABS(D45))
D48    (D47+1.421413741)*1/(1+.3275911117*ABS(D45))
```

D49	(D48-.284496736)*1/(1+.327591117*ABS(D45))
D50	(D49+.254829592)*1/(1+.327591117*ABS(D45))
D51	IF(D45>10,0,IF(D45<-10,2,IF(D45<0, 2-EXP(-D45*D45)*D50,EXP(-D45*D45)*D50)))

Copy the formulas in cells D45:D51 into cells E45:K45.

53-59   Copy the formula in cell D45 into cell D53; edit the formula in cell D53 to change the minus sign to a plus sign. Copy the formulas in cells D46:D51 into cell D54. Copy the formulas in cells D53:D59 into cells E53:K53.

61   Enter the following formula in the cell shown:

D61	D40/2*EXP(D12/2/F8)* (D51*EXP(-D12*D41/2/F8)-D59*EXP(D12*D41/2/F8))

Copy the formula in cell D61 into cells E61:K61.

64-70   Enter the following formulas in the cells shown:

A65	A15
D65	(D12^2+F$8*$A15^2/$F$9)*$F$4/(4*$F$8*$F$3*$K$9)

Copy the formula in cell A65 into cells A66:A70 and the formula in cell D65 into cells D65:K70.

71-77   Enter the following formulas in the cells shown:

A72	A15
D72	SQRT(D12^2+F$8*$A15^2/$F$9)*$D$41/2/$F$8

Copy the formula in cell A72 into cells A73:A77 and the formula in cell D72 into cells D72:K77.

78-90   Enter the following formulas in the cells shown:

A79	A15
D79	IF(OR(ISERROR(D79)=1,AND(D79=0,ITER>1)),0, EXP(-EXP(LOG(D65)+(2*ITER-3)* (10-LOG(D65))/24)-D72^2/4/EXP(LOG(D65)+ (2*ITER-3)*(10-LOG(D65))/24)))
A85	A15
D85	IF(ITER=1,0,D85+D79)

Copy the formula in cell A79 into cells A80:A84 and the formula in cell A85 into cells A86:A90. Copy the formula in cell D79 into cells D79:K84 and the formula in cell D85 into cells D85:K90.

91-103   Enter the following formulas in the cells shown:

```
A92    A15
D92    IF(OR(ISERROR(D92)=1,AND(D92=0,ITER>1)),0,
           EXP(-EXP(LOG(D65)+(2*ITER-2)*
           (10-LOG(D65))/24)-D72^2/4/EXP(LOG(D65)+
           (2*ITER-2)*(10-LOG(D65))/24)))
A98    A15
D98    IF(ITER=1,0,D98+D92)
```

Copy the formula in cell A92 into cells A93:A97 and the formula in cell A98 into cells A99:A103. Copy the formula in cell D92 into cells D92:K97 and the formula in cell D98 into cells D98:K103.

104-10   Enter the following formulas in the cells shown:

```
A105    A15
D105    ((10-LOG(D65))/24)*(EXP(-EXP(LOG(D65))-
            D72^2/4/EXP(LOG(D65)))+4*D85+2*D98+
            IF(ISERROR(D105)=1,0,
            EXP(-EXP(10)-D72^2/4/EXP(10))))/3
```

Copy the formula in cell A105 into cells A106:A110 and the formula in cell D105 into cells D105:K110.

111-17   Enter the following formulas in the cells shown:

```
A112    A15
D112    $D$40/(4*PI*SQRT($F$8*$F$9))*
            EXP(D$12/2/$F$8)*D105
```

Copy the formula in cell A112 into cells A113:A117 and the formula in cell D112 into cells D112:K117.

Protection of the Model

To safeguard the model from inadvertent changes, you should "protect" the entire worksheet and then "unprotect" the following cells: F3:F9, K4:K5, K7, K9, D12:K12, and A15:B20.

Graphs

To obtain the figures shown in chapter 2, above, select the bar type of graph. Select the following data blocks as variables: D15:K15, D16:K16, D17:K17, D18:K18, D19:K19, and D20:K20. Label the points on the x-axis with the contents of cells D12:K12. Label the variables with the contents of cells A15:A20. Use the contents of cell A1 as the title and cell K9 for the sub-title. Label the x-axis with cell F11 and the y-axis with cell F14.

Recipe 3: PRESERVING A SPECIES

BEARS uses 18 columns and 199 rows, although about half of this space is used as a scratch-pad for recording the results of model runs so that they can subsequently be displayed in a graph. The model itself is about 19 Kb in size. It contains 11 different equations, which together perform about 280 discrete arithmetic or logical operations (exclusive of those involved in graphing). There are two macros to help run the model through repeated iterations.

In addition to the four standard arithmetic operations, the model uses inequalities and the following spreadsheet functions:

    SUM(), ROUND(), RANDOM, and VLOOKUP().

Note that VLOOKUP() can be abbreviated to VLU() in SuperCalc. Note too that cells that contain (or are dependent on other cells that contain) the RANDOM function may not show the same values in your model as they do in Listing 3.1, even though you have entered the formulas correctly.

Layout

Set the default column width to 5; then reformat columns A, D, E, F, M, N, O, and R to 3, 6, 2, 1, 2, 2, 3, and 6 respectively.

Format cells B3:D3, P3:R3, P25:R25, B27:D27, and D99 as text-right (i.e., right-justified). Set the default numerical format to integer; then format cell K8 to show percentages and cells B28:D48 to two decimal places.

In cell A2, enter a continuous horizontal line by using a repeating hyphen or, if your spreadsheet accepts character graphics, Alt-196. Copy the contents of cell A2 into cell F12, F18, A21, A49, and A93.

In cell E3, enter a vertical line by using the vertical-bar key or, if your spreadsheet accepts character graphics, Alt-179. Copy the contents of cell E3 into cells E4:E20 and N3:N20.

Text and Numerical Entries

Referring to the following rows in Listing 3.1 (pages 71-72),

1   Enter the title and other information shown in row 1.

3-20   Enter the text shown in cells A3:D3, A4, A20, G3, G4:G8, G10:G11, G13:G14, G16:G17, G19:G20, O3:R3, O4, and O20. Enter the values shown in cells A5:A19, B4:B19, C4:C19, K4:K8, L4:L7, K10:K11, and O5:O19. Leave cells K13:K14 and K16:K17 blank.

24-48   Enter the text shown in cells A24:A25, O24, O25:R25, B27:D27, O27, and O42:O43. Enter the values shown in cells B28:D48 and O28:O41.

51   Enter the text shown in cell A51.

94-96   Enter the source citation shown in rows 94-96.

98-199   Enter the text shown in cells A98, G98 (note the leading spaces), D99, A100, and A102. Enter the values shown in cells G99:P99, B101, and B103. Enter the numbers from 1 to 100 in cells D100:D199 respectively.

Formulas

Referring to the following rows in Listing 3.1 (pages 71-72),

3-20   Enter the following formulas in the cells shown:

```
D4     B4+C4
B20    SUM(B4:B19)
K19    SUM(P9:P19)
K20    SUM(Q9:Q19)
```

           Copy the formula in cell D4 into cells D5:D19 and the formula in cell B20 into cells C20:D20.

24-48   Enter the following formulas in the cells shown:

```
P27   ROUND(K20*(K6+L6*VLU(RAN,C28:D48,1))*
        (K7+L7*VLU(RAN,C28:D48,1))*(1-K8))*(K19>0)
Q27   ROUND(K20*(K6+L6*VLU(RAN,C28:D48,1))*
        (K7+L7*VLU(RAN,C28:D48,1)*L7)*K8)*(K19>0)
R27   P27+Q27
P28   ROUND(P4*(1-(K$4+L$4*(VLU(RAN,C$28:D$48,1)))))
Q28   ROUND(Q4*(1-(K$5+L$5*(VLU(RAN,C$28:D$48,1)))))
P42   ROUND(P18*(1-(K$4+L$4*(VLU(RAN,C$28:D$48,1))))+
        RAN*P19*(1-(K$4+L$4*(VLU(RAN,C$28:D$48,1)))))
Q42   ROUND(Q18*(1-(K$5+L$5*(VLU(RAN,C$28:D$48,1))))+
        RAN*Q19*(1-(K$5+L$5*(VLU(RAN,C$28:D$48,1)))))
P43   SUM(P27:P42)
```

Copy the formula in cell R27 into cells R28:R42, the formulas in cells P28:Q28 into cells P29:P41, and the formula in cell P43 into cells Q43:R43.

Macros

Enter the following macro, using column B for the name, column C for the instructions, and column L for the comments (which are optional):

```
53   {\S}   {home}{paneloff}                          home cursor; dialog off
54          {let K16,0}{let K17,0}                    zero extinction counters
55          {let K13,1}                               initialize cycles
56          {let K14,1}                               [C56] initialize years
57          /c B4:D20,P4,v                            copy initial pop to current
58          !                                         [C58] calc next pop
59          {if K14=K11}{let K13,K13+1}{branch C56}   if max years, reinitialize
60          {if K13>K10}{branch C69}                  if max cycles, quit
61          {let K14,K14+1}                           increment year
62          {if P20<1}{branch C66}                    if no males left....
63          {if Q20<1}{branch C66}                    if no females left....
64          /c P27:R43,P4,v                           copy next pop to current
65          {branch C58}                              loop to next year
66          {let K17,(K17*K16+K14)/(K16+1)}           [C66] adjust mean time
67          {let K16,K16+1}{beep}                     increment times extinct
68          {let K13,K13+1}{branch C56}               reinitialize
69          {beep}{panelon}                           [C69] dialog on
```

Using the commands appropriate to your spreadsheet, link the macro name in cell B53 to the instructions in cells C53:C69.

Enter the following macro, using column B for the name, column C for the instructions, and column L for the comments (which are optional):

```
71   {\T}   {home}{paneloff}                          home cursor; dialog off
72          {if K10>10}{let K10,10}                   limit max cycles
73          {if K11>100}{let K11,100}                 limit max years
74          /b G100:P199 ~                            blank records
75          {let K16,0}{let K17,0}{let K13,1}         initialize counters
```

```
76            {let K14,1}                              [C76] initialize years
77            /c B4:D20,P4,v                           copy initial pop to current
78            !                                        [C78] calc next pop
79            /c R20,[B101+K13;B103+K14],v             update records
80            {if K14=K11}{let K13,K13+1}{branch C76}  if max years, reinitialize
81            {if K13>K10}{branch C91}                 if max cycles, quit
82            {let K14,K14+1}                          else increment year
83            {if P20<1}{branch C87}                   if no males left....
84            {if Q20<1}{branch C87}                   if no females left....
85            /c P27:R43,P4,v                          copy next pop to current
86            {branch C78}                             loop to next year
87            {let K17,(K17*K16+K14)/(K16+1)}          [C87] adjust mean time
88            {let K16,K16+1}{beep}                    increment times extinct
89            {if K14<K11}{let [B101+K13;B103+K14+1],0}    add zero pop to records
90            {let K13,K13+1}{branch C76}              reinitialize
91            /b [B101+K13;B103+K14]~                  [C91] blank the last cycle
92            {beep}{panelon}                          dialog on
```

Using the commands appropriate to your spreadsheet, link the macro name in cell B71 to the instructions in cells C71:C92.

Protection of the Model

To safeguard the model from inadvertent changes, you should "protect" the entire worksheet and then "unprotect" the following cells: B4:D20, K4:L7, K8, K10:K11, K13:K14, K16:K17, P4:R20, P27:R43, and G100:R199.

Graphs

To obtain the figures shown in chapter 3, above, select the line type of graph. Select the following data: G100:G199, H100:H199, I100:I199, J100:J199, K100:K199, L100:L199, M100:M199, N100:N199, O100:O199, and P100:P199. Label the points on the x-axis with the contents of cells D100:D199. Label the variables with the contents of cells G99:P99. Use the contents of cells A1 and G3 as title and sub-title respectively. Label the x-axis with cell G11.

Recipe 4: SUSTAINABLE YIELD

TREES uses 15 columns and 245 rows and is about 110 Kb in size. It contains 17 different formulas, which together perform about 12,300 discrete arithmetic or logical operations (exclusive of those involved in graphing). There are two macros to help run the model through repeated iterations.

In addition to the four standard arithmetic operations, the model uses inequalities, powers, and the following spreadsheet functions:

     `SUM(), AV(), AND(), OR(), MAX(), LN(), IF(), ISERROR()`, and `RANDOM`

Note that cells that contain (or are dependent on other cells that contain) the RANDOM function may not show the same values in your model as they do in Listing 4.1, even though you have entered the formulas correctly.

Layout

Set the default column width to 6. Then reformat columns A, H, and I to 4; columns B, C, K, and L to 5; and column G to 1.

Format the following cells as text-right: D3:F3, K9:O9, J10:O10, B11:F11, J11:M11, B12:F12, B13:D13, E81, J81, N81, and E137. Set the default numerical format to integer. Then format cell K4 as two-decimal percent, K5:K6 as three decimal places, and K7 as integer percent.

In cell A2, enter a continuous horizontal line by using a repeating hyphen or, if your spreadsheet accepts character graphics, Alt-196. Copy the contents of cell A2 into cell A8, A10, H59, A75, A83, A131, A139, A187, M238, and A242.

In cell G3, enter a vertical line by using the vertical-bar key or, if your spreadsheet accepts character graphics, Alt-179. Copy the contents of cell G3 into cells G4:G7, L3:L7, G9:G74, H80:H82, L80:L82, H84:H130, L84:L130, H136:H138, L136:L138, H140:H186, L140:L186, and L188:L241. Note that some of these cells will be edited later on.

Text and Numerical Entries

Referring to the following rows in Listing 4.1 (pages 95-98),

1   Enter the title and other information shown in row 1.

3-7   Enter the text shown in cells A3:A7, D3:F3, and H3:H7 (note the leading space in the last three cells). Edit the contents of cells L3:L7 to include the text shown. Enter the values shown in cells D4:F4, F6:F7, K4:K7, O3, O5, and O7. Note that, when entering percentages, you enter them as decimal values without the percent sign.

9   Enter the text shown in cells A9 and E9.

9-74   Enter the text shown in cells H9:H13, J10:J11, K9:N11, O9:O10, A12:A13, B11:E13, F11:F12, H61, H63:H66, H68, M68, J70:J73, L70, and A74. Enter the values shown in cells A14:A73, and H14:I58.

77-82   Enter the text shown in cells A77, B78, D80 (with a leading space), E81, J81, N81, and B82. Edit the contents of cells H80 and L80 to include the text shown. Enter the values shown in cells D82:F82, I82:K82, and M82:O82.

84-130 Enter the text shown in cells A84:A85. Enter the values shown in cells A86:B130.

133-38 Copy the contents of cells A77:O82 into cell A133.

140-86 Enter the text shown in cells A140:A141. Copy the contents of cells A86:B130 into cell A142.

189   Enter the text shown in cell M189.

243-45 Enter the source citation shown in rows 243-45.

Formulas

Referring to the following rows in Listing 4.1 (pages 95-98),

9   Enter the following formula in the cell shown:

```
F9      5*K4+(K4=0)*(K5*(-LN(1-RAN))^(1/K6)+
        K5*(-LN(1-RAN))^(1/K6)+K5*(-LN(1-RAN))^(1/K6)+
        K5*(-LN(1-RAN))^(1/K6)+K5*(-LN(1-RAN))^(1/K6))
```

9-74   Enter the following formulas in the cells shown:

```
J13     SUM(J14:J58)
J14     SUM(D86:F86)
K14     5*SUM(I86:K86)
L14     5*SUM(M86:O86)
M14     5*SUM(D142:F142)
N14     SUM(I142:K142)
O14     5*(SUM(M142:O142))
B74     IF(ISERROR(AV(B14:B73))," ",AV(B14:B73))
```

Copy the formula in cell J13 into cells K13:O13, the formulas in cells J14:O14 into cells J15:J58, and the formula in cell B74 into cells C74:F74.

84-130   Enter the following formulas in the cells shown:

```
D84    SUM(D86:D130)
D86    M240*(RAN+.5)
I86    $F$9/5*D86*5*(.0284+.5047/10^11*$I14^5-
       .9224/10^4*$I14^2+.0007492*LN($I14)-
       .3353/10^8*$I14^4+.003540*$I14-
       .4139/10^8*$I14^3*I$82+
       .8554/10^6*$I14^3+.7035/10^4*$I14*I$82)
M86    I86*$K$7
```

Copy the formula in cell D84 into cells E84:F84, cells I84:K84, and cells M84:O84. Copy the formula in cell D86 into cells D86:F130, the formula in cell I86 into cells I86:K130, and the formula in cell M86 into cells M86:O130.

140-86   Enter the following formulas in the cells shown:

```
D140    SUM(D142:D186)
D142    ($O$3>0)*(OR($I14>=$O$5,
        AND($O$3>1,AND($I14>$O$7-5,$I14<$O$7+5)))))
        *(D86-I86)
I142    D86/10^3*MAX(0,-265.8+3.381*$I14+.7876*$I14*D$82-
        .3519/10^6*$I14^4+156.4*D$82-319.4*LN(D$82)-
        .4719*$I14*D$82^2+.001176*$I14^2*D$82^2)
M142    IF(D86=0,0,(D142+M86)/D86*I42)
```

Copy the formula in cell D140 into cells E140:F140, cells I140:K140, and cells M140:O140. Copy the formula in cell D142 into cells D142:F186, the formula in cell I142 into cells I142:K186, and the formula in cell M142 into cells M142:O186.

Macros

Enter the following macro, using column A for the name, column B for the instructions, and column H for the comments (which are optional):

```
189   {\S}   {comment} SIMULATION MACRO
190          {paneloff}                        dialog off
191          {home} =A20 ~                      move to see
192          {let C9,0}                         set year
193          {if F7=C9} {branch B206}           exit if finished
194          {branch B202}                      skip growth
195          /c D86:F130,M192,v                 [B195] copy to pad
196          /c I86:K130,M192,-                 less annual burn
197          /c D142:F186,M192,-                less annual cut
198          /c M236:O236,M191,v                copy dead into new
199          /c I84:K84,M191,+                  add total burn
```

```
200              /c D140:F140,M191,+              add total harvest
201              /c M191:O235,D86,v              copy back from scratchpad
202              {let C9,C9+5}                   [B202] increment yr
203              ! {down}                        recalc; move down
204              /c K13:O13,[2;13+C9/5],v        copy to summary
205              {if C9<F7} {branch B195}        repeat
206              {panelon} {home} {beep}         [B206] dialog on
```

Using the commands appropriate to your spreadsheet, link the macro name in cell A189 to the instructions in cells B189:B206.

Enter the following macro, using column A for the name, column B for the instructions, and column H for the comments (which are optional):

```
208   {\I}     {comment} INITIALIZATION MACRO
209              {paneloff} {home}              dialog off
210              = A80 ~                        move to see
211              {let D86,0}                    zero site 1
212              {let E86,0}                    zero site 2
213              {let F86,0}                    zero site 3
214              /c D86:F86,D87:D130,v          copy to all
215              !                              recalc
216              {let D86,5*D4/D5}              set site 1
217              {let E86,5*E4/E5}              set site 2
218              {let F86,5*F4/F5}              set site 3
219              /c D86,D87:[4;85+D5/5],v       copy
220              /c E86,E87:[5;85+E5/5],v       copy
221              /c F86,F87:[6;85+F5/5],v       copy
222              {if F6<>1} {branch B236}       skip randomizing
223              /c D86:F86,M240,v              store exact values
224              =A80~ =D86~                    setup
225              M240*(RAN+.5)                  [B225] randomize
226              {down}                         down and check okay
227              {if crow<=(85+D5/5)} {branch B225}
228              =A80~ =E86~                    setup
229              N240*(RAN+.5)                  [B229] randomize
230              {down}                         down and check okay
231              {if crow<=(85+E5/5)} {branch B229}
232              =A80~ =F86~                    setup
233              O240*(RAN+.5)                  [B233] randomize
234              {down}                         down and check okay
235              {if crow<=(85+F5/5)} {branch B233}
236              /b B14:F73 ~                   [B236] blank last
237              {let C9,0}                     zero year
238              {panelon}                      dialog on
239              {home}                         ready for options
240              {beep}                         signal
```

Using the commands appropriate to your spreadsheet, link the macro name in cell A208 to the instructions in cells B208:B240.

Protection of the Model

To safeguard the model from inadvertent changes, you should "protect" the entire worksheet and then "unprotect" the following cells: D4:F5, F6:F7, K4:K7, O3, O5, O7, J13:O13, B14:F73, D86:F130, M191:O236, and M240:O240.

Graphs

To obtain the figures shown in chapter 4, above, the following graphs are required:

- Select the stacked-bar type of graph. Select the following data ranges as variables: D86:D130, E86:E130, and F86:F130. Label the points on the x-axis with the contents of cells I14:I58. Label the variables with the contents of cells D3:F3. Use the contents of cells A1 and H63 as title and sub-title respectively. Label the x-axis with cell J73 and the y-axis with cell J11.

- Select the line type of graph. Select the following data ranges as variables: B14:B73 and D14:D73. Label the points on the x-axis with the contents of cells A14:A73. Label the variables with the contents of cells J70 and L70. Use the contents of cells A1 and H64 as title and sub-title respectively. Label the x-axis with cell M68 and the y-axis with cell A4.

- Select the line type of graph. Select the following data ranges as variables: E14:E73 and F14:F73. Label the points on the x-axis with the contents of cells A14:A73. Label the variables with the contents of cells E12 and F12. Use the contents of cells A1 and H65 as title and sub-title respectively. Label the x-axis with cell M68 and the y-axis with cell E13.

Recipe 5: HERE COMES THE SUN

SOLAR uses 17 columns and 106 rows and is about 72 Kb in size. It contains 26 different formulas, which together perform about 9,530 discrete arithmetic or logical operations (exclusive of those involved in graphing).

In addition to the four standard arithmetic operations, the model uses powers, the constant PI, and the following spreadsheet functions:

```
MAX(), SUM(), AV(), STD(), SQRT(), EXP(), SIN(), and COS().
```

Layout

Set the default column width to 5; then reset the width of columns A, B, C, D, and O to 4, 4, 1, 4, and 2 respectively. If columns A to Q do not fit within the width of your screen, you can reduce the width of columns C and O until they do.

Set the default numerical format to general. Format the block D8:P20 as integer and column Q to show two decimal places. Format row 6 and column O as text-right (i.e., right-justified) and then format cell A6 as text-left (i.e., left-justified).

In cell A5, enter a continuous horizontal line by using a repeating hyphen or, if your spreadsheet accepts character graphics, Alt-196. Copy the contents of cell A5 into cells A21 and A108.

In cell C6, enter a vertical line by using the vertical-bar key or, if your spreadsheet accepts character graphics, Alt-179. Copy the contents of cell C6 into cells C7:C20 and O6:O20.

Text and Numerical Entries

Referring to the following rows in Listing 5.1 (pages 129-30),

1-4   Enter the title and other information shown in row 1. Enter the text shown in cells A2:A3, E2:E4, and O3:O4. Enter the values shown in cells A4 and M2:M4.

6-20   Enter the text shown in cells A6, B6, D6:N6, P6, Q6, B7, E7, and A8:A19. Enter the values shown in cells A7 and B8:B19.   Enter the text shown in cell A20.

27-36   Enter the text shown in cells A27:A31, E28:E31, and J30:J31. Enter the text shown in cells A34:A35 and G34:G36.

39-52   Enter the text shown in cells A39 and A41:A52.

55-71   Enter the text shown in cells A55 and A57:A59. Enter the values shown in cells D57:N57.

74-87   Enter the text shown in cells A74:A75.

90-106   Enter the text shown in cell A90. Copy the contents of cells A57:N59 into cell A92.

109-10   Enter the source citation shown in rows 109-10.

Formulas

Referring to the following rows in Listing 5.1 (pages 129-30),

6-20   Enter the following formulas in the cells shown:

```
Q7     IF(AND(ISERROR(STD(P8:P19))=0,AV(P8:P19)>=1),
           STD(P8:P19)/AV(P8:P19))
D8     MAX(0,$A$4*(1-$B8)*116.4*
           (1+.034*COS(360*$A60/365*PI/180))*D95*.56*
           (IF(ISERROR(EXP(-.65*D76)+EXP(-.095*D76))=0,
           EXP(-.65*D76)+EXP(-.095*D76))))
P8     SUM(D8:√8)
Q8     IF(AND(ISERROR(STD(D8:N8))=0,AV(D8:N8)>=1),
           STD(D8:N8)/AV(D8:N8))
```

Copy the formula in cell D8 into cells D8:N19, the formula in cell P8 into cells P9:P19, and the formula in cell Q8 into cells Q9:Q19.

Enter the following formulas in the cells shown:

```
B20    (B8*31+B9*28+B10*31+B11*30+B12*31+B13*30+B14*31+
           B15*31+B16*30+B17*31+B18*30+B19*31)/365
D20    AV(D8:D19)
```

Copy the formula in cell D20 into cells E20:N20 and into cells P20:Q20.

34-36   Enter the following formulas in the cells shown:

```
I34    M2*PI/180
I35    M3*PI/180
I36    M4*PI/180
```

39-52   Enter the following formulas in the cells shown:

```
B41    $A$7
B42    B41+31
B43    B42+28
B44    B43+31
B45    B44+30
B46    B45+31
B47    B46+30
B48    B47+31
```

```
B49    B48+31
B50    B49+30
B51    B50+31
B52    B51+30
D41    23.45*SIN(360*(284+B41)/365*PI/180)*PI/180
```

Copy the formulas in cells B41:B52 into cell A60 and cells A60, A76, and A95; copy the formula in cell D41 into cells D42:D52.

55-71   Enter the following formulas in the cells shown:

```
D58    D57*PI/180
D60    SIN($I$34)*SIN($D41)+
       COS($I$34)*COS($D41)*COS(D$58)
```

Copy the formula in cell D58 into cells E58:N58 and the formula in cell D60 into cells D60:N71.

74-87   Enter the following formula in the cell shown:

```
D76    SQRT(1229+(614*D60)^2)-(614*D60)
```

Copy the formula in cell D76 into cells D76:N87.

90-106   Enter the following formula in cell D95:

```
D95    SIN($D41)*
       (SIN($I$34)*COS($I$36)-
       COS($I$34)*COS($I$35)*SIN($I$36))+
       COS($D41)*COS(D$93)*(COS($I$34)*COS($I$36)+
       SIN($I$34)*COS($I$35)*SIN($I$36))+
       COS($D41)*SIN($I$35)*SIN($I$36)*SIN(D$93)
```

Copy the formula in cell D95 into cells D95:N106.

Conversion to Imperial Units

The following modifications are required to convert the model from metric to Imperial units of measurement:

- Edit the text in cell E7 to read "(BTU/sq ft/hour on day shown for each month)".

- Edit the formula in cell D8 by changing the number 116.4 to 429, and copy the amended formula in cell D8 into cells D8:N19.

- Edit the text in cells E31 and J31 to read "BTU/sq ft/hour".

Protection of the Model

To safeguard the model from inadvertent changes, you should "protect" the entire worksheet and then "unprotect" the following cells: A4, M2:M4, O4, A7, and B8:B19.

Graphs

To obtain the figures shown in chapter 5, above, the following graphs are required:

- Select the bar type of graph. Select the following data range as a variable: P8:P19. Label the graph with the text in cells E28 and N1 as title and sub-title respectively. Label the x-axis and y-axis with cells E30 and E31 respectively. Label the x-axis points with cells A8:A19.

- Select the bar type of graph. Select the following data ranges as variables: D8:D19, E8:E19, F8:F19, G8:G19, H8:H19, I8:I19, J8:J19, K8:K19, L8:L19, M8:M19, and N8:N19. Label the graph with the text in cells E28 and N1 as title and sub-title respectively. Label the x-axis and y-axis with cells E30 and E31 respectively. Label the x-axis points with cells A8:A19.

- Select the bar type of graph. Select the following data ranges as variables: D8:N8, D9:N9, D10:N10, D11:N11, D12:N12, D13:N13, D14:N14, D15:N15, D16:N16, D17:N17, D18:N18, and D19:N19. Label the graph with the text in cells E28 and N1 as title and sub-title respectively. Label the x-axis and y-axis with cells J30 and J31 respectively. Label the x-axis points with cells D6:N6.

Recipe 6: THE KLEIN MODEL

KLEIN uses 9 columns and 43 rows and is about 10 Kb in size. It contains 12 formulas, which together perform about 240 discrete arithmetic or logical operations (exclusive of those involved in graphing). Because the model is "circular", it is solved by iteration. With a delta value of .001 for iteration control, the model in chapter 6 typically requires about 15 iterations to reach a new solution.

The model uses only the four standard arithmetic operations and no functions.

Layout

Set the default column width to 8; then reset the width of columns A and B to 16 and 3 respectively. If columns A to I do not fit within the width of your screen, you can reduce the width of column A until they do.

Set the default numerical format to three decimal places, or whatever is appropriate for the monetary units in use. Format cell I15 as general, and row 22 and cell C17 as integer. Format the following cells as text-right: D8:D11, F8:F11, H8:H10, I17, and B24:B32. Finally, format cells I18:I20 as percentages to one decimal place. Note that, when entering values in cells formated with percent (%) signs, you do not normally enter the symbol directly: for example, to get "9%", you enter ".09" (without the quotation marks).

In cell A2, enter a continuous horizontal line by using a repeating hyphen or, if your spreadsheet accepts character graphics, Alt-196. Copy the contents of cell A2 into cells A16, A21, and A33.

Text and Numerical Entries

Referring to the following rows in Listing 6.1 (page 157),

1   Enter the title and other information shown in row 1.

3-15   Enter the text shown in cells A3:A15 (note the use of three leading spaces in cells A3, A7, and A12 to increase readability), B4:B6, B8:B11, D8:D11, F8:F11, H8:H10, B13:B15, and H13:H15. Note that the "formulas" in cells B4:B6 and B13:B15 are not operational: that is, they are not part of the model but are provided solely to remind users of the structure of the model. Enter the value shown in cell I15. If your spreadsheet allows you to set a delta value for controlling the convergence, then specify cell I15 as the parameter.

17-20   Enter the text shown in cells C17:C20, E17:E20, and I17. Enter the values shown in cells C17:C20, G17:G20, and I18:I20.

22-32   Enter the text shown in cells A22, A24:A32, and B24:B32.

34-43   Enter the source citation shown in rows 34-43.

Formulas

Referring to the following rows in Listing 6.1 (page 157),

22-32   Enter the following formulas in the cells shown:

C22   C17
C24   C18
C25   C19
C26   G18
C27   C20
C28   G19
C29   G20
C30   C24+C25+C26−C31
C31   C27+C28+C29
C32   G17

Enter the following formulas in the cells shown:

D22   C22+1
D24 $C$8+$E$8*D29+$G$8*(D27+D28)+$I$8*C29+$C$11
D25 $C$9+$E$9*D29+$G$9*C29+$I$9*C32+$E$11
D26   C26*(1+$I$18)
D27 $C$10+$E$10*(D31+D30−D28)+$G$10*(C31+C30−C28)+
       $I$10*(D22−1931)+$G$11
D28   C28*(1+$I$19)
D29   C29*(1+$I$20)
D30   D24+D25+D26−D31
D31   D27+D28+D29
D32   C32+D25

Copy the formulas in cells D22:D32 into cells E22:I22.

Protection of the Model

To safeguard the model from inadvertent changes, you should "protect" the entire worksheet and then "unprotect" the following cells: I15, C8:C11, E8:E11, G8:G11, C17:C20, G17:G20, and I18:I20.

Graphs

To obtain the figures shown in chapter 6, above, the following graphs are required:

●   Select the line type of graph. Select the following eight data ranges as variables: C24:I24, C25:I25, C26:I26, C27:I27, C28:I28, C29:I29, C30:I30, and C31:I31. (Variable C32:I32 is omitted because it is shown in a separate graph, below.) Label the variables with the text in cells A32:A39. Title the graph with the text in cell A1. Label the variables with the text in cells C22:I22.

- Select the bar-chart type of graph. Select the following data range as a variable. Title the graph with the text in cell A1; use the text in cell A32 as a sub-title. Label the variables with the text in cells C22:I22.

Recipe 7: THE LOWRY MODEL

LOWRY uses 11 columns and 48 rows and is about 11 Kb in size. It contains 12 different formulas, which together perform about 290 discrete arithmetic or logical operations (exclusive of those involved in graphing). Because the model is "circular" in nature, the spreadsheet automatically solves by iteration. With a delta value for iteration of .01 (the default in my spreadsheet), this model typically iterates about 15 times to reach a new solution.

In addition to the four standard arithmetic operations, the model uses only powers and the SUM() function.

Layout

Set the default column width to 6; then reset the width of columns A, B, C, D, and K to 17, 4, 9, 2, and 7 respectively. If columns A to K do not fit within the width of your screen, you can reduce the width of column A until they do.

Set the default numerical format to general. Format rows 14, 15, 17, and 20 as integers; and rows 32-44 to two decimal places. Format cell C5 and rows 12 and 30 as text-right (right-justified).

In cell A2, enter a continuous horizontal line by using a repeating hyphen or, if your spreadsheet accepts character graphics, Alt-196. Copy the contents of cell A2 into cells A11, A21, and A46.

Text and Numerical Entries

Referring to the following rows in Listing 7.1 (page 173),

1   Enter the title and other information shown in cell A1.

3-10   Enter the text shown in cells A3, A4, E4, A6, A7, A9, and A10. Enter the text shown in cells E3:J3, cell C5, and cells D5:D10. Enter the values shown in cells B4, B7, B10, and E5:J10.

12-20   Enter the text shown in cells E12:K12 and A13:A20. Enter the values shown in cells E13:J13.

24-28   Enter the text shown in cells A24:A28 and B25:B28.

30-37   Enter the text shown in cells E30:K30, E31, A32:A36, and D32:D37.

39-44   Enter the text shown in cells A39:A43 and D39:D44.

46-48   Enter the source citation shown in rows 46-48.

Formulas

Referring to the following rows in Listing 7.1 (page 173),

12-20   Enter the following formulas in the cells shown:

```
K13    SUM(E13:J13)
E14    $E$13*E39+$F$13*E40+$G$13*E41+$H$13*E42+
           $I$13*E43+$J$13*E44
E15    E14*$B$7
E16    E20*$B$10
E17    $E$16*E39+$F$16*E40+$G$16*E41+$H$16*E42+
           $I$16*E43+$J$16*E44
E18    E17*$B$7
E19    E13+E16
E20    E15+E18
```

Copy the formula in cell K13 into cells K14:K20 and the formulas in cells E14:E20 into cells F14:J14.

30-37   Enter the following formulas in the cells shown:

```
E32    1/E5^$B$4
K32    SUM(E32:J32)
```

Copy the formula in cell E32 into cells E32:J37 and the formula in cell K32 into cells K33:K37.

39-44   Enter the following formulas in the cells shown:

```
E39    E32/$K32
K39    SUM(E39:J39)
```

Copy the formula in cell E39 into cells E39:J44 and the formula in cell K39 into cells K40:K44.

Protection of the Model

To safeguard the model from inadvertent changes, you should "protect" the entire worksheet and then "unprotect" the following cells: B4, B7, B10, E5:J10, and E13:J13.

Graphs

To obtain the figures shown in chapter 7, above, select the bar type of graph. Select the following two data ranges as variables: E13:J13 and E20:J20. Label the graph with the text in cells B25:B28 as title, sub-title, and variable names respectively.

Recipe 8: THE BERTAUD MODEL

BERTAUD uses 10 columns and 77 rows and is about 16 Kb in size. It contains 50 different formulas, which together perform about 250 discrete arithmetic or logical operations (exclusive of those involved in graphing).

In addition to the four standard arithmetic operations, the model uses only powers and the SUM() and IF() functions.

Layout

Set the default column width to 8; then reset the width of columns A, B, and D to 9, 7, and 3 respectively. If columns A to J do not fit within the width of your screen, you can reduce the width of column C and/or column J until they do.

The optimum formating of the worksheet depends on the currency units used. In the version shown in chapter 8, the default format is set to integer and the following cells are formated in the manner shown below:

Integer & Commas	Integer & Percent	Integer $ & Commas	Two Decimal Places	Two Decimals & Dollar ($)
I3:I11	C5:C11	C14:C20	B3:B11	C23:C25
C32:C38	J5:J11	G15:J17	G5:H7	F24:J24
	E20:J20	E19:J19	F15:F17	J25:J26
	J27	G34:J37	C26:C28	J28

C43:C48	I44:J48	C31	G39:J39
F37	C63, H63	C39:C42	C66, H66
F39:F40	C69:C70		C68, H68
	H69:H70		C75, H75
	C72, H72		C77, H77
	C82:C83		C79, H79
	H80		C81

Note that, when entering values in cells formated with percent (%) or dollar ($) signs, you do not normally have to enter either symbol directly. Thus, to get 9%, you enter .09; similarly, to get $5.10, you enter 5.1.

Format the following cells as text-right (right-justified): F4:H4, F9:I9, E10, F14:J14, E18:J18, F23:J23, J30, G32:J32, J42, and I43:J43.

In cell A2, enter a continuous horizontal line by using a repeating hyphen or, if your spreadsheet accepts character graphics, Alt-196. Copy the contents of cell A2 into cells A12, A21, A29, E41, A49, A52, A61, and A74.

In cell D3, enter a vertical line by using the vertical-bar key or, if your spreadsheet accepts character graphics, Alt-179. Copy the contents of cell D3 into cells D4:D11, D13:D20, D22:D28, D30:D48, D53:D60, and D62:D73.

Text and Numerical Entries

Referring to the following rows in Listing 8.1 (pagea 191-92),

1	Enter the title and other information shown in row 1.
3-11	Enter the text shown in cells A3, C3, E3, J3, F4:H4, A5, A7:A11, E5:E11, and F9:I9; for increased readability, insert a space before the text in cells C3 and J3. Enter the values shown in cells B3, G3, H3, F5:H7, F10:H10, and B7:B10.
13-20	Enter the text shown in cells A13, E13, G13, A14:A20, E14:J14, and E18:J18. Note that, for graphing purposes, the text that appears in cell E13 is entered partly in cell E13 and partly in cell G13. Enter the values shown in cells C14:C20, F15:J17, and I19.
22-28	Enter the text shown in cells A22:A28, E22, F23:J23, and E24:E28. Enter the values shown in cells C23:C25 and J27.

30-48   Enter the text shown in cells A30:A48, E30, G32:J32, E34:E40, E42, J42, I43, J43, and E44:E48. Enter the values shown in cells C31 and F37:F40.

50-73   Enter the text shown in cells A50 and A51. Enter the text shown in cells A53, A55:A60, E53, E55:E60, and I53:I60. Enter the values shown in cells C53, C55:C59, H53, and H55:H59. Enter the text shown in cells A62, A64:A73, E62, E64:E70, and I62:I73. Enter the values shown in cells C62, C64:C72, H62, and H64:H69.

75-77   Enter the source citation shown in rows 75-77.

Formulas

Referring to the following rows in Listing 8.1 (pages 191-92),

3-11   Enter the following formulas in the cells shown:

```
I3      G3*H3
B5      I3/10000
B11     B3-SUM(B5:B10)
C5      B5/$B$3
F8      SUM(F5:F7)
I5      F5*G5*H5
J5      I5/$I$3
I10     SUM(F10:H10)
I11     I3-SUM(I5:I7)-I10
```

Copy the formula in cell C5 into cells C7:C11, the formula in cell F8 into cell I8, the formula in cell I5 into cells I6:I7, and the formula in cell J5 into cells J6:J8 and cells J10:J11.

13-20   Enter the following formulas in the cells shown:

```
E19     C70
F19     C83
G19     H70
H19     H80
J19     SUM(E19:I19)
E20     E19/$J19
```

Copy the formula in cell E20 into cells F20:J20.

22-28   Enter the following formulas in the cells shown:

```
C26    J8*(B5+B11-J10*SUM(B7:B10))
C27    SUM(B7:B9)
C28    C26+C27
F24    C14/B3*.1
G24    C15/B3*.1
H24    SUM(C16:C20)/B3*.1
I24    J19/I3
J24    SUM(F24:I24)
J25    J24*B3/C28
J26    (C28*J25-B7*C23-B8*C24-B9*C25)/C26
J28    J26*I8/(I5+(1+J27)*I6+(1+2*J27)*I7)
```

30-48   Enter the following formulas in the cells shown:

```
C32    C$26*10000/I$8*F5
C35    SUM(C32:C34)
C36    C35*C31/B3
C37    C35/B3
C38    F8/I3*10000
C39    B7*10000/(C$36*B$3)
```

Copy the formula in cell C32 into cells C33:C34. Copy the formula in cell C39 into cells C40:C41. Enter the following formulas in the cells shown:

```
C42    (B10*10000+I11*(C26*10000/I8))/(C36*B3)
C43    C26/B3
C44    C7
C45    C8
C46    C9
C47    C42*C36/10000
C48    1-SUM(C43:C47)
```

Enter the following formulas in the cells shown:

```
G34    F15*G15+H15+I15+J15
H34    F16*G16+H16+I16+J16
I34    F17*G17+H17+I17+J17
J34    (G3*F$5+H34*F$6+I34*F$7)/SUM(F$5:F$7)
```

Copy the formula in cell J34 into cells J35:J36 and into J39. Enter the following formulas in the cells shown:

```
G35    J28*G5*H5
H35    (1+J27)*J28*G6*H6
I35    (1+2*J27)*J28*G7*H7
G36    G34+G35
```

Copy the formula in cell G36 into cells H36:I36. Enter the following formulas in the cells shown:

```
G37    G36*$F37
G39    G36*((1-$F37)*((1+$F39)^$F38)*($F39-$F40))/
       (12*(((1+$F39)^$F38)-(1+$F40)^$F38))
```

Copy the formula in cell G37 into cells H37:J37 and the formula in cell G39 into cells H39:I39. Enter the following formulas in the cells shown:

```
I44    C14
I45    SUM(C16:C20)
I46    (G24+J24)*B3*10
I47    (C32*G34+C33*H34+C34*I34)/1000
I48    SUM(I44:I47)
J44    1000*I44/C$35
```

Copy the formula in cell J44 into cells J45:J48.

50-73   Enter the following formula in the cell shown:

```
C60    IF(C53=0,C55*C56+C57*C58+C59,C53)
```

Copy the formula in cell C60 into cell H60. Enter the following formulas in the cells shown:

```
C73    IF(C62=0,C64*C65+C66*C67+C68*C69+C70*C71+C72,C62)
H70    IF(H62=0,H64*H65+H66*H67+H68*H69,H62)
```

Protection of the Model

To safeguard the model from inadvertent changes, you should "protect" the entire worksheet and then "unprotect" the following cells: B3, G3, H3, F5:H7, F10:H10, B7:B10, C14:C20, F15:J17, I19, C23:C25, J27, C31, F37:F40, C53, C55:C59, H53, H55:H59, C62, C64:C72, H62, and H64:H69.

Graphs

To obtain the figures shown in chapter 8, above, the following graphs are required:

● Select the pie type of graph. Select the following data variable: B5:B11. Label the graph with the text in cells A1 and A3 as title and sub-title respectively. Label the data points with cells A5:A11.

- Select the bar type of graph. Select the following data variables: I5, I6, I7, F10, G10, H10, and I11. Label the graph with the text in cells A1 and E3 as title and sub-title respectively. Label the y-axis with cell J3. Label the data points with cells E5, E6, E7, F9, G9, H9, and E11 respectively.

- Select the pie type of graph. Select the following data range as the variable: C14:C20. Label the graph with the text in cells A1 and A13 as title and sub-title respectively. Label the data points with cells A14:A20.

- Select the pie type of graph. Select the following data range as the variable: E19:I19. Label the graph with the text in cells A1 and G13 as title and sub-title respectively. Label the data points with cells E18:I18.

- Select the pie type of graph. Select the following data range as the variable: C26:C27. Label the graph with the text in cells A1 and A22 as title and sub-title respectively. Label the data points with cells A26:A27.

- Select the pie type of graph. Select the following data range as the variable: F24:I24. Label the graph with the text in cells A1 and E22 as title and sub-title respectively. Label the data points with cells F23:I23.

- Select the pie type of graph. Select the following data range as the variable: C43:C48. Label the graph with the text in cells A1 and A30 as title and sub-title respectively. Label the data points with cells A43:A48.

- Select the bar type of graph. Select the following data ranges as variables: G34:J34, G35:J35, G36:J36, G37:J37, and G39:J39. Label the graph with the text in cells A1 and E30 as title and sub-title respectively. Label the y-axis with cell J30. Label the x-axis points with the text in cells G32:J32. Label the variables with cells E34, E35, E36, E37, and E39 respectively.

- Select the pie type of graph. Select the following data range as the variable: I44:I47. Label the graph with the text in cells A1 and E42 as title and sub-title respectively. Label the data points with cells E44:E47.

Recipe 9: TRAFFIC ON THE ROADS

TRAFFIC uses 14 columns and 172 rows and is about 46 Kb in size. It contains 63 different formulas, which together perform more than 2,650 discrete arithmetic or logical operations (exclusive of those involved in graphing). Because the model is "circular" in nature, the spreadsheet automatically solves by iteration. With a delta value for iteration of .01 (the default value in my spreadsheet), this model typically iterates about 70 times to reach a new solution.

In addition to the four standard arithmetic operations, the model uses powers, inequalities, and the SUM() function.

Layout

Set the default column width to 6. Then reset the widths of columns A and B to 4; D, E, and F to 5; G to 3; and N to 7.

Set the default format to integer and make the following adjustments:

General Format	Text Right	Numbers Left	2 Decimal Places	4 Decimal Places
E17:E19	N3	G5:G10	D27:F31	H107:N112
C27:C32	C23	G15:G20	D46:F56	H129:N134
D35:F35	C24:F25	G107:G112	F99:F102	H151:N156
E38:F38	N24	G116:G121		
C46:C56	C43:F44	G129:G134		
	N43	G138:G143		
	N73, N82	G151:G156		
	N105, F107	G160:G165		
	N114, F116			
	N127, F129			
	N136, F138			
	N149, F151			
	N158, N160			

In cell A2, enter a continuous horizontal line by using a repeating hyphen or, if your spreadsheet accepts character graphics, Alt-196. Copy the contents of cell A2 into cells A12, A21, A40, A71, A80, A103, A125, A147, and A169.

Text and Numerical Entries

Referring to the following rows in Listing 9.1 (pages 215-17),

1   Enter the title and other information shown in row 1.

3-11   Enter the text shown in cells G3, N3, A4, G4, F5, A6, A7, A9, and G11. Enter the values shown in cells H3:M3 and G5:G10.

13-20   Enter the text shown in cells A13, G13, G14, F15, B16, and D17:A19. Enter
             the values shown in cells H13:M13, G15:G20, H15:M20, and E17:E19.

22-39   Enter the text shown in cells A22, C23, K23, A24:G25, H24:N24, A26, A27:
             A32 (which are all text entries), H26 (with three leading spaces), H33 (with
             three leading spaces), A34, A35, A37, and A38. Enter the values shown in
             cells C27:F31 and H27:M31. Enter the values shown in cells D35:F35 and
             E38:F38.

41-70   Enter the text shown in cells A41, K42, A43:G44, H43:N43, A45, H45 (with
             four leading spaces), and H58 (with four leading spaces). Enter the text
             shown in cells A59:A70 and then copy the contents of those cells into cells
             A46:A57. Enter the values shown in cells C46:F56 and H46:M56.

72-79   Enter the text shown in cells A72, K72, H73:N73, H74, and D75:D78.

81-102   Enter the text shown in cells A81, K81, H82:N82, A83, H83, D84:D87, A88,
             D89:D92, A93, D94:D97, A98, and D99:D102. Enter the value zero in cell
             H84 and copy it into cells I84:N84.

104-24   Enter the text shown in cells A104, A105, G105:N105, G106, F107,
             G107:G112, A114, G114:N114, G115, F116, G116:G122, and A124.

170-7   Enter the source citation shown in rows 170-72.

Formulas

Referring to the following rows in Listing 9.1 (pages 215-17),

3-11   Enter the following formulas in the cells shown:

             H5   H116+H138+H160
             N5   SUM(H5:M5)
             H11   SUM(H5:H10)

             Copy the formula in cell H5 into cells H5:M10, the formula in cell N5 into
             cells N6:N11, and the formula in cell H11 into cells I11:M11.

22-39   Enter the following formulas in the cells shown:

             G27   SUM(D27:F27)
             N27   SUM(H27:M27)

```
C32    SUM(C27:C31)
H32    SUM(H27:H31)
```

Copy the formula in cell G27 into cells G28:G31, the formula in cell N27 into cells N28:N31, and the formula in cell H32 into cells I32:N32.

Enter the following formulas in the cells shown:

```
H34    H27*$C27
N34    SUM(H34:M34)
H39    SUM(H34:H38)
```

Copy the formula in cell H34 into cells H34:M38, the formula in cell N34 into cells N35:N38, and the formula in cell H39 into cells I39:N39.

41-70   Enter the following formulas in the cells shown:

```
G46    SUM(D46:F46)
N46    SUM(H46:M46)
C57    SUM(C46:C56)
```

Copy the formula in cell G46 into cells G47:G56, the formula in cell N46 into cells N47:N56, and the formula in cell C57 into cells H57:N57.

Enter the following formulas in the cells shown:

```
H59    H46*$C46
N59    SUM(H59:M59)
H70    SUM(H59:H69)
```

Copy the formula in cell H59 into cells H59:M69, the formula in cell N59 into cells N60:N69, and the formula in cell H70 into cells I70:N70.

72-79   Enter the following formulas in the cells shown:

```
H75    (H34*$D27+H35*$D28+H36*$D29+H37*$D30+H38*$D31)/
       $D35
H76    (H34*$E27+H35*$E28+H36*$E29+H37*$E30+H38*$E31)/
       $E35
H77    (H34*$F27+H35*$F28+H36*$F29+H37*$F30+H38*$F31)/
       $F35
N75    SUM(H75:M75)
H78    SUM(H75:H77)
```

Copy the formula in cell H75 into cells I75:M75, the formula in cell H76 into cells I76:M76, the formula in cell H77 into cells I77:M77, the formula in cell N75 into cells N76:N77, and the formula in cell H78 into cells I78:N78.

81-102   Enter the following formulas in the cells shown:

```
H85     $E38*H32
N85     SUM(H85:M85)
H86     $F38*H32
H87     SUM(H84:H86)
```

Copy the formula in cell H85 into cells I85:M85, the formula in cell N85 into cell N86, the formula in cell H86 into cells I86:M86, and the formula in cell H87 into cells I87:N87.

Enter the following formulas in the cells shown:

```
H89     H59*$D46+H60*$D47+H61*$D48+H62*$D49+H63*$D50+
        H64*$D51+H65*$D52+H66*$D53+H67*$D54+
        H68*$D55+H69*$D56
N89     SUM(H89:M89)
H90     H59*$E46+H60*$E47+H61*$E48+H62*$E49+H63*$E50+
        H64*$E51+H65*$E52+H66*$E53+H67*$E54+
        H68*$E55+H69*$E56
H91     H59*$F46+H60*$F47+H61*$F48+H62*$F49+H63*$F50+
        H64*$F51+H65*$F52+H66*$F53+H67*$F54+
        H68*$F55+H69*$F56
H92     SUM(H89:H91)
```

Copy the formula in cell H89 into cells I89:M89, the formula in cell N89 into cells N90:N91, the formula in cell H90 into cells I90:M90, the formula in cell H91 into cells I91:M91, and the formula in cell H92 into cells I92:N92.

Enter the following formulas in the cells shown:

```
H94     H84+H89
F99     N75/N94
H99     H94*$F99
```

Copy the formula in cell H94 into cells H94:N97, the formula in cell F99 into cells F100:F102, and the formula in cell H99 into cells H99:N102.

104-24   Enter the following formulas in the cells shown:

```
H107   (H15>0)/H15^$E$17
N107   SUM(H107:M107)
H116   $H75*H107*H124/
          ($H107*$H124+$I107*$I124+$J107*$J124+
          $K107*$K124+$L107*$L124+$M107*$M124)
H117   $I75*H108*H124/
          ($H108*$H124+$I108*$I124+$J108*$J124+
          $K108*$K124+$L108*$L124+$M108*$M124)
H118   $J75*H109*H124/
          ($H109*$H124+$I109*$I124+$J109*$J124+
          $K109*$K124+$L109*$L124+$M109*$M124)
H119   $K75*H110*H124/
          ($H110*$H124+$I110*$I124+$J110*$J124+
          $K110*$K124+$L110*$L124+$M110*$M124)
H120   $L75*H111*H124/
          ($H111*$H124+$I111*$I124+$J111*$J124+
          $K111*$K124+$L111*$L124+$M111*$M124)
H121   $M75*H112*H124/
          ($H112*$H124+$I112*$I124+$J112*$J124+
          $K112*$K124+$L112*$L124+$M112*$M124)
H122   SUM(H116:H121)
H124   (H99>0)*H99^2/H122
```

Copy the formula in cell H107 into cells H107:M112, the formula in cell N107 into cells N108:N112 and N116:N122, the formulas in cells H116:H121 into cells I116:M116, the formula in cell H122 into cells I122:M122, and the formula in cell H124 into cells I124:N124.

126-46   This part of the model is similar to the part we have just completed. So begin by copying the contents of cells A104:N124 into cell A126. Then edit the text in cell A126 to read as shown in Listing 9.1 (on page 217); and edit the formulas in the following cells to read as shown:

```
H129   (H15>0)/H15^$E$18
N129   SUM(H129:M129)
H138   $H76*H129*H146/
          ($H129*$H146+$I129*$I146+$J129*$J146+
          $K129*$K146+$L129*$L146+$M129*$M146)
H139   $I76*H130*H146/
          ($H130*$H146+$I130*$I146+$J130*$J146+
          $K130*$K146+$L130*$L146+$M130*$M146)
H140   $J76*H131*H146/
          ($H131*$H146+$I131*$I146+$J131*$J146+
          $K131*$K146+$L131*$L146+$M131*$M146)
H141   $K76*H132*H146/
          ($H132*$H146+$I132*$I146+$J132*$J146+
          $K132*$K146+$L132*$L146+$M132+$M146)
```

```
H142    $L76*H133*H146/
        ($H133*$H146+$I133*$I146+$J133*$J146+
        $K133*$K146+$L133*$L146+$M133*$M146)

H143    $M76*H134*H146/
        ($H134*$H146+$I134*$I146+$J134*$J146+
        $K134*$K146+$L134*$L146+$M134*$M146)
H144    SUM(H138:H143)
H146    (H100>0)*H100^2/H144
```

Copy the formula in cell H129 into cells H129:M134, the formula in cell
N129 into cells N130:N134 and N138:N144, the formulas in cells H138:H143
into cells I138:M138, the formula in cell H144 into cells I144:M144, and the
formula in cell H146 into cells I146:N146.

148-68   Once again, this part of the model is similar to the part we have just completed. So begin by copying the contents of cells A126:N146 can be copied into
cell A148. Then edit the text in cell A148 to read as shown in Listing 9.1 (on
page 217); and edit the formulas in the following cells to read as shown:

```
H151    (H15>0)/H15^$E$19
N151    SUM(H151:M151)
H160    $H77*H151*H168/
        ($H151*$H168+$I151*$I168+$J151*$J168+
        $K151*$K168+$L151*$L168+$M151*$M$168)
H161    $I77*H152*H168/
        ($H152*$H168+$I152*$I168+$J152*$J168+
        $K152*$K168+$L152*$L168+$M152*$M168)
H162    $J77*H153*H168/
        ($H153*$H168+$I153*$I168+$J153*$J168+
        $K153*$K168+$L153*$L168+$M153*$M168)
H163    $K77*H154*H168/
        ($H154*$H168+$I154*$I168+$J154*$J168+
        $K154*$K168+$L154*$L168+$M154*$M168)
H164    $L77*H155*H168/
        ($H155*$H168+$I155*$I168+$J155*$J168+
        $K155*$K168+$L155*$L168+$M155*$M163)
H165    $M77*H156*H168/
        ($H156*$H168+$I156*$I168+$J156*$J168+
        $K156*$K168+$L156*$L168+$M156*$M163)
H166    SUM(H160:H165)
H168    (H101>0)*H101^2/H166
```

Copy the formula in cell H151 into cells H151:M156, the formula in cell
N151 into cells N152:N156 and N160:N166, the formulas in cells H160:H165

into cells I160:M160, the formula in cell H166 into cells I166:M166, and the formula in cell H168 into cells I168:N168.

Protection of the Model

To safeguard the model from inadvertent changes, you should "protect" the entire worksheet and then "unprotect" the following cells: H15:M20, E17:E19, C27:F31, D35:F35, C46:F56, and H46:M56.

Graphs

To obtain the figures shown in chapter 9, above, the following graphs are required:

- Select the bar-chart type of graph. Select the following data ranges as variables: H5:H10, I5:I10, J5:J10, K5:K10, L5:L10, and M5:M10. Label the points along the x-axis with the text in cells H3:M3. Label the variables with the text in cells H3, I3, J3, K3, L3, and M3. Title the graph with the text in cell A4; use cell A6 as the subtitle, cell K72 for the x-axis, and cell A9 for the y-axis.

- Select the bar-chart type of graph. Select the following data ranges as variables: H5:M5, H6:M6, H7:M7, H8:M8, H9:M9, and H10:M10. Label the points along the x-axis with the text in cells H3:M3. Label the variables with the text in cells H3, I3, J3, K3, L3, and M3. Title the graph with the text in cell A4; use cell A7 as the subtitle, cell K72 for the x-axis, and cell A9 for the y-axis.

Recipe 10: THROWING THINGS AWAY

WASTE uses 18 columns and 83 rows and is about 19 Kb in size. It contains 20 different formulas, which together perform about 560 discrete arithmetic or logical operations (exclusive of those involved in graphing).

In addition to the four standard arithmetic operations, the model uses powers, inequalities, and the following spreadsheet functions:

SUM(), AV(), MAX(), MIN(), EXP(), RANDOM, and VLOOKUP().

Note that VLOOKUP() can be abbreviated to VLU() in SuperCalc. Note too that cells that contain (or are dependent on other cells that contain) the RANDOM function may not show the same values in your model as they do in Listing 10.1, even though you have entered the formulas correctly.

Layout

Set the default column width to 7; then reset the width of columns E, F, and H to 2, 4, and 6 respectively. If columns A to L do not fit within the width of the screen, you can reduce the width of columns K or L. If the last column (column M) is visible on the screen, it can be hidden or moved further to the right.

Set the default numerical format to general. Format column D as integer and format cell D19 to show percentages. Format rows 3, 39, and 40 as text-right (right-justified). Then format cell A3 as text-left (left-justified).

In cell A2, enter a continuous horizontal line by using a repeating hyphen or, if your spreadsheet accepts character graphics, Alt-196. Copy the contents of cell A2 into cells A8, A20, A35, and A81.

In cell E3, enter a vertical line by using the vertical-bar key or, if your spreadsheet accepts character graphics, Alt-179. Copy the contents of cell E3 into cells E4:E34.

Text and Numerical Entries

Referring to the following rows in Listing 10.1 (pages 239-40),

1	Enter the title and other information shown in row 1.
3-34	Enter the text shown in cells A3:A7, A9:A15, and A17:A19. Enter the values shown in cells D4:D7. Enter the text shown in cells F3:M3. Enter the values shown in cells F4:F34.
38-80	Copy the text shown into cells A38, J39, L39, F40, G40, I40, J40, and L40. Enter the values shown in cells F41:F80, G41, and L41:L80.
82-83	Enter the source citation shown in rows 82-83.

Formulas

Referring to the following rows in Listing 10.1 (pages 239-40),

3-34   Enter the following formulas in the cells shown:

```
D10    AV(G4:G34)
D11    AV(H4:H34)
D12    MIN(H4:H34)
D13    MAX(H4:H34)
D14    SUM(L4:L34)
D15    MAX(F4:F34)
D17    SUM(K4:K34)
D18    D17/D7
D19    D18/SUM(H4:H34)
```

Enter the following formulas in the cells shown:

```
G4    D4/2
H4    VLU(M4,$J$41:$L$80,2)
I4    H4*$D$7
J4    MIN($D$4-G4,I4)
K4    I4-J4
L4    G4>=$D$5
M4    RANDOM
G5    G4+J4-L4*$D$5
```

Copy the formula in cell G5 into cells G6:G34. Copy the formulas in cells H4:M4 into cells H5:H34.

38-80   Enter the following formulas in the cells shown:

```
I41    ($D$6^F41)/(EXP($D$6)*G41)
J41    I41+J40
G42    F42*G41
```

Copy the formula in cell G42 into cells G43:G80. Copy the formulas in cells I41:J41 into cells I42:I80.

Protection of the Model

To safeguard the model from inadvertent changes, you should "protect" the entire worksheet and then "unprotect" the following cells: D4:D7.

Graphs

To obtain the figures shown in chapter 10, above, select the line type of graph. Select the following two data ranges as variables: cells H4:H34 and K4:K34. Label the graph with the text in cells A1 and A9 as title and sub-title respectively. Label the x-axis and y-axis with the text in cells F3 and I3 respectively. Label the x-axis points with the text in cells F4:F34.

Recipe 11: MULTI-CRITERIA ANALYSIS AND EIA

EIA uses 23 columns and 186 rows and is about 170 Kb in size. It contains 19 different formulas, which together perform about 13,600 discrete arithmetic or logical operations (exclusive of those involved in graphing). A macro is used to run the model 20 times and display a graph of the results.

In addition to the four standard arithmetic operations, the model uses the following spreadsheet functions:

> SUM( ), AV( ), STD( ), MAX( ), MIN( ), ABS( ), IF( ), ISERROR( ), and RANDOM.

Note that cells that contain (or are dependent on other cells that contain) the RANDOM function may not show the same values in your model as they do in Lising 11.1, even though you have entered the formulas correctly.

Layout

Set the default column width to 5; then reset the width of column A to 20; the widths of columns B, D, and E to 4; the widths of columns C and X to 1; and the widths of columns F and O to 2. Set the global format to general.

Format row 3 as text-right (right-justified); then format cells A3, D3, P3, T3, and U3 as text-left (left-justified). Format cells P8:S9 and U9:W10 as text-right. Format row 11 as text-right; then format cells A11, T11, and U11 as text-left. Format row 69 as text-right. Format row 127 as text-right; then format cell A127 as text-left. Format cells G133:N133 as text-right.

In cell A2, enter a continuous horizontal line by using a repeating hyphen or, if your spreadsheet accepts character graphics, Alt-196. Copy the contents of cell A2 into cells A8, A10, P7, P10, U5, U8, U11, A68, A126, A132, A161, and A184.

In cell C3, enter a vertical line by using the vertical-bar key or, if your spreadsheet accepts character graphics, Alt-179. Copy the contents of cell C3 into cells O3:O67, T3:T67, C9, F9, C11:C67, F11:F67, F69:F125, O69:O125, and O133:O183.

Text and Numerical Entries

Referring to the following rows in Listing 11.1 (pages 260-62),

1   Enter the title and other information shown in row 1. Enter the first three words in cell A1 (so they can be used for graph titles) and the rest in cell F1 and other cells, as appropriate.

3-9   Enter the text shown in cells A3:A7, A9, B4, B6, D3, D5:D7, D9, F3, G3:N3, G9 (note the leading space), L9 (note the three leading spaces), P3:P6, P8:S9, U3:U4, U6:W6, U7, and U9:W10. Enter the values shown in cells B5, B7, B9, E9, K9, and N9.

11-67   Enter the text shown in cells A11:A67, B11:B54, D11:E11, G11:N11. Enter the values shown in cells D12:E67 and G12:N67.

69-125   Enter the text shown in cells A72, A74, A76:A77, A79:A82, A84, A88, A90:A91, A93:A98, G69:N69, and P69:W69.

127-83   Enter the text shown in cells A127, A129:A131, P127:W127, A133:A135, D134:D154, D156:D160, and G133:N133. Enter the values shown in cells E135:E154.

162   Enter the text shown in cell A162.

185-86   Enter the source citation shown in rows 185-86.

Formulas

Referring to the following rows in Listing 11.1 (pages 260-62),

3-9   Enter the following formulas in the cells shown:

```
G5      IF(ISERROR(G156)," ",G156)
V7      MAX(U12:U67)
W7      ABS(SUM(U12:U67))
```

Copy the formula in cell G5 into cells G5:N7.

11-67   Enter the following formulas in the cells shown:

```
P12     MAX(G12:N12)
Q12     MIN(G12:N12)
R12     SUM(G12:N12)
S12     STD(G12:N12)
U12     (D12*$K$9+E12*$N$9)/($K$9+$N$9)
V12     IF($B$5=1,U12/V$7,U12/W$7)
```

```
W12    $V12*(1-$E$9+2*$E$9*RANDOM)
```

Copy the formulas in cells P12:S12 into cells P13:P67 and the formula in cells U12:W12 into cells U13:U67.

69-125   Enter the following formulas in the cells shown:

```
G70    IF(ISERROR(1/$P12/$R12/($P12-$Q12)/$S12)=1," ",
       IF($B$7=1,G12/$P12,
       IF($B$7=2,G12/$R12,
       IF($B$7=3,(G12-$Q12)/($P12-$Q12),
       (G12-AV($G12:$N12))/$S12))))
P70    IF(G70=" "," ",G70*(1-$E$9+2*$E$9*RANDOM))
```

Copy the formula in cell G70 into cells G70:N125 and the formula in cell P70 into cells P70:W125.

127-83   Enter the following formula in the cell shown:

```
P128   IF(P70=" "," ",P70*$W12)
G134   SUM(P128:P183)
G156   AV(G137:G154)
G157   STD(G137:G154)
G158   G157/SQRT(COUNT(G137:G154))
G159   G156+2.093*G158
G160   G156-2.093*G158
```

Copy the formula in cell P128 into cells P128:W183. Copy the formula in cell G134 into cells H134:N134. Copy the formulas in cells G156:G160 into cells H156:N156. Do not enter anything into cells G135:N154.

Macros

Enter the following macro, using column D for the name, column E for the instructions, and column K for the comments (which are optional):

```
164   \s    {home}{paneloff}                    home cursor
165         /b G135:N154 ~                      blank previous run
166         {let B9,0}                          zero counter
167         /c [5;B9+135],B9,v                  [E167] increment
168         /c G134:N134,[7;B9+135],v           copy current results
169         !                                   recalc
170         {if B9<20}{branch E167}             if counter<20, loop
171         {panelon}                           dialog on
172         {beep}{view}                        signal and graph
```

Using the commands appropriate to your spreadsheet, link the macro name in cell D164 to the instructions in cells E164:E172.

Protection of the Model

To safeguard the model from inadvertent changes, you should "protect" the entire worksheet and then "unprotect" the following cells: B5, B7, G5:N7, B9, E9, K9, N9, D12:E67, G12:N67, G134:N160, and E135:E154.

Graphs

To obtain the figures shown in chapter 11, above, select what is called in most spreadsheets the "high-low" type of graph. (This type of graph is often used for plotting stock prices.) Select the following data ranges: G159:N159, G160:N160, and G156:N156. Label the points on the x-axis using the text in cells P127:W127. Use the text in cells A1 and A127 as title and sub-title; use the text in cell D3 for the x-axis.

Recipe 12: THE GAME OF LIFE

LIFE uses 26 columns and 86 rows and is about 45 Kb in size. It contains five different formulas, which together perform about 3,200 discrete arithmetic or logical operations. There are three macros in the model to help set it up and run it through repeated iterations.

In addition to the standard arithmetic operations, the model uses inequalities and the following spreadsheet functions:

    MAX(), COUNT(), AND(), OR(), and IF().

Also, the RANDOM function is used in one of the macros; but it does not affect the model itself.

Layout

Set the default column width to 2; then reset the width of columns A, B, C, D, and E to 12, 6, 6, 6, and 1 respectively. If columns A to Z do not fit within the width of your screen, you can reduce the width of column A until they do.

Set the default numerical format to general. Format column Z as text-right (right-justified).

In cell A3, enter a continuous horizontal line by using a repeating hyphen or, if your spreadsheet accepts character graphics, Alt-196. Copy the contents of cell A3 into cells A15, A20, A23, and A53.

In cell E1, enter a vertical line by using the vertical-bar key or, if your spreadsheet accepts character graphics, Alt-179. Copy the contents of cell E1 into cells E2:E20, Z1:Z20, E31:E50, and Z31:Z50.

Text and Numerical Entries

Referring to the following rows in Listing 12.1 (pages 284-85),

1-20   Enter the text shown in cells A1, A2, A5, A8, A11, A12, and A16:A19. Enter the value shown in cell C5. In cell F1, enter (as text) a space followed by a dot. Copy cell F1 into cells F1:Y20.

27-50   Enter the text shown in cells A27, A29, and A31.

59   Enter the text shown in cells A59, B59, and H59.

Formulas

Referring to the following rows in Listing 12.1 (pages 284-85),

1-20   Enter the following formulas in the cells shown:

```
C8     COUNT(F1:Y20)
C11    IF(C5=1,C8,MAX(C8,C11))
C12    IF(C5=1,0,IF(C11>C8,C12,C5))
```

27-50   Enter the following formulas in the cells shown:

```
C31    C5+1
G32    IF(OR(AND(COUNT(F1:H3)=4,G2=1),COUNT(F1:H3)=3),
       1," .")
```

Note that there is a single space before the dot at the end of the second formula. Copy the formula in cell G32 into cells G32:X49.

Copy the formula in cell G32 into cell G31. Edit the formula in cell G31 to change the argument of both COUNT functions from (F1:H3) to

(F20,F1,F2,G20,G1,G2,H20,H1,H2)

and copy the formula in cell G31 into cells H31:X31.

Copy the formula in cell G49 into cell G50. Edit the formula in cell G50 to change the argument of both COUNT functions from (F1:H3) to

(F19,F20,F1,G19,G20,G1,H19,H20,H1)

and copy the formula in cell G50 into cells H50:X50.

Copy the formula in cell G32 into cell F32. Edit the formula in cell F32 to change the argument of both COUNT functions from (F1:H3) to

(Y1,Y2,Y3,F1,F2,F3,G1,G2,G3)

and copy the formula in cell F32 into cells F33:F49.

Copy the formula in cell X32 into cell Y32. Edit the formula in cell Y32 to change the argument of both COUNT functions from (F1:H3) to

(X1,X2,X3,Y1,Y2,Y3,F1,F2,F3)

and copy the formula in cell Y32 into cells Y33:Y49.

Copy the formula in cell G31 into cell F31. Edit the formula in cell F31 to change the argument of both COUNT functions to

(Y20,Y1,Y2,F20,F1,F2,G20,G1,G2)

Copy the formula in cell X31 into cell Y31. Edit to formula in cell Y31 to change the argument of both COUNT functions to

(X20,X1,X2,Y20,Y1,Y2,F20,F1,F2)

Copy the formula in cell G50 into cell F50. Edit the formula in cell F50 to change the argument of both COUNT functions to

(Y19,Y20,Y1,F19,F20,F1,G19,G20,G1)

Copy the formula in cell X50 into cell Y50. Edit the formula in cell Y50 to change the argument of both COUNT functions to

$$(X19,X20,X1,Y19,Y20,Y1,F19,F20,F1)$$

Macros

Enter the following macro, using column A for the name, column B for the instructions, and column H for the comments (which are optional):

```
61   {\S}   {home}                     home cursor
62          {paneloff}                 turn dialog line off
63          !                          [B63] manual recalc
64          {if C8=0}{branch B68}      exit if population reaches zero
65          /c C31,C5,v                update number of cycles
66          /c F31:Y50,F1,v            copy results of calculation to display
67          {branch B63}               loop until stopped
68          {beep} {panelon}           [B68] turn dialog line on again
```

Using the commands appropriate to your spreadsheet, link the macro name in cell A61 to the instructions in cells B61:B68.

Enter the following macro, using column A for the name, column B for the instructions, and column H for the comments (which are optional):

```
70   {\I}   {home}                     home cursor
71          {let C5,1}                 set number of cycles to one
72          {let F1," ."}              assign value " ." to f1
73          /c F1,F1:Y20,v             copy " ." throughout display
74          !                          manual recalc
75          =010 ~ {beep}              goto middle of display
```

Using the commands appropriate to your spreadsheet, link the macro name in cell A70 to the instructions in cells B70:B75.

Enter the following macro, using column A for the name, column B for the instructions, and column H for the comments (which are optional):

```
77   {\Z}   {home}                        home cursor
78          {let C5,1}                     set number of cycles to one
79          {let F1," ."}                  assign value " ." to f1
80          /c F1,F1:Y20,v                 copy " ." throughout display
81          =F1 ~                          goto upper left corner
82          {if ran>.5} 1                  [B82] replace with "1" at random
83          {if currow<20} {down} {branch B82}
84          {if curcol<25} {up 20} {right} {branch B82}
85          ~{home}
86          ! {beep}                       manual recalc
```

Using the commands appropriate to your spreadsheet, link the macro name in cell A77 to the instructions in cells B77:B86.

Protection of the Model

To safeguard the model from inadvertent changes, you should "protect" the entire worksheet and then "unprotect" the following cells: C5, C8, C11:C12, F1:Y20, C31, and F31:Y50.

Graphs

There are no graphs associated with this model.

Recipe 13: THE GAME OF CHOICE

CHOICE uses 26 columns and 88 rows and is about 79 Kb in size. It contains six different formulas, which together perform about 6,800 discrete arithmetic or logical operations. There are three macros in the model to help set it up and run it through repeated iterations.

In addition to the four standard arithmetic operations, the model uses inequalities and the following spreadsheet functions:

   MAX( ), COUNT( ), AND( ), IF( ), and RANDOM.

Note that cells that contain (or are dependent on other cells that contain) the RANDOM function may not show the same values in your model as they do in Listing 13.1, even though you have entered the formulas correctly.

Layout

The layout of this model is similar to that of the previous model (LIFE); so, if you have already entered that model, you may want to start with a copy and convert it into CHOICE.

If you are starting from scratch, set the default column width to 2; then reset the width of columns A, B, C, D, and E to 12, 6, 6, 6, and 1 respectively. If columns A to Z do not fit within the width of your screen, you can reduce the width of column A until they do. Set the default numerical format to general. Format cells B7:C8 and column Z as text-right (right-justified).

In cell A3, enter a continuous horizontal line by using a repeating hyphen or, if your spreadsheet accepts character graphics, Alt-196. Copy the contents of cell A3 into cells A15, A20, A23, and A53.

In cell E1, enter a vertical line by using the vertical-bar key or, if your spreadsheet accepts character graphics, Alt-179. Copy the contents of cell E1 into cells E2:E20, Z1:Z20, E31:E50, and Z31:Z50.

Text and Numerical Entries

Referring to the following rows in Listing 13.1 (pages 296-97),

1-20   Enter the text shown in cells A1, A2, A5, B7:C7, B8:C8, A9:A10, A12:A13, and A16:A19. Enter the values shown in cells C5 and B13:C13.

                 In cell F1, enter (as text) a space followed by a dot. Copy the contents of cell F1 into cells F1:Y20.

27-50   Enter the text shown in cells A27, A29, and A31.

59   Enter the text shown in cells A59:B59 and H59.

Formulas

Referring to the following rows in Listing 13.1 (pages 296-97),

1-20   Enter the following formulas in the cells shown:

```
B9    COUNT(F1:Y20)
C9    400-B9
B10   IF(C5=1,B9,MAX(B9,B10))
C10   IF(C5=1,C9,MAX(C9,C10))
```

27-50   Enter the following formulas in the cells shown:

```
C31   C5+1
G32   IF(G2=1,
         IF(((8-$B$13-COUNT(F1,F2,F3,G1,G3,H1,H2,H3))/
         (8-$B$13))>RAN," .",1),
         IF(((COUNT(F1,F2,F3,G1,G3,H1,H2,H3)-$C$13)/
         (8-$C$13))>RAN,1," ."))
```

In the second formula, note that there is meant to be a space before the dot at the end of both nested IF() functions. Copy the formula in cell G32 into cells G32:X49.

Copy the formula in cell G32 into cell G31. Edit the formula in cell G31 to change the argument of both COUNT functions to

    (F20,F1,F2,G20,G2,H20,H1,H2)

and copy the formula in cell G31 into cells H31:X31.

Copy the formula in cell G49 into cell G50. Edit the formula in cell G50 to change the argument of both COUNT functions to

    (F19,F20,F1,G19,G1,H19,H20,H1)

and copy the formula in cell G50 into cells H50:X50.

Copy the formula in cell G32 into cell F32. Edit the formula in cell F32 to change the argument of both COUNT functions to

    (Y1,Y2,Y3,F1,F3,G1,G2,G3)

and copy the formula in cell F32 into cells F33:F49.

Copy the formula in cell X32 into cell Y32. Edit the formula in cell Y32 to change the scope of both COUNT functions to

    (X1,X2,X3,Y1,Y3,F1,F2,F3)

and copy the formula in cell Y32 into cells Y33:Y49.

Copy the formula in cell G31 into cell F31. Edit the formula in cell F31 to change the scope of both COUNT functions to

    (Y20,Y1,Y2,F20,F2,G20,G1,G2)

Copy the formula in cell X31 into cell Y31. Edit the formula in cell Y31 to change the scope of both COUNT functions to

    (X20,X1,X2,Y20,Y2,F20,F1,F2)

Copy the formula in cell G50 into cell F50. Edit the formula in cell F50 to change the scope of both COUNT functions to

(Y19,Y20,Y1,F19,F1,G19,G20,G1)

Copy the formula in cell X50 into cell Y50. Edit the formula in cell Y50 to change the scope of both COUNT functions to

(X19,X20,X1,Y19,Y1,F19,F20,F1)

Macros

Enter the following macro, using column A for the name, column B for the instructions, and column H for the comments (which are optional):

```
61   {\S}   {home}                    home cursor
62          {paneloff}                turn dialog line off
63          !                         [B63] manual recalc
64          {if B9=0}{branch B69}     exit if one population reaches zero
65          {if C9=0}{branch B69}     exit if dots population reaches zero
66          /c C31,C5,v               update number of cycles
67          /c F31:Y50,F1,v           copy results of calculation to display
68          {branch B63}              loop until stopped
69          {beep} {panelon}          [B69] turn dialog line on again
```

Using the commands appropriate to your spreadsheet, link the macro name in cell A61 to the instructions in cells B61:B69.

Enter the following macro, using column A for the name, column B for the instructions, and column H for the comments (which are optional):

```
71   {\I}   {home}                    home cursor
72          {let C5,1}                set number of cycles to one
73          {let F1," ."}             assign value " ." to f1
74          /c F1,F1:Y20,v            copy " ." throughout display
75          !                         manual recalc
76          =010 ~ {beep}             goto middle of display
```

Using the commands appropriate to your spreadsheet, link the macro name in cell A71 to the instructions in cells B71:B76.

Enter the following macro, using column A for the name, column B for the instructions, and column H for the comments (which are optional):

```
78   {\Z}   {home}                    home cursor
79          {let C5,1}                set number of cycles to one
80          {let F1," ."}             assign value " ." to f1
```

```
81          /c F1,F1:Y20,v                    copy " ." throughout display
82          =F1 ~                             goto upper left corner
83          {if ran>.5} 1                     [B83] replace with "1" at random
84          {if currow<20} {down} {branch B83}
85          {if curcol<25} {up 20} {right} {branch B83}
86          ~{home}
87          !                                 manual recalc
88          =B13 ~ {beep}                     go to tolerance settings
```

Using the commands appropriate to your spreadsheet, link the macro name in cell A78 to the instructions in cells B78:B88.

Protection of the Model

To safeguard the model from inadvertent changes, you should "protect" the entire worksheet and then "unprotect" the following cells: C5, B13:C13, F1:Y20, C31, and F31:Y50.

Graphs

There are no graphs associated with this model.

Recipe 14: THE GAME MACHINE

GAMACH uses 33 columns and 123 rows and is about 67 Kb in size. It contains 11 different formulas, which together perform about 8,400 discrete arithmetic or logical operations. There are three macros in the model to help set it up and run it through repeated iterations.

In addition to the four standard arithmetic operations, the model uses powers, inequalities and the following spreadsheet functions:

IF() and LOOKUP().

Also the RANDOM function is used in one of the macros, but not in the model itself. Note that LOOKUP() can be abbreviated to LU() in SuperCalc.

Layout

The layout of this model is similar to that of the two previous models; so, if you have already created either of them, you may want to start with a copy and convert it into this model.

If you are starting from scratch, set the default column width to 2. Then reset the width of columns A, D, G, and J to 5, and the width of columns C, F, I, and L to 1. If columns A to AG do not fit within the width of your screen, you can reduce the width of columns M and AG by 1 each.

Set the default numerical format to general. Format column AG as text-right (right-justified).

In cell A3, enter a continuous horizontal line by using a repeating hyphen or, if your spreadsheet accepts character graphics, Alt-196. Copy the contents of cell A3 into cells A5, A15, A20, A23, and A53.

In cell L1, enter a vertical line by using the vertical-bar key or, if your spreadsheet accepts character graphics, Alt-179. Copy the contents of cell L1 into cells L2:L20, AG1:AG20, C7:C14, F7:F14, I7:I14, L31:L50, and AG31:AG50.

Text and Numerical Entries

Referring to the following rows in Listing 14.1 (pages 308-9),

1-20   Enter the text shown in cells A1, A2, A4, G4, A6:A14, D6:D14, G6:G14, J6:J14, and A16:A19. Enter the values shown in cells B7:B14, E7:E14, H7:H14, and K7:K14. In cell M1, enter (as text) a space followed by a dot. Copy cell M1 into cells M1:AF20.

27-50   Enter the text shown in cells A27, A29, and A31.

59   Enter the text shown in cells A59, J59, and L59.

61-92   Enter the values shown in cells J61:J92.

96   Enter the text shown in cells A96, G96, and O96.

Formulas

Referring to the following rows in Listing 14.1 (pages 308-9),

1-20   Enter the following formula in the cell shown:

    J4 COUNT(M1:AF20)

27-50   Enter the following formulas in the cells shown:

```
J31    D4+1
N32    IF(LU((N2=1)*2^5+(M2=1)*2^4+(N1=1)*2^3+
       (N3=1)*2^2+(O2=1)*2,$J$61:$J$92)>0,1," .")
```

Note that there is meant to be a space before the dot at the end the second formula. Copy the formula in cell N32 into cells N32:AE49.

Copy the formula in cell N32 into cell N31. Edit the formula to

```
N31    IF(LU((N1=1)*2^5+(M1=1)*2^4+(N20=1)*2^3+
       (N2=1)*2^2+(O1=1)*2,$J$61:$J$92)>0,1," .")
```

and copy the formula in cell N31 into cells O31:AE31.

Copy the formula in cell N49 into cell N50. Edit the formula to

```
N50    IF(LU((N20=1)*2^5+(M20=1)*2^4+(N19=1)*2^3+
       (N1=1)*2^2+(O20=1)*2,$J$61:$J$92)>0,1," .")
```

and copy the formula in cell N50 into cells O50:AE50.

Copy the formula in cell N32 into cell M32. Edit the formula to

```
M32    IF(LU((M2=1)*2^5+(AF2=1)*2^4+(N1=1)*2^3+
       (N3=1)*2^2+(O2=1)*2,$J$61:$J$92)>0,1," .")
```

and copy the formula in cell M32 into cells M33:M49.

Copy the formula in cell AE32 into cell AF32. Edit the formula to

```
AF32   IF(LU((AF2=1)*2^5+(AE2=1)*2^4+(AF1=1)*2^3+
       (AF3=1)*2^2+(M2=1)*2,$J$61:$J$92)>0,1," .")
```

and copy the formula in cell AF32 into cells AF33:AF49.

Copy the formula in cell N31 into cell M31. Edit the formula as follows:

```
M31    IF(LU((M1=1)*2^5+(AF1=1)*2^4+(M20=1)*2^3+
       (M2=1)*2^2+(N1=1)*2,$J$61:$J$92)>0,1," .")
```

Copy the formula in cell AE31 into cell AF31. Edit the formula as follows:

```
AF31   IF(LU((AF1=1)*2^5+(AE1=1)*2^4+(AF20=1)*2^3+
       (AF2=1)*2^2+(M1=1)*2,$J$61:$J$92)>0,1," .")
```

Copy the formula in cell N50 into cell M50. Edit the formula as follows:

```
M50    IF(LU((M20=1)*2^5+(AF20=1)*2^4+(M19=1)*2^3+
       (M1=1)*2^2+(N20=1)*2,$J$61:$J$92)>0,1," .")
```

Copy the formula in cell AE50 into cell AF50. Edit the formula as follows:

```
AF50   IF(LU((AF20=1)*2^5+(AE20=1)*2^4+(AF19=1)*2^3+
       (AF1=1)*2^2+(M20=1)*2,$J$61:$J$92)>0,1," .")
```

61-92   Enter the following formulas in the cells shown:

```
K61    B7
K69    E7
K77    H7
K85    K7
```

Copy the formula in cell K61 into cells K62:K68, the formula in cell K69 into cells K70:K76, the formula in K77 into cells K78:K84, and the formula in cell K85 into cells K86:K92.

Macros

Enter the following macro, using column A for the name, column G for the instructions, and column Q for the comments (which are optional):

```
98    {\S}    {home}                    home cursor
99            {paneloff}                turn dialog line
100           !                         [G100] manual recalc
101           {if J4=0}{branch G105}    exit if population reaches zero
102           /c J31,D4,v               update number of cycles
103           /c M31:AF50,M1,v          copy results of calculation to display
104           {branch G100}             loop until stopped
105           {beep} {panelon}          [G105] turn dialog line on again
```

Using the commands appropriate to your spreadsheet, link the macro name in cell A98 to the instructions in cells G98:G105.

Enter the following macro, using column A for the name, column G for the instructions, and column Q for the comments (which are optional):

```
107   {\I}    {home}                    home cursor
108           {let D4,1}                set number of cycles to one
109           {let M1," ."}             assign value " ." to f1
110           /c M1,M1:AF20,v           copy " ." throughout display
111           !                         manual recalc
112           =V10 ~ {beep}             goto middle of display
```

Using the commands appropriate to your spreadsheet, link the macro name in cell A107 to the instructions in cells G107:G112.

Enter the following macro, using column A for the name, column G for the instructions, and column Q for the comments (which are optional):

```
114   {\Z}   {home}                                  home cursor
115          {let D4,1}                               set number of cycles to one
116          {let M1," ."}                            assign value " ." to f1
117          /c M1,M1:AF20,v                          copy " ." throughout display
118          =M1 ~                                    goto upper left corner
119          {if ran>.5} 1                            [G119] replace with "1" at random
120          {if currow<20} {down} {branch G119}
121          {if curcol<25} {up 20} {right} {branch G119}
122          ~{home}
123          ! {beep}                                 manual recalc
```

Using the commands appropriate to your spreadsheet, link the macro name in cell A114 to the instructions in cells G114:G123.

Protection of the Model

To safeguard the model from inadvertent changes, you should "protect" the entire worksheet and then "unprotect" the following cells: D4, B7:B14, E7:E14, H7:H14, K7:K14, M1:AF20, J31, and M31:AF50.

Graphs

There are no graphs associated with this model.

Index

ABS() 365, 402
AND() 102, 351, 374, 376, 405, 409
APL 277
ASCII 345
AV(), AVERAGE() 135, 263, 340, 374, 378, 399, 402, 404
Age-group(s) 90, 91, 93, 94, 99, 101, 103, 110
Age-class 91, 96, 101, 104, 106, 108, 110, 112, 119
Algorithm(s) 52, 59, 247, 279, 291, 334
Analog 4
Approximation 28, 60, 155
Artificial 2, 11, 12, 112, 275, 278, 288, 291, 302, 304, 322, 323, 328
Automata 278, 279, 281, 282, 288-92, 301, 303, 305, 306, 312, 318, 321-23, 331
Average 63, 64, 67, 68, 80, 94, 99, 114, 122, 124, 135, 169, 171, 186, 191, 193-96, 198, 210, 231-34, 236, 237, 241, 246, 256, 263, 340, 354

BASIC 5, 6, 16, 33, 45, 47, 50, 55, 59, 62, 64, 122-24, 130, 134, 149-51, 163, 166-68, 170, 171, 173-78, 184, 186, 190, 197, 203, 212, 251, 252, 287, 293, 300, 314, 321, 329, 355
Bar-chart 385, 399
Boolean 6, 103, 220, 221, 242, 311, 331, 350
Box and Muller method 81, 82, 84
Branching 25, 277, 286, 288, 298
Brute force 281

COS() 134, 135, 141, 378, 381
COUNT() 277, 282, 286, 294, 298, 299, 340, 405, 409
Cell size 341
Cellular 2, 8, 278-82, 288-92, 301, 303, 305, 306, 312, 318, 321-23, 331, 332
Chaos 15, 32, 40, 203, 229, 277, 289, 291, 301, 303, 304, 322, 332, 333
Circular, circularity 169, 175, 341, 342

Coefficient(s) 25-27, 38, 48, 59, 135-38, 150-53, 155-59, 161, 268, 271
Cohort 12, 64, 69, 74, 75, 80
Complexity 3, 11, 13, 252, 289, 329, 332
Components 44, 170, 252, 255-57, 259, 260, 268, 331
Computer(s) 1, 4, 5, 9, 11, 15, 16, 66, 69, 74, 80, 82, 85, 90, 113, 122, 142, 149, 156, 162, 163, 172, 184, 190, 193, 195, 201, 225, 247, 250, 273, 277, 279, 282, 283, 288, 289, 291, 292, 295, 302, 303, 307, 310, 321-23, 329, 331-34, 339, 354-59
Confidence limit 264
Cost-benefit 251
Counter-intuitive 245
Cumulative probability 236
Cursor 71, 72, 99, 101, 262, 282, 285, 287, 297, 298, 309, 372, 404, 408, 412, 416, 417
Cycle 24, 69-73, 77, 94, 100, 101, 110, 119, 148, 165, 278-81, 283, 284, 286-88, 294-96, 298, 299, 306-8, 311, 312, 315, 316, 318-320, 349, 373
Cyclical 110, 232, 278, 283, 298, 307

DCOUNT() 83
Data range 83, 382, 385, 392
Database 83, 339, 353
Decimal 100, 127, 352, 356-58, 365, 370, 374, 379, 383, 385, 387, 393
Decision making 2, 262, 274, 289, 293, 294
Delta value 155, 157, 158, 365, 382, 383, 385, 393
Developing countries 12, 15, 40, 62, 89, 150, 161, 163, 180, 181, 185, 202, 204, 272
Development 12, 15, 19, 42, 90, 94, 113, 114, 118, 148, 165, 166, 173, 177, 178, 184-86, 188, 191, 193-96, 198, 203-5, 229, 249, 250, 272, 274, 290, 328, 331, 334
Digit(s) 86, 100, 248, 310, 356, 358
Digital 288, 289, 331
Dimension(s) 20-22, 24, 44-47, 49-51, 53, 57, 58, 65, 74, 255, 288, 289, 301, 303, 319, 332-33, 353

Dimensional 7, 45, 48, 51-57, 59-61, 251, 278, 279, 291, 303-5, 320, 321, 332, 355

Display area 131, 134, 282, 283, 286, 287, 295, 298, 307, 309, 319, 346

Diversity 85, 89, 293, 307, 355

EXP() 37, 52, 134, 238, 347, 362-65, 367-69, 378, 380, 399, 401

Ecological 16, 67, 87, 89, 142

Ecology 16, 85-87, 253, 262, 274

Economics 40, 148, 163, 251, 253, 328

Education 205

Energy 12, 14, 121-24, 126, 127, 130, 131, 135-44, 204, 205, 230

Environment 4, 6, 15, 23, 24, 42, 66, 73, 119, 142, 149, 185, 202, 205, 224-26, 249, 251, 253, 256, 272, 273, 277, 292, 294, 306, 318, 320, 321, 323, 343

Environmental 1, 8, 11, 12, 15, 19, 23, 24, 40-44, 47, 62, 65, 79, 86, 121, 141, 181, 229, 232, 247-51, 256, 260, 262, 268, 272-74, 289, 303, 304, 328, 329, 333, 334

Equation(s) 21, 28, 37, 65, 92, 115, 116, 119, 124-26, 134, 141, 147-55, 157-59, 162, 164, 238, 359, 370

Error checking 341

Error(s) 3, 6, 59, 60, 103, 104, 115-16, 134-36, 256, 259, 263-65, 281, 341, 342, 344-47, 349, 354, 356-58

Evaluation 36, 37, 58, 80, 90, 112, 118, 140, 162, 178, 179, 201, 224, 246, 251, 252, 268, 271-74, 288, 300, 319

Excel 82, 234, 333, 352, 357, 359

Experiment 6, 110, 271, 283, 317, 328

Expert(s) 7, 8, 28, 67, 80, 122, 329, 333

Exponent 37, 59, 134, 233, 356

Factorial 233

Forecast 64, 165, 252

Forecasting 2, 3, 150, 163, 166, 227

Format 9, 341, 343, 349, 351, 352, 356, 357, 362, 365, 370, 374, 379, 383, 385, 387, 388, 393, 400, 402, 405, 409, 414

Formatting 118, 356

Formula(s) 5, 6, 8, 9, 11, 13, 14, 25, 28, 38, 39, 48, 51-53, 61, 73-76, 81-83, 85, 90, 99, 101-4, 112-16, 134, 135, 141, 155, 157, 159, 172, 174, 175, 194-96, 219-22, 233,

238, 241, 242, 259, 263, 264, 267, 269, 270, 283, 286, 287, 295, 298, 299, 311, 337, 338, 341-47, 349-52, 355, 357-59, 361-76, 378, 380-87, 389-92, 394-406, 401, 403, 404, 406-12, 414-16

Fortran 6, 20, 46, 91, 98, 150, 225

Fractal(s) 303-5, 313, 316, 319-23

Function(s) 6, 7, 13, 14, 22, 37, 59-61, 83, 85, 92, 93, 101-5, 112, 114, 115, 117, 134, 135, 137, 138, 142, 149, 152, 153, 168, 173, 179, 185, 207, 208, 218-20, 234, 238, 241, 259, 277, 282, 286, 288, 294, 295, 298, 299, 306, 311, 331, 332, 337, 340, 341, 347, 348, 350-55, 359, 362, 365, 370, 374, 378, 382, 385, 387, 393, 399, 402, 405-9, 411-13

Garden of Eden 281

Gaussian 20-23, 25-27, 39, 40

Government 8, 16, 148, 152, 153, 161, 167, 272

Graph(s) 6, 25-28, 55, 61, 73, 77, 82, 84, 85, 94, 96, 108, 113, 128, 129, 131, 159, 161, 173, 176, 198, 226, 234, 242, 243, 246, 255, 258, 262, 264, 342, 343, 352-54, 359, 361, 365, 370, 373, 378, 382, 384, 385, 387, 391, 392, 399, 401, 402-5, 409, 413, 417

Graphing 7, 20, 25, 53, 176, 184, 332, 352, 354, 355, 362, 365, 370, 373, 378, 382, 385, 387, 388, 393, 399, 402

Gravity model 167, 208

HLOOKUP() 348, 352

HLU() 351, 352

Hantush function 61

IF() 51, 103, 104, 263, 286, 288, 298, 299, 347, 350, 362, 365, 367, 374, 387, 402-5, 409-11, 413

ISERROR() 103, 341, 347, 365, 374, 402

ITER 341, 365, 368, 369

Imagination 185

Impact(s) 11, 12, 20, 35-37, 40-44, 47, 58, 62, 79, 90-92, 105, 112-16, 161, 176, 184, 198, 205, 249-53, 255-57, 259, 260, 267, 268, 271-74, 328, 329, 334

Infinite 3, 46, 278, 282, 294, 304, 306, 316

Initialization 14, 94, 98, 288, 340, 377
Initialize 10, 71, 72, 108, 154, 284, 287, 296, 298, 308, 311, 318, 372, 373
Integer(s) 233, 238, 241, 303, 304, 362, 365, 370, 374, 379, 383, 385, 387, 393, 400
Integration 61, 252, 279, 300
Intuitive 245
Inverse power function 179, 208, 220
Iterate 162, 221, 348, 349
Iteration 8, 13, 14, 53, 149, 154, 155, 157, 158, 162, 169, 221, 342, 348-50, 354, 365, 382, 385, 393

Kb, Kilobyte 11, 337, 339, 341, 355, 362, 365, 370, 373, 378, 382, 385, 387, 392, 399, 402, 405, 409, 413
Knowledge 113, 330, 333

LN() 37, 101, 347, 362, 374, 376
LOG() 352, 365
LOOKUP(), LU() 277, 311, 348, 351, 413
Label(s) 5, 6, 25-28, 83, 84, 94, 96, 128, 129, 172, 173, 338, 352, 361, 365, 370, 373, 378, 382, 384, 385, 387, 391, 392, 399, 401, 405
Logarithm 37, 113
Lookup table(s) 25, 70, 73-76, 81, 82, 92, 237-39, 241, 307, 309, 310
Looping 8, 24, 32, 77, 287, 337
Lotus 1-2-3 8, 59, 134, 155, 158, 164, 226, 333, 347, 350-52, 357, 359

MAX() 134, 374, 378, 399, 402, 405, 409
MIN() 241, 362, 399, 401, 402
Macro(s) 6-8, 10, 11, 13, 14, 25, 70, 71, 73, 76, 77, 90, 94, 98-101, 104, 108, 165, 166, 246, 258, 262, 264, 277, 282, 283, 285, 287, 295, 297, 298, 307, 309, 311, 332, 340-42, 349-352, 354, 355, 361, 370, 372, 373, 376, 377, 402, 404, 405, 408, 409, 412, 413, 416, 417
Management 2, 6, 7, 11, 12, 15, 36, 40, 41, 47, 62, 72, 83, 85-87, 90, 91, 93-95, 98, 100, 105, 106, 110, 112, 118, 119, 173, 180, 181, 204, 210, 211, 213, 217, 226, 229, 232, 239, 247, 272, 273, 333, 339, 353
Managing 33, 35, 36, 86, 89, 118, 226, 244, 272

Matrix, Matrices 7, 25, 90, 122, 169-75, 180, 181, 187, 209, 210, 214, 215, 219, 225, 250, 251, 254, 262, 271, 274, 278, 283, 286, 287, 295, 298, 311, 319, 332, 348, 354
Measurement 11, 91, 171, 190, 213, 236, 252, 381
Microcomputer(s) 1, 4, 5, 15, 16, 36, 61, 66, 90, 122, 149, 150, 162, 173, 180, 206, 210, 211, 213, 217, 225-27, 272, 290, 293, 327, 329, 332-34, 356, 359
Model(s) 2-4, 7-14, 16, 20-25, 27, 28, 32-37, 40-48, 51-53, 57-59, 62, 64-66, 69, 70, 72-74, 77-82, 85, 86, 90-95, 98, 100, 101, 103-5, 108, 112-15, 118, 119, 122-24, 126-28, 130, 131, 134-36, 138-43, 147-81, 183-92, 194-98, 201-9, 212, 214, 215, 218-22, 224-26, 229, 231, 232, 233, 236-40, 242, 243, 245-53, 255-59, 262-65, 267, 268, 271, 278, 279, 282, 283, 286-89, 291-95, 298-300, 302-7, 310-12, 314, 316, 319-21, 323, 327-32, 335, 337-47, 350-56, 361, 362, 364, 365, 369, 370, 373, 374, 378, 381, 382-85, 387, 391, 393, 397-402, 405, 409, 413, 417
Modeling 1-4, 6-8, 10, 16, 17, 20, 22, 28, 39-42, 44-48, 50, 57, 61, 62, 68, 86, 122, 145, 147, 150, 151, 158, 162-64, 166, 169, 185, 205, 206, 225, 226, 231, 232, 247, 248, 251, 275, 292, 303, 304, 318, 320, 323, 327-30, 332, 333, 339
Monte Carlo (simulation technique) 65, 66, 86, 156, 165, 231, 232, 239, 247, 248
Moore neighborhood 282, 294, 299, 305, 306, 313, 314, 320
Multiplier 168-71, 173-75, 178

Natural number (*e*) 37
Neighborhood(s) 2, 231, 282, 287, 293-95, 299, 303, 305-7, 310-16, 319, 320
Network 5, 178, 192, 251, 333, 339
Normal distribution 69, 74, 81, 82, 84
Normalization 270, 271
Normalized 267, 268, 270

OR() 102, 351, 374, 376, 405
Overflow 28, 345, 346
Overlay 251

Parameter(s) 3, 7, 10, 11, 23, 25, 27, 28, 34, 37, 44, 48, 58, 65, 69, 76, 77, 114, 93, 94, 99, 104, 105, 112-14, 128, 148, 155, 156, 162, 172, 175-78, 185, 201, 211, 222, 227, 237, 239, 245, 246, 258, 264, 267, 295, 342, 352, 383

Parity 307, 315, 316

Pascal 6, 20, 26, 225

Percentage 68, 70, 99, 100, 140, 184, 187, 190, 193, 211, 213

Perception 3, 339

Planner(s) 1, 8, 19, 20, 32, 33, 36, 37, 41, 42, 44, 143, 164-66, 179-81, 183-86, 201, 204-8, 251

Planning 10, 12, 15, 32, 36, 37, 40, 118, 119, 143, 161, 163-65, 173, 180, 181, 184, 186, 187, 201, 203, 204, 210, 211, 213, 215, 217, 224-26, 251, 252, 271-74, 289, 291, 301, 304, 322, 330, 333, 334

Plonski yield function 92, 115, 117, 119

Poisson distribution 233-39, 241, 352

Policy making 161

Politics 39-41, 164, 230

Pollutant 28, 41, 43-47, 51, 53, 55, 57-61

Pollution 2, 12, 19-24, 26-28, 32, 33, 35-37, 39-45, 51, 55, 57, 58, 126, 142, 255, 332

Population 12, 63-81, 85, 87, 90, 156, 165-68, 170-78, 185, 187, 281-90, 297, 298, 305, 307, 309, 311, 315, 354, 358, 408, 412, 416

Power function, power law 28, 173, 179, 208, 220

Predator-prey 66

Predictability 277, 278, 304

Predictable 278, 328

Prediction 36, 39, 42, 179, 181, 251

Probabilistic linear vector analysis 250, 260

Probability 64, 65, 80, 170, 172-75, 179, 180, 232-36, 241, 256, 295, 358

Program(s) 2, 4, 5, 7-10, 26, 40, 44, 46, 47, 50, 61, 62, 70, 90, 112, 118, 131, 149, 164, 174, 211, 227, 246, 264, 289, 290, 293, 327, 330, 333, 334, 337-41, 351, 353-55, 359, 361

Programming 6-10, 14, 16, 20, 41, 119, 155, 225, 247, 277, 283, 287, 306, 334, 337, 343, 351, 352

Programming language(s) 6-8, 20, 225, 283, 287, 337, 352

Pseudorandom 231, 232, 340

Quattro Pro 351, 352, 359

RAN, RAND 5, 13, 75, 79, 81, 83, 98, 99, 181, 226, 260, 285, 297, 299, 309, 352, 372, 375-77, 408, 410, 413, 417

RANDOM 13, 14, 63, 65, 69, 70, 74-76, 81, 82, 84, 86, 95, 99, 100, 104, 105, 110, 112, 114, 231, 232, 234, 236, 238, 239, 242, 247, 248, 250, 256, 259, 261, 263, 282, 285, 287, 294, 295, 297-300, 305, 306, 309, 311, 315, 317, 328, 340, 341, 352, 356, 358, 359, 370, 374, 399, 401, 402, 404, 405, 408, 409, 413, 417

ROUND() 370

Randomization 14, 256-259, 264-66, 268

Randomize 98, 284, 296, 308, 318, 377

Reality 2, 3, 36, 45, 58, 81, 171, 206, 329, 330, 333

Recalculation 14, 243, 286, 287, 298, 349, 355, 362

Recipe(s) 2, 14, 335, 338, 343, 351, 352, 355, 361, 362, 365, 370, 373, 378, 382, 385, 387, 392, 399, 402, 405, 409, 413

Regression 7, 92, 101, 151-53, 155, 225

Research 1, 15, 16, 32, 40, 44, 62, 72, 85, 87, 118, 119, 142, 143, 148, 163, 173, 185, 203, 204, 210, 211, 213, 217, 226, 227, 247, 272, 273, 279, 291, 301, 322, 328, 330-33

SIN() 134, 135, 141, 378, 381

SQRT() 52, 53, 81, 83, 362, 365, 367, 369, 378

STD() 378, 402

SUM() 135, 218, 219, 340, 370, 374, 378, 385, 387, 393, 399, 402

Science 8, 40, 61, 62, 66, 85, 118, 143, 144, 204, 226, 231, 291, 292, 302, 321-23, 329, 331, 359

Scientific 62, 85, 142, 144, 273, 291, 292, 304, 322, 323, 327, 329, 334

Scores 255-63, 267, 270, 271

Scratch pad 4, 25, 48, 77, 94, 98, 101

Sensitivity 3, 37, 79, 161, 178, 222, 267

Sensitivity analysis 3, 37

Simpson's rule 61

Simulation model(s), modeling 1-4, 10, 32, 36, 37, 41, 47, 65, 69, 82, 118, 122, 142, 247, 248, 251, 327-29, 333

Spreadsheet 1, 2, 4-14, 16, 20, 27, 36, 40, 47, 53, 66, 73, 81-83, 86, 90, 104, 113, 114, 118, 122, 128, 131, 141, 149, 155, 162, 168, 174, 175, 176, 179, 184, 185, 198, 220, 221, 224, 225, 234, 246, 255, 277, 279, 282, 283, 287, 288, 294, 325, 332, 337-43, 345-47, 349-51, 353-58, 361, 362, 365, 366, 370-74, 377-79, 383, 385, 388, 393, 399, 400, 402, 405, 406, 408, 409, 410, 412-414, 416, 417

Standard deviation 22, 38, 39, 69, 70, 74-76, 80-82, 136-38, 259, 264, 270, 271

Standard error 264

Standardization 255, 256, 258, 260, 263, 267, 268, 271

Standardized 127, 256, 261-63, 269-71

Stochastic 41, 63, 65, 69, 74, 76, 112, 149-53, 155, 157, 158, 231

Stochasticity 65, 79

String 6, 7, 354

SuperCalc 2, 82, 83, 241, 338, 340, 341, 348, 349, 351, 352, 359, 361, 362, 370, 399, 413

Sustainable 89, 108, 110, 373

Symmetrical 169, 286, 298

Symmetry 307

Teach 328

Template(s) 6, 8, 16

Time 3, 5, 6, 11, 14, 19, 21, 22, 43-45, 49, 51, 53, 55, 59-61, 63-65, 68-72, 74, 77-80, 89, 90, 92, 104, 105, 108, 110, 114, 122, 124-26, 129, 131, 147-49, 151, 152, 153-156, 162, 165, 167, 169-72, 176, 179, 180, 183, 184, 195, 196, 202, 207-10, 214, 215, 218, 219, 231-34, 236, 237, 245, 247, 255, 256, 267, 277, 278, 289, 293, 298, 300, 305, 311, 313, 314, 316, 317, 319, 321, 329, 333, 337, 339, 342, 354, 358, 359, 366, 372, 373

Time-distance matrix 209, 210, 214, 215, 219

Toroid, Toroidal 282, 294, 316

Transition rules 278, 283, 288, 293, 298, 305-7, 310-18, 320

Trial and error 3, 281

Turing machine 279

Unity ()1) 8, 151, 169, 171, 175, 178, 261, 269-71, 348, 350, 352

User-friendly 9, 332

VLOOKUP(), VLU() 348, 351, 352, 362, 364, 370, 399

Variable(s) 3, 6, 27, 60, 61, 65, 68, 69, 73, 96, 142, 147-50, 152, 154-59, 162, 167, 173, 198, 201, 202, 253, 255, 331, 332, 349, 352, 354, 365, 370, 373, 378, 382, 384, 385, 387, 391, 392, 399, 401

Vector 174, 175, 250, 260, 348

Virtual reality 333

Visicalc 4

Von Neumann neighborhood 305, 313, 314, 320

Weibull function 92, 93, 95, 99, 100, 105, 112, 114

Weighted average 263

Weights 253, 255-59, 262, 263, 265, 267, 271, 315

What-if 3, 6, 7, 20, 185, 198, 339

Window 5, 76, 104, 198

Worksheet 5, 6, 9, 10, 14, 43, 53, 62, 69, 70, 73, 75, 76, 82, 99, 101, 103, 104, 128, 131, 156, 157, 190, 196, 219, 242, 245, 263, 264, 277, 284, 287, 296, 308, 310, 338-40, 343, 345, 346, 348, 353, 357, 364, 369, 373, 378, 382, 384, 387, 391, 399, 401, 405, 409, 413, 417

Zero 8, 23, 25, 39, 71, 72, 76, 81, 82, 92, 93, 98-100, 102, 103, 108, 114, 123, 124, 127, 134, 136, 155, 157, 172, 180, 194, 196, 219, 234-36, 241, 242, 257, 259, 262, 264, 265, 268-70, 285, 297, 309-14, 316, 347, 350, 352, 357, 358, 372, 373, 377, 394, 404, 408, 412, 416

Zone(s) 43, 45, 46, 58, 167-80, 173-78, 180, 206-16, 218, 219, 221-25, 274